FOCUSING
OF CHARGED PARTICLES

VOLUME I

CONTRIBUTORS

M. Y. Bernard

P. Durandeau

C. Fert

J. C. Francken

C. Germain

M. E. Haine

K. J. Hanszen

P. W. Hawkes

R. Lauer

D. Linder

T. Mulvey

E. Regenstreif

C. Weber

Focusing
of Charged Particles

Edited by
ALBERT SEPTIER

INSTITUT D'ÉLECTRONIQUE FONDAMENTALE
FACULTÉ DES SCIENCES D'ORSAY
UNIVERSITÉ DE PARIS
91-ORSAY, FRANCE

VOLUME I

1967

ACADEMIC PRESS   New York and London

CHEMISTRY

ACADEMIC PRESS INC.
111 Fifth Avenue, New York, New York 10003

United Kingdom Edition published by
ACADEMIC PRESS INC. (LONDON) LTD.
Berkeley Square House, London W.1

LIBRARY OF CONGRESS CATALOG CARD NUMBER: 67-17596

PRINTED IN THE UNITED STATES OF AMERICA

LIST OF CONTRIBUTORS

Numbers in parentheses refer to the pages on which authors' contributions begin.

M. Y. Bernard, *Conservatoire National des Arts et Metiers, Paris, France* (3).

P. Durandeau, *Faculté des Sciences, Toulouse, France* (309).

C. Fert, *Faculté des Sciences, Toulouse, France* (309).

J. C. Francken, *University of Groningen, The Netherlands* (101).

C. Germain, CERN, *Geneva, Switzerland* (163).

M. E. Haine, *A.E.I. Research Laboratory, Rugby, Warwickshire, England* (233).

K. J. Hanszen, *Physikalisch-Technische, Bundesanstalt, Braunschweig, Germany* (251).

P. W. Hawkes, *The Cavendish Laboratory, Cambridge and Peterhouse, Cambridge, England* (411).

R. Lauer, *Physikalisch-Technische, Bundesanstalt, Braunschweig, Germany* (251).

D. Linder, *A.E.I. Instrumentation Division, Harlow, Essex, England* (233).

T. Mulvey, *Department of Physics, University of Aston, Birmingham, England* (469).

E. Regenstreif, *Faculté des Sciences, Rennes, France* (353).

C. Weber, *Philips Research Laboratories N. V. Philips' Gloeilampenfabrieken, Eindhoven, The Netherlands* (45).

EDITOR'S PREFACE

That it was possible to focus beams of charged particles was experimentally proved by 1930. Since that time a number of treatises on electronic and ionic geometrical optics have been published. The early books made a point of investigating in the utmost detail the optical properties of lenses and prisms in the first-order approximation, then considering the aberrations in order to improve the first devices using low-density beams of particles: cathod-ray tubes, electron microscopes, and mass spectrographs.

The need to solve new and extremely complex problems accompanied the development of microwave generators, isotope separators employing intense ionic current and high-energy particle accelerators of the nuclear physics centers. For high-intensity beams, new optical devices were required to compensate for the defocusing force due to the space charge along the trajectory of particles. For very high-energy particles, the classical lenses in use since the beginning of corpuscular optics were much too weak; new focusing principles were needed to guide the beams of fast particles over long distances and to increase the internal efficiency of particle accelerators.

Few books have dealt with these latter aspects of corpuscular optics. To meet the needs of students, researchers, and engineers working with charged particles, we thought it useful to collect a number of articles written by specialists and dealing with the various aspects of the problems we have just mentioned. We did not intend this work to be a mere collection of formulas, or a purely theoretical work. We attempted to define what is now known in geometrical corpuscular optics. The reader may not necessarily find in it the exact solution to his particular problem, but we hope the many results concerning lenses and prisms, the methods for computing fields, and the potentials and optical properties of numerous systems may prove valuable.

In the first chapter of Volume I we discuss the general properties of potentials, fields and trajectories, the latest methods for resolving Laplace's and Poisson's equations and computing trajectories with or without space charge, and a description of the methods used for the measurement of magnetic fields.

The second chapter concerns the optics of straight axis systems for producing and focusing low-intensity beams: high-brightness electron guns, electrostatic and magnetic electron lenses, and strong focusing lenses for high-energy beams. Whatever symmetry is used in their making, the aberrations of these lenses are all treated together. The chapter ends with the particular and very delicate problem of the production of electron microprobes.

Volume II contains three chapters dealing, respectively, with the production and focusing of very high-intensity beams, the optics of systems with a cylindrical symmetry and curved axis (prisms), and the very special problems of focusing particles inside accelerators (electrostatic, circular, and linear particle accelerators) and when they leave them.

The editor wishes to express his sincere gratitude to Professor Grivet who generously consented to write a preface and to the contributors of the various chapters who kindly accepted to share their vast experimental knowledge with the reader and managed to summarize the utmost number of experimental or theoretical results within a necessarily strictly limited space.

Orsay, France   A. SEPTIER
June, 1967

PREFACE

These volumes attempt to give a view of the two modern aspects of particle optics—the classical electron and the ion optical domains, and the new accelerator field. A well-balanced and up-to-date review of both fields is given with the links between the two developments as well as the techniques for tailoring the common theories to the special application in view being presented.

The electron was discovered nearly one hundred years ago and the following century has been a very active period of development for modern physics with the electron and ion beams playing the most important role of preferred instrumentation for generations of physicists. The beams have been put to a great variety of uses. One finds them everywhere and at every epoch —probing the first quantum pecularities of the atom (Franck and Hertz), analyzing the mass isotopes in Mendeleev's table (Aston), searching for nuclear reactions in accelerators, and now investigating the still deeper secrets of elementary and strange particles in the giant machines of recent years.

Although the principles of the laws of propagation of beams have been understood for a long time, they were not investigated in detail. In the twenties, however, the rapid development of the cathode-ray tube as a practical oscillograph of unrivaled speed and accuracy gave rise to a true science of electron optics. It was at this time that Busch, working with magnetic fields, coined the expression "electron lens," and soon afterward, Davisson and Calbick introduced the concept of electrostatic lenses.

The field of electron and ion optics then developed rapidly, mainly through the work of German physicists, as evidenced by the publication in the thirties of the first treatise in the field by Scherzer, which is now a classic. It appeared that particle optics had become a favored field for the transposition of the laws of glass optics, because the beams exhibited just the small-angle properties necessary for the application of the optical theory in the Gauss approximation. Moreover, one could often employ the next approximation—for example, the aberration theory—to obtain a very precise treatment. A clever use of these ideas and a great experimental effort by a few research teams culminated, just before World War II, in achieving supermicroscopic resolving power in the electron microscope. On the industrial side, very fine oscillograph and television tubes were put at the disposal of the public.

This remarkable scientific development continued after the war, though initially only along the same lines, that is, within the realm of the classical laws of electron optics. A new collection of fine instruments reached theoretical perfection during this period; they ranged from the shadow x-ray microscope to the metallurgical sonde of Professor Castaing. Resolving power that overstepped the bounds determined by theory some twenty years earlier was actually realized in the electron microscope. Furthermore, development has by no means reached its end as, for example, indicated by the recent progress in Professor Dupouy's laboratory with the high-tension electron microscope. Superconductive lenses, or quadrupole lenses, also seem to open new possibilities within the classical theory of the instrumentation and may well lead to significant improvements in the present record in experimental resolving power (3 Å by H. Fernandez Moran).

In this postwar period, however, a new and original development has progressively taken shape. Although the channeling of particles was already of great interest in a large variety of machines and devices, which we may broadly call "accelerators," the close analogy between an accelerator and the classical electron optical instrument was not widely recognized. In the former, the beam appears very broad, the velocities are not homogeneous, and the "spots" may reach a diameter of 1 centimeter or more. This could not be directly compared with the delicate beams of classical electron optics, a domain in which the Ångstrom appears as a natural unit and high-density beams are able to bore holes in a solid that are well under 1 micron in diameter.

Recognition, however, that similar procedures could be followed in optimizing the two systems and that many of the ultimate goals were similar has led to a fruitful incorporation of accelerators into the science of optical devices.

It is to be hoped that these volumes will contribute to the development of a new generation of young researchers in this fascinating field; for any experimental progress offers the added interest of exerting an immediately beneficial influence on the development of the most "advanced" part of physics, the discovery of new and strange particles. Large accelerators are the modern microscope through which young physicists are probing the secrets of the subatomic world.

Orsay, France   PIERRE GRIVET
June, 1967

CONTENTS

CONTENTS OF VOLUME II

1
Potential, Fields, Trajectories

CHAPTER 1.1

PARTICLES AND FIELDS: FUNDAMENTAL EQUATIONS

Michel Y. Bernard

Conservatoire National des Arts et Métiers
Paris, France

3

1.1.1. The Electrostatic Field

A. ELECTRODE SYSTEM

We shall first consider the electrostatic fields that are created to focus beams of charged particles. These fields are obtained with stainless steel or brass electrodes that have been brought to constant potentials. In so far as possible, the values required are obtained with a potentiometer and a single voltage supply. The circuit has the substantial advantage of producing ratios of potential differences independent of voltage supply variations. This is important, since in many cases, particle motion is solely dependent on ratio of potential differences (Fig. 1)

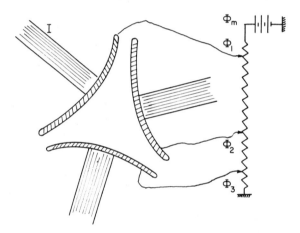

FIG. 1. General structure of electrodes system: *I* is the isolator.

The electrodes are held in insulators (plexiglas, araldite, and so on) that should be placed as far as possible from the particle beam. The designers of electrostatic lenses must in fact respect the rule that an "observer" following the beam should not see the insulators, but see only the electrodes. If this rule is not respected, particles or positives ions are deposited on the insulators and disturb the field. Our considerations are therefore limited to those electrostatic fields in vacuum that have been formed by charges on perfectly conductive electrodes. The dielectrics are not involved in calculation of the field, although they figure, in as far as the "breakdown" performance is concerned, in the design of lenses.

We now situate the beam in the electrode structure. We shall ignore the influence of its charges on those of the electrodes; the focusing fields can thus be calculated independently of the beam it is to focus; this is an approximation that may be accepted in the case of low-intensity particle focusing beams. Although it is not always possible to ignore the influence of one particle in the beam on another, the space-charge effect will be studied independently (Section 1.1.4,C) as a factor for final correction of particle path calculation.

B. LAWS GOVERNING ELECTROSTATIC FIELDS IN VACUUM: CALCULATION METHODS

The electrostatic field equations are, first,

$$\boldsymbol{\nabla} \wedge \mathbf{E} = 0 \tag{1.1}$$

which expresses in fact that $\mathbf{E}$ (in volts/meter) depends on a scalar potential Φ (in volts).

$$\mathbf{E} = -\boldsymbol{\nabla}\Phi \tag{1.2}$$

and secondly, the Gauss theorem, which relates the electrostatic field to charge density ϱ (in C/m³).

$$\boldsymbol{\nabla} \cdot \mathbf{E} = \varrho/\varepsilon_0 \tag{1.3}$$

ε_0 is a fundamental constant, equal to $8{,}85434 \cdot 10^{-12}$ (farad/meter). For a superficial electrode charge density σ (in C/m²) we obtain the value of the field normal to the electrode:

$$| \mathbf{E} | = \sigma/\varepsilon_0 \tag{1.4}$$

The elimination of $\mathbf{E}$ between Eqs. (1.3) and (1.2) gives the Poisson equation:

$$\nabla^2\Phi = -\varrho/\varepsilon_0 \tag{1.5}$$

If we know the repartition of space charge [namely, the function $\varrho(x, y, z)$], the calculation of electric potential Φ is possible; but, generally the space charge is created by particles and the motion of which is conditioned by the fields, that is, by the repartition of potential. The problem is very complicated. However, if we ignore the space charge in calculating the fields created by the electrodes, we have $\varrho = 0$. In this simple case we obtain the Laplace equation:

$$\nabla^2\Phi = 0 \tag{1.6}$$

The potential is a harmonic function; it is known that such a function has no "extremum" elsewhere other than on electrodes. It is worthwhile observing that Eq. (1.6) expresses also the fact that the integral

$$J = \iiint (\nabla \Phi)^2 \, dv \tag{1.7}$$

is "extremal"; the integration volume can have any value. We must therefore find a harmonic function Φ accepting the imposed values at the electrode surfaces. Equation (1.4) will then give us the surface density, and —by integration— the electrode charges. Further, it is evident that knowing the function Φ makes it possible for us to calculate the electrical field at all points by derivations.

Let us first stress the linearity of the Laplace equation. If two harmonic solutions are known, for example, Φ_1 and Φ_2, then any linear combination $\lambda_1 \Phi_1 + \lambda_2 \Phi_2$ is another solution. We can thus find any series capable of satisfying limit conditions. All this will be applied later. A further interesting consequence is the singleness of solution for given limit conditions. If two solutions (for example, those obtained by different methods) are known for the same problem, both satisfying the same limit conditions, it will be readily understood that the difference between these two solutions $\Phi - \Phi'$ must be 0 on all the electrodes. This demonstrates that the solution $\Phi - \Phi'$ must be identically nil, since the potential cannot assume "extremum" values elsewhere than on the electrodes. Hence, $\Phi \equiv \Phi'$. The two solutions are identical, and for practical purposes, the simpler method can be employed for calculation.

There are three major calculation methods. The first consists of applying the variables separation process to the Laplace equation. Because of the linearity of the equations, it is thus possible to express the solution as a series of orthonormalized functions, whose coefficients can be determined more or less easily. We shall see a few examples of this. The second method proceeds in the inverse sense; it consists of first selecting a function that satisfies limit conditions and is dependent on certain parameters. These parameters are then determined to bring the solution as close as possible to the Laplace equation. The criteria adopted for approximation of the solution may pertain to the "extremum" method, or the finite differences equation. The reader is referred to Chapter 1.2 for the application of these finite differences equations. We shall not discuss the "extremum" method which is of relatively slight practical utility. In fact, the function Φ must be situated in the integral (1.7), and the parameters must be determined so as to find J extremum. The third method cannot be overlooked, although it is somewhat artificial, because it is useful. This is the "conformal-mapping" system, which is no more than a way of producing harmonic functions. With patience, the suitable function can be found. Unfurtunately, the method is only applicable to problems with two coordinates; we shall see an example of this.

All of these calculations are lenghty and tiresome and contain many causes of error. It is frequently valuable to have an idea of the form of the equipotentials, particularly in the successive approximation method, which takes as its basis an assumed solution that should be as close as possible to the real solution. Let us take the case of a problem in plane geometry (the only really convenient case). The electrodes are drawn on a sheet of paper, and the potentials are indicated (Fig. 2). We know that

(a) The electrodes are equipotentials;

(b) The lines of force are perpendicular to the equipotentials and hence to the electrodes;

(c) The potential varies regularly along the lines of force (the potential cannot be "extremum");

(d) Two lines of force cannot cross (any more than equipotentials can do so) at any point other than where the field is nil, and so on.

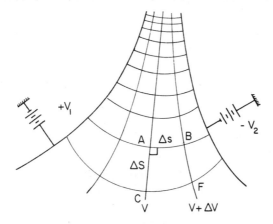

FIG. 2. Lehmann's method for the determination of equipotential curves and field lines.

All of the above rules, judiciously used, make the rapid drawing of an approximate topography of the field quite possible, according to the process devised by Th. Lehman. In order to do this, a number of potential values are selected at regular intervals of ΔV, as also a conventional value $\Delta \psi$ is

8   MICHEL Y. BERNARD

selected for the flux of the field through a tube of force lines. The tube is
defined by two lines of force and any height (for instance, unit of length),
perpendicular to the plane of the figure. We now consider the "pseudo-
rectangle" $ABCF$. Along AB and CF, the "work" of the field is equal to ΔV.
We write

$$E \cdot \Delta s = \Delta V \qquad (\Delta s \simeq AB \simeq CF) \tag{1.8}$$

Furthermore, the flux of $\mathbf{E}$ across AC or FB is the same; it is equal to

$$\Delta \psi = E \cdot \Delta S \qquad (\Delta S \simeq AC \simeq FB) \tag{1.9}$$

from which we deduce

$$\frac{\Delta V}{\Delta \psi} \simeq \frac{\Delta s}{\Delta S} \simeq \frac{AB}{AC} = \text{const} \tag{1.10}$$

If the equipotentials and lines of force are correctly drawn, all the "pseudo-
rectangles" will be similar. In particular, it is preferable to have them rather
as "pseudosquares," choosing $\Delta V = \Delta \psi$; thus $AB \simeq AC$. In this
case, the problem simply becomes one of tracing "curvilinear squares"
to fill the entire plane. In practice, a little experience is required, and the
structure of the field becomes apparent in a few minutes.

C. VARIABLES-SEPARATION METHOD

The method consists of attempting to find *one* solution of the Laplace
equation in the form of the product of three functions, each function depending upon one variable only. In order to give an example in cylindrical
coordinates Orz, we assume revolution symmetry, so that θ is not involved.
The Laplace equation is written

$$\frac{\partial^2 \Phi}{\partial r^2} + \frac{1}{r} \frac{\partial \Phi}{\partial r} + \frac{\partial^2 \Phi}{\partial z^2} = 0 \tag{1.11}$$

We stipulate $\Phi(r, z) = R(r)Z(z)$ and obtain

$$\frac{1}{R} \left(\frac{d^2 R}{dr^2} + \frac{1}{r} \frac{dR}{dr} \right) = -\frac{1}{Z} \frac{d^2 Z}{dz^2} \tag{1.12}$$

The first member depends solely on r, and the second solely on z. Since
these are independent variables, the only way of meeting this requirement
is to write

$$\frac{1}{R}\frac{d^2R}{dr^2} + \frac{1}{rR}\frac{dR}{dr} = k^2 \qquad (1.13)$$

$$\frac{1}{Z}\frac{d^2Z}{dz^2} = -k^2 \qquad (1.14)$$

whereby k is a constant, undetermined at the moment. Evidently, we could have written $-k^2$ in Eq. (1.13) and k^2 in Eq. (1.14); we shall return to this later. From Eqs. (1.13) and (1.14) we get

$$R = AI_0(kr) + BK_0(kr) \qquad (1.15)$$

$$z = C\cos k(z - z_0) \qquad (1.16)$$

in which A, B, C, and z_0 are arbitrary constants. I_0 and K_0 are modified Bessel functions of the first and second kind.

Unfortunately, not all systems of coordinates permit the separation of variables. This method of calculation is thus suitable only if the form of the electrodes permits selection of a system of coordinates allowing this separation (as is fortunately quite frequently the case). The reader will find full details in the bibliography. The following is a useful equation for the rapid determination of the Laplacian in a system of *orthogonal* coordinates q_1, q_2, q_3, the only really practically utilizable case. The element of length is calculated

$$ds^2 = e_1{}^2 dq_1{}^2 + e_2{}^2 dq_2{}^2 + e_3{}^2 dq_3{}^2 \qquad (1.17)$$

The Laplace equation is then written

$$\nabla^2\Phi = \frac{1}{e_1 e_2 e_3}\left[\frac{\partial}{\partial q_1}\left(\frac{e_2 e_3}{e_1}\frac{\partial\Phi}{\partial q_1}\right) + \frac{\partial}{\partial q_2}\left(\frac{e_1 e_3}{e_2}\frac{\partial\Phi}{\partial q_2}\right) \right.$$
$$\left. + \frac{\partial}{\partial q_3}\left(\frac{e_1 e_2}{e_3}\frac{\partial\Phi}{\partial q_3}\right)\right] \qquad (1.18)$$

In spherical coordinates, for example, the coordinates are r, θ, φ and it is quite easy to find

$$ds^2 = dr^2 + r^2 d\theta^2 + r^2 \sin^2\theta \, d\varphi^2 \qquad (1.19)$$

which gives us

$$e_1 = 1, \qquad e_2 = r, \qquad e_3 = r\sin\theta \qquad (1.20)$$

and for the Laplace equation

$$\nabla^2\Phi = \frac{1}{r^2}\frac{\partial}{\partial r}\left(r^2\frac{\partial\Phi}{\partial r}\right) + \frac{1}{r^2\sin\theta}\frac{\partial}{\partial\theta}\left(\sin\theta\frac{\partial\Phi}{\partial\theta}\right) + \frac{1}{r^2\sin^2\theta}\frac{\partial^2\Phi}{\partial\varphi^2}. \quad (1.21)$$

Like the cylindrical coordinates, spherical coordinates also permit the separation of variables using Legendre functions.

D. Calculation of Potential with Separation of Variables

The problem is *relatively* simple if a system of coordinates is available permitting separation of the variables, and such that *the electrodes entirely cover a coordinated surface without a gap of finite width.* Let us consider, for example, an electrostatic focusing system consisting of two coaxial cylinders of radius R. Figure 3 is a view in perspective and is a cross section

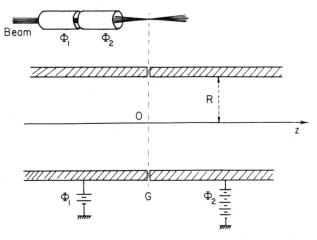

Fig. 3. Two-cylinder electron lens. G is a gap of zero width.

showing the significance of the parameter. An *infinitely narrow* interval separates the semicylinder of potential Φ_1 from the other, of potential Φ_2. Thus, the coordinate's surface $r = R$ is *entirely covered* by the electrodes, with the exception of a gap of negligible width, and there is *no electrode anywhere else.* Conditions correspond to application of the method.

Let us take a solution obtained by separation of variables:

$$AI_0(kr) \sin k(z - z_0) \tag{1.22}$$

We set aside the Bessel function of the second kind, because it is infinite on the axis, which is unsuited to the physical conditions of the problem. Let us take the origin of the axis of the z at the break and calculate

$$y = \Phi - \frac{\Phi_1 + \Phi_2}{2} \tag{1.23}$$

This modification of the origin of the potentials allows us to manipulate odd functions only, since the function y is odd, going from $-(\Phi_2 - \Phi_1)/2$ to $(\Phi_2 - \Phi_1)/2$ when z goes from $-\infty$ to $+\infty$.

Let us therefore attempt to find a solution of the form

$$y = \Phi - \frac{\Phi_1 + \Phi_2}{2} = \sum_k A(k)I_0(kr) \sin kz \qquad (1.24)$$

The function $A(k)$ is determined by trial and error, so as to satisfy limit conditions for the function Φ. For this, transformation of this sum into an integral is indicated, with use of the Fourier integral formula:

$$\Phi = \frac{\Phi_1 + \Phi_2}{2} + \int_0^\infty A(k)I_0(kr) \sin kz \, dk \qquad (1.25)$$

Since the potential is known for $r = R$, we reverse the equation. Thus,

$$A(k)I_0(kR) = \frac{1}{\pi} \int_0^\infty \left(\Phi(z,R) - \frac{\Phi_1 + \Phi_2}{2} \right) \sin kz \, dz \qquad (1.26)$$

Since $\Phi(z, R) = \Phi_2$ for z positive, we immediately obtain

$$A(k) = \frac{\Phi_2 - \Phi_1}{\pi} \frac{1}{kI_0(kR)} \qquad (1.27)$$

which brings us to

$$\Phi(z,r) = \frac{\Phi_1 + \Phi_2}{2} + \frac{\Phi_2 - \Phi_1}{\pi} \int_0^\infty \frac{I_0(kr)}{kI_0(kR)} \sin kz \, dk \qquad (1.28)$$

All that remains to be done, is numerical calculation.

We conclude by indicating another, absolutely parallel method. We have expressed the constant as k^2 for the first equation when separating the variables. Nothing prevents us from writing $-k^2$, which results in exponential (and not sinusoidal) functions and in "normal" Bessel functions for the second equation. Taking symmetry into account, we obtain for $z > 0$ only a form of solution.

$$\Phi = \sum_k A_k J_0(kr) \exp(-kz) \qquad (1.29)$$

eliminating functions $N_0(kr)$ and e^{+kz}, which become infinite in the area studied. It is evident that symmetry will make it possible for us to obtain the solution for $z < 0$. Retaining serial form, we determine the coefficients A_k to satisfy boundary conditions. For $r = R$, the potential is Φ_2, except for $z = 0$ at the break. By taking the solutions of the equation, $J_0(kR) = 0$ is indicated for the various values of k, with the addition of Φ_2 to the series. We obtain thereby

$$\Phi(r,z) = \Phi_2 + \sum_{n=1}^{\infty} A_n J_0 \left(\mu_n \frac{r}{R} \right) \exp\left[- \mu_n \frac{z}{R} \right] \qquad (1.30)$$

which satisfies boundary conditions for $r = R$. μ_n being the succession of the roots of $J_0(x)$ given in the tables.

Let us now state that for $z = 0$, the potential at the plane of symmetry is $\frac{1}{2}(\Phi_1 + \Phi_2)$. Using a conventional property of the Bessel functions integrals, we find

$$A_n = \left(\frac{\Phi_1 + \Phi_2}{2} - \Phi_2 \right) \int_0^R r J_0 \left(\mu_n \frac{r}{R} \right) d\dot{r} \qquad (1.31)$$

The integral is easily calculated, using the function $J_1(x)$; for $z > 0$:

$$\Phi = \Phi_2 + (\Phi_1 - \Phi_2) \sum_{n=1}^{\infty} \frac{J_0(\mu_n r/R)}{\mu_n J_1(\mu_n)} \exp\left[- \mu_n \frac{z}{R} \right] \qquad (1.32)$$

This mathematical series, equivalent to the integral form (1.28), is more convenient for numerical calculation. The reader can try it out, for example, in calculating the axial potential. He will observe, perhaps with surprise, that the function is practically identical with

$$\varphi(z) = \frac{\Phi_1 + \Phi_2}{2} + \frac{\Phi_2 - \Phi_1}{2} \tanh \omega z \qquad (1.33)$$

with $\omega = 1.318$. This coincidence, pointed out by Bertram (1942), is very useful for numerical calculation. It is in fact essential not to forget that calculations that can be completed without approximation have to be accepted. It is recommendable to do this at the end of numerical calculation, taking a simple function, to approximate results; Eq. (1.33) is an example of this method.

The conditions of validity of the method described can be slightly less stringent, in one case; when the surface of coordinates is not entirely covered by the electrodes, but where a gap of finit length is present. It is frequently possible to find a function correctly representing the potential at the gap and thus to complete the calculation. We shall take the same example (two coaxial cylinders) at the same radius R, but shall now assume an interval $2d$ between the two cylinders (which is closer to practice, if break-

down of potential is to be avoided). We can no longer use the general method, since the potential is not completely known at the cylinder. We can
however complete calculation if we assume that it varies linearily between
the two "edges" of the break, as shown in Fig. 4. Thus we write

$$\Phi(z, R) = \frac{\Phi_1 + \Phi_2}{2} + \frac{\Phi_2 - \Phi_1}{2d} z \qquad (-d < z < d) \qquad (1.34)$$

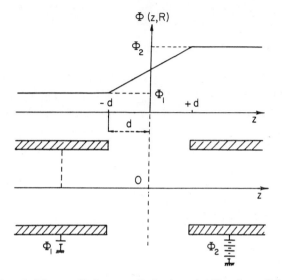

FIG. 4. More realistic two-cylinder lens (width of gap $2d$).

In this case, complete calculation gives us

$$\Phi = \frac{\Phi_1 + \Phi_2}{2} + \frac{\Phi_2 - \Phi_1}{\pi} \int_0^\infty \frac{I_0(kr)}{kI_0(kR)} \frac{\sin kd}{kd} \sin kz \, dk \qquad (1.35)$$

as the reader can verify. He may also also attempt to use the second method,
which will result in a series. Finally, numerical calculation will convince
the reader that the function

$$\varphi = \frac{\Phi_1 + \Phi_2}{2} + \frac{\Phi_2 - \Phi_1}{4\omega d} \log \frac{\cosh \omega(z + d)}{\cosh \omega(z - d)} \qquad (1.36)$$

supplies, by a fortunate coincidence, a really excellent approximation of the
axial potential in this type of electrode.

Evidently, if we are less concerned with rigorous precision as in the earlier
examples, and if we are looking merely for a suitable approximation, we

can use the method of superimposing the solutions derived from the separation of variables with much greater freedom. Let us give one example to terminate the discussion. The equation stating the potential created by a circular hole of radius R in a metal plate at potential Φ_0 separating two uniform field regions E_1 an E_2 can be precisely demonstrated (the fields E_1 and E_2 are perpendicular to the plate). We find

$$\Phi(r, z) = \Phi_0 - \frac{E_1 + E_2}{2} z + \frac{1}{\pi} (E_1 - E_2) |z| \left(\arctan \frac{a}{R} + \frac{R}{a} \right)$$

$$a = \{\tfrac{1}{2}(z^2 + r^2 - R^2) + \tfrac{1}{2}[4R^2z^2 + (z^2 + r^2 - R^2)^2]^{1/2}\}^{1/2}$$

(1.37)

If we add three functions of this type, with correct offset of the abscissa z, and a suitable radius R, we obtain a harmonic function "roughly" representing the potential produced by three flat circular plates, that is to say, by the electrodes of a three-diaphragm electron lens. For the field values, we take those between the electrodes, distant from the axis. Evidently, the method is not stringently precise, and boundary conditions are scarcely satisfied, but the experiment shows nonetheless that the axial potential is fairly well represented. An article by Regenstreif (1951) gives details concerning this type of approximation.

E. Conformal Mapping Method

This method is solely applicable to problems involving two variables, x and y. If an analytical function is available,

$$Z = f(z) = X(x, y) + jY(x, y) \tag{1.38}$$

in which functions $X(x, y)$ and $Y(x, y)$ are harmonic functions identifiable with a potential distribution. It is thus possible to compile an atlas of potential distribution charts for all the analytical functions that can be supposed. Then, one of the charts may be found to supply the solution to the problem that is being studied (a matter of pure luck), that is to say, in the fortunate case when the charted equipotentials happen to be those of the electrodes studied.

It is clear that everything would be very simple if we knew how to construct the function whose chart suits our requirements. It is also clear that this is generally a matter of integrating the Laplace equation. The reader will find indications in the bibliography (1) concerning the Schwarz method, which solves the problem in a particular case. We shall limit ourselves to a few examples.

First consider the function

$$z = k(Z + e^Z) \qquad (1.39)$$

which implicitly defines Z as a function of z. Now let us extract the real and complex parts. We get

$$x + jy = kX + jkY + k\,e^X(\cos Y + j\sin Y) \qquad (1.40)$$

$$x = kX + ke^X \cos Y \qquad y = kY + ke^X \sin Y \qquad (1.41)$$

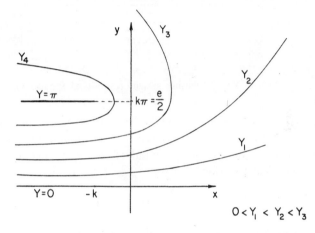

FIG. 5. An example of conformal mapping. Equipotential lines of a flat capacitor.

Figure 5 shows the lines $Y = $ const, which are the most interesting, for our example. We have evidently encountered the "edge effect" problem of the thin electrode flat capacitor. The distance between the electrodes e allows us to determine the coefficient k. Thus

$$k = e/2\pi \qquad (1.42)$$

Take, as a further example, a beam of particles of velocity v parallel with Ox. The beam is infinite at the Oy negative side, as also along Oz, perpendicular to the plane of the figure (Fig. 6). The potential distribution along such a beam is known. If it is assumed zero at the cathode, then the potential is V at the anode, at a distance L, and varies according to

$$\Phi = V(x/L)^{4/3} \qquad (1.43)$$

The equipotential surfaces are planes perpendicular to Ox. The field is

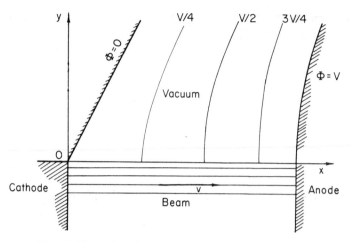

FIG. 6. Pierce focusing system (cylindrical symmetry).

merely a component, following Ox. The beam will be confined if we can fill the upper space ($y > 0$) with an electrical field connecting with boundary conditions suiting the internal potential of the beam. Let us try to find a suitable analytical function. We try

$$Z = kz^{4/3} \qquad (Z = X + jY) \qquad (1.44)$$

which may be broken down into real and complex parts, X signifying the potential. For $y = 0$, the two potentials connect, subject to choosing

$$k = V/L^{4/3} \qquad (1.45)$$

The potential, which is the real part of Z, is then

$$\Phi = (V/L^{4/3}) (x^2 + y^2)^{2/3} \cos[\tfrac{4}{3} \arctan (y/x)] \qquad (1.46)$$

Calculation shows that the electrical fields meet perfectly on either side of the plane $y = 0$. It is then possible to construct the equipotentials, and more particularly $\Phi = 0$, which will be the electrode extending the cathode, and $\Phi = V$, which extends the grid acting as anode. Figure 6 shows their form. All this brings us to the Pierce gun (which the figure shows) and which is capable of delivering a well-focused, intense beam. There is a variant possessing axial symmetry, which is discussed by Pierce (1941).

We conclude by mentioning that the "conformal mapping" resulting from the functions,

$$Z = kz^n \qquad (1.47)$$

make it possible to study multipole fields of the order $2n$. In particular, $n = 2$ gives us the quadrupole lenses with the potential

$$\Phi = k(x^2 - y^2) \tag{1.48}$$

In conclusion, when faced with the calculation of an electrostatic field, separation of the variables should be attempted if the conditions of validity are satisfied. Otherwise, analogue and numerical analysis methods (as discussed in Chapter 1.2) should be adopted. If the problem is a matter of plane geometry, a look through a "conformal-mapping"-chart atlas (Kober, 1952) may be worthwhile, before resigning oneself to the adoption of these methods.

1.1.2. The Magnetic Field

A. MAGNETIC FOCUSING STRUCTURES

Although certain analogies may be found to exist between electrical and magnetic systems (analogies from which we shall profit later), it is of importance to recognize the substantial differences between the two types of assemblies.

(a)   The "iron-free coils," the first means of obtaining a desired magnetic field, are crossed by currents whose distribution is known. Whereas it has never been possible to use the integral method in electrostatics (based on the Coulomb law), owing to lack of knowledge concerning the location of the charges, the same method (based on the Biot-Savart law) will certainly become practice in "iron-free" magnetism.

(b)   In electrostatics, the dielectrics holding the electrodes (quite indispensable) have to be concealed from the beam. It has never been possible to use their dielectric properties to produce the field. In magnetism there are highly permeable bodies that are conductors, and that can therefore be located close to the beam without danger (if badly focused charges strike them, they flow to ground without disturbing the field). We can therefore profit from the interesting properties of ferromagnetic bodies in the technology of magnetic focusing structures.

(c)   The measurement of electrical fields is very difficult. On the contrary, magnetic measurements are relatively easy. If, therefore, the theoretical methods we shall review involve excessively long calculations, there should be no hesitation in producing a model and simply taking measurements. There is little point in seeking to establish sophisticated calculation processes or complex analogue methods in the case of magnetic structures.

(d)   Finally, and except in the relatively rare case of permanent magnets, magnetic assemblies consume power. Before undertaking any focusing project, it should therefore first be ascertained whether or not the necessary power can be supplied in the form of electricity and evacuated as heat.

B. FIELDS OF IRON-FREE COILS

The Biot and Savart equation is used in the case of iron-free magnetic assemblies. It obtains the magnetic induction **B** at all points in space. We write

$$\mathbf{B} = \frac{\mu_0}{4\pi} I \oint \frac{d\mathbf{s} \wedge \mathbf{r}}{r^3} \tag{2.1}$$

Figure 7 illustrates the parameters. In MKSA units $\mu_0 = 12.56 \times 10^{-7}$ B is measured in teslas ($= 10,000$ gauss). It is always possible to obtain magnetic fields by a numerical computation of integral (2.1).

It will be remembered that the very long solenoid produces an internal

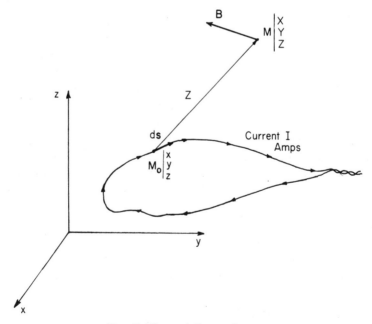

FIG. 7. Biot and Savart theorem.

magnetic induction which is substantially uniform, parallel to the axis, and of the value

$$B = \mu_0 n I \tag{2.2}$$

where n is the linear density of the turns carrying intensity I. If the solenoid is of finite length, it can be shown that axial induction is equal to

$$B = \tfrac{1}{2} \mu_0 n I (\cos \theta_2 - \cos \theta_1) \tag{2.3}$$

θ_2 and θ_1 being the aperture angles of the cones, whose summits are at the point studied, and whose bases rest on the initial and final turns of the coil. A rectilinear wire of length L produces the magnetic induction whose direction is shown in Fig. 8 at point M. Figure 8 also illustrates the parameters.

$$B = \frac{\mu_0 I}{2\pi a} (\sin \alpha_2 - \sin \alpha_1) \tag{2.4}$$

The foregoing equation permits calculation of the fields produced by circuits consisting of straight sections, for instance, in the case of iron-free quadrupole lenses.

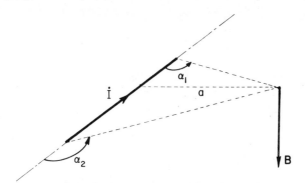

FIG. 8. Magnetic field produced by a rectilinear wire.

A circular turn of radius R, in which a current of intensity I flows, creates an induction at its axis of

$$B = \frac{\mu_0 I}{2\,R[1 + (z/R)^2]^{3/2}} \tag{2.5}$$

where z is the axial distance measured from the center of the turn. For the points not on the axis, the field is expressed by using elliptical integrals. The section of the conductors is not always negligible, as assumed by the preceding equations. Complete calculation, which consists of dividing the cross section into filiform elements and summating raises no difficulties of principle, but usually results in complicated formulas, which are difficult to use. Usually filiform coils are preferred, which produce the same field in the effective area and tolerate a degree of approximation. For example, Lyle (1902) demonstrated that the field produced at its axis by a circular coil of square section is substantially the same as that produced by a coil of filiform section, in the axis of symmetry of the system and of the radius:

$$R = R_0[1 + (a^2/24R_0^2)] \tag{2.6}$$

Figure 9 explains the notations. It also indicates a second case of equivalence, where the image of the winding obtained with two coils, also in the symmetrical plane of the assembly, and traveled by currents equal to $I/2$. Radii are $R + e$ and $R - e$, with

$$R = R_0 \left[1 + \frac{a^2}{24 R_0^2} \right] \qquad e^2 = \frac{b^2 - a^2}{12} \tag{2.7}$$

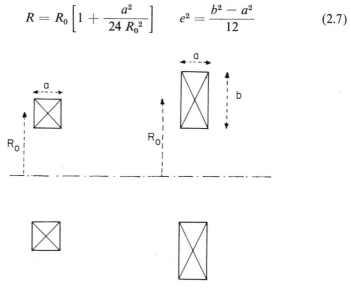

FIG. 9. Iron-free coils of square or rectangular cross section.

Iron-free windings do not produce very intense fields, unless superconductors are used, thereby eliminating the difficulties due to heat evacuation. Fields of about 100,000 gauss are possible, but in using them in focusing particles we encounter difficult problems of technology. In fact, with normal conductors, iron-free coils can be used if the field is not expected to be greater than a few hundred gauss in continuous operation. If the current density is about 3 A/mm², radiation cooling is sufficient. Beyond this density, forced cooling becomes necessary.

C. MAGNETIC CIRCUIT OF AN IRON-CORE COIL

Figure 10 is a general illustration of an iron-core circuit with flat, parallel poles. Calculation of the fields in the air gap is possible, allowing some approximation. The fundamental laws governing the magnetic field are described by the Ampere and Gauss theorems. We shall express the first in its integral form

$$\oint \mathbf{H} \cdot d\mathbf{s} = I \tag{2.8}$$

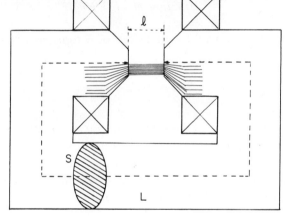

FIG. 10. Magnetic circuit with iron: $\hbar$ is the gap width; L is the mean length of the flux line in iron; S is the cross-sectional area of the yoke; $B_1 B_2$ are the coils ($N/2$ turns in each coil).

where I is the total intensity flowing through the closed path along which the integral is calculated (conventional orientation, allows us to use the terms "positive" and "negative" intensity). We express the Gauss magnetic law in its differential form,

$$\boldsymbol{\nabla} \cdot \mathbf{B} = 0 \tag{2.9}$$

recalling that it expresses the flux of $\mathbf{B}$ from a closed surface as zero. Care must be taken not to confuse $\mathbf{B}$ and $\mathbf{H}$, which are no longer proportional, as in the case of the iron-free coils, but which are now related by a complicated law dependent on the material, and frequently on the history of the occurrences. As a rule, it is necessary to know the initial magnetization curve, the hysteresis loop from saturation, and the demagnetization curve. The manufacturers of magnetic materials supply these data.

We shall consider a ferromagnetic circuit of average length L and section S. The air gap of area s and length l is connected to the core by truncated-cone poles. We shall assume that the lines of force are totally guided in the air gap (no leakage), and that the field and magnetic induction are practically uniform. B_f and H_f are their value in the core and B and H, their value in the air. The Gauss theorem, which expresses the conversation of the flux of B, demands

$$B_f \cdot S = B \cdot s \qquad (2.10)$$

and the Ampere law

$$H_f \cdot L + H \cdot l = NI \qquad (2.11)$$

if I is the intensity flowing through the coil of N turns providing the magnetization. We obtain

$$H_f L + \frac{B}{\mu_0} l = H_f L + \frac{B_f S}{\mu_0 s} l = NI \qquad (2.12)$$

the relation of B_f to H_f. But these quantities are also related by the magnetization curve; graphical solution of the problem is evident (Fig. 11). Our theory is valid only if the field remains weak; thus the core permeability remains high. Otherwise, the lines of force escape the core in the truncated-cone region, which is contrary to our assumption.

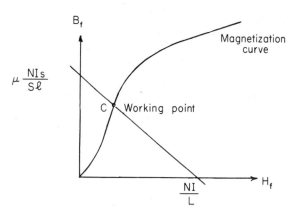

FIG. 11. Determination of the working point by graphical method, for iron-core circuit.

Let us now assume that the ferromagnetic material possesses hysteresis. As the current diminishes, the operating point descends the cycle (Fig. 12), and for zero current, we obtain the operating point D, where the field in the core is "demagnetizing." We have produced a permanent magnet. As the product of Eqs. (2.10) and (2.11) in this case (where $I = 0$) we obtain

$$| B_f \cdot H_f \cdot L \cdot S | = | B \cdot H \cdot s \cdot l | = \mu_0 H^2 v \qquad (2.13)$$

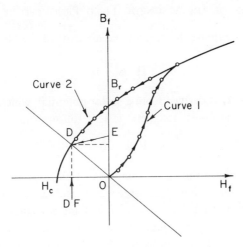

Fig. 12. Graphical method, for a permanent magnet. Curve 1 shows the displacement of the working point with increasing magnetizing current I. Curve 2 shows the displacement of the working point with decreasing magnetizing current. D is the working point when $I = 0$. $D.F.$ is the demagnetizing field; B_r is the remanent induction and H_c is the coercitive field.

If therefore it is desired to obtain a given field in an air-gap of given volume v, the product $B_f \cdot H_f$ must be maximum, for a minimum core volume $S \cdot L$. This is the optimum point of operation indicated by manufacturers. When the magnetic circuit is closed with a piece of soft iron, l is reduced practically to zero. Although the line DO steepens (since its slope is $\mu_0 sL/Sl$), it must not be thought that the operating point returns to the initial cycle. Because of hysteresis, the material follows a new, almost rectilinear curve, referred to as the "return line," as far as point E. Induction is the less than B_f. When the circuit is open, the point of operation returns to D if the magnet is well designed, since the "go" and "return" in such a small cycle are practically the same.

This method makes it possible to determine the circulation of the magnetic field between the poles ($B \cdot l/\mu_0 = Hl$), which we shall use. If the permeability of the core were infinite, $H_f = 0$ (since B_f is finite) and we would have

$$H \cdot l = \frac{B}{\mu_0} \cdot l = NI \tag{2.14}$$

In fact, the "field circulation" is slightly less, and we have discovered a means of estimating this correction.

D. CALCULATION OF THE MAGNETIC FIELD PRODUCED BY THE POLES

We shall now examine the case of the air gap. In the absence of current, we have

$$\oint \mathbf{H} \ ds = 0 \tag{2.15}$$

which shows that **H** is derived from a scalar potential, which we shall call ψ. On the basis of the Gauss magnetic law, it follows that

$$\nabla^2 \psi = 0 \tag{2.16}$$

The scalar potential is thus a harmonic function, and the calculation processes described in Section 1.1.1. are valid. But we must not forget the "multiform" nature of ψ. If we leave the point considered and after long travel return, conventional calculation will show that we will not generally find the same value of ψ, unless the path followed does not surround a current. If it does, ψ becomes $\psi + nI$, n being the number of times that the path taken goes round the conductor crossed by current I. This is an immediate consequence of the Ampere law.

The "multiform" character of ψ is not a serious matter, since only the gradient of ψ has a physical sense. What is more serious, is the fact that ψ cannot be defined other than in certain regions of space, namely those containing current. This raises very delicate problems in the logical definition of boundary conditions, since these conditions must be known, if Eq. (2.16) is to be integrated. A reasonable assumption is to regard the poles as "magnetic equipotentials." The assumption is justified if the permeability of the core is high, the flow of H_f between one point of the pole and the neighboring point is very slight, H_f being small and the distance always short. Then, boundary conditions being known, the problem can be solved by the method discussed in Section 1.1.1, or the analogue methods mentioned in Chapter 1.3. The following is an example.

Figure 13 is a cross section of a magnetic-lens circuit, symmetrical in revolution around Oz. We assume that the lines of force guided by the core pass through the lateral wall and come near the axis, as shown in the figure.

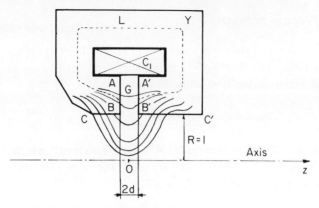

FIG. 13. Magnetic lens with iron yoke: C_1 is the coil (N turns, $N\,I$ A.t.); L is the mean length of the iron; G is the gap of width $2d$; R is taken as unit of length.

Let us now attempt to calculate the scalar potential ψ near the air gap, with the assumption that the core is highly permeable. On the surface ABC the potential value will be ψ_1, and on the surface $A'B'C'$ it will be ψ_2, The difference $\psi_2 - \psi_1$ represents the flow of H in the air, between the two electrodes. By virtue of the Ampere law, the difference is also

$$\psi_2 - \psi_1 = NI - H_f L \qquad (2.17)$$

In Section 1.1.1,C we discussed a process for obtaining $H_f L$, a slight correction if the material remains far from saturation.

It is then sufficient to compare this problem with the one studied in Section 1.1.1,D to recognize their identity of form; the formula (1.35), or better, its approximate expression (1.36) states the value of the potential. In fact, it is the axial field which is of principal interest. The induction, obtained by simple derivation is

$$B = \frac{\psi_2 - \psi_1}{4d} \; \frac{\sinh 2\omega d}{\cosh \omega(z + d) \cdot \cosh \omega(z - d)} \qquad (2.18)$$

Experimental measurements are in "relatively good agreement" with this type of equation. However, there should be no illusions as to the possibilities of this type of theory, based on approximation. In effect, they are good solely for the drafting of the design of a coil with a core and for definition of its main characteristics. But if the exact distribution of the field is to be known, particularly in the case of strong fields, where core saturation begins to be felt, the coil must be produced and measurements made (see Chapter 1.4).

E. The Vector Potential

This is a form of representation that is of little interest for the calculation of static fields, but that is indispensable for the calculation of paths. The Gauss theorem makes it possible to pose

$$\mathbf{B} = \boldsymbol{\nabla} \wedge \mathbf{A} \tag{2.19}$$

where $\mathbf{A}$ is a field of vectors, the "vector potential" which is evidently defined within a gradient field. We add the complementary condition, referred to as "Maxwell's condition,"

$$\boldsymbol{\nabla} \cdot \mathbf{A} = 0 \tag{2.20}$$

and the Ampere theorem leads us to the equation

$$\boldsymbol{\nabla}^2 \mathbf{A} = -\mu_0 \mathbf{i} \tag{2.21}$$

in which $\mathbf{i}$ is the current-density vector. Let us retain simply the relation between $\mathbf{A}$ and $\mathbf{B}$ (2.19) and the fact that $\mathbf{A}$ is a harmonic function in a vacuum.

1.1.3. Determination of Trajectory

A. Relativistic-Corpuscular Dynamics

A particle of mass m is characterized by its three Cartesian rectangular coordinates x, y, z. The velocity vector $\mathbf{v}$ whose components are $\dot{x}$, $\dot{y}$, $\dot{z}$ obeys the fundamental relativistic dynamics law:

$$\frac{d}{dt} \left[\frac{m\mathbf{v}}{[1 - (v/c)^2]^{1/2}} \right] = \mathbf{F} \tag{3.1}$$

$\mathbf{F}$ is the force on the particle and c is a constant that is equal to 2.99×10^8 meters/s.

The force $\mathbf{F}$ is created by the electrical field $\mathbf{E}$ and magnetic field $\mathbf{B}$, in which the particle moves. If Φ and $\mathbf{A}$ are the scalar and vector potentials, we may write either

$$\mathbf{F} = e[\mathbf{E} + (\mathbf{v} \wedge \mathbf{B})] \tag{3.2}$$

$$\mathbf{F} = e[-\boldsymbol{\nabla} \Phi + \mathbf{v} \wedge (\boldsymbol{\nabla} \wedge \mathbf{A})] \tag{3.3}$$

The second expression carries an advantage. By simple transformation we may write

$$\mathbf{F} = - d(e\mathbf{A})/dt - \boldsymbol{\nabla}\,(e\Phi - e\mathbf{v} \cdot \mathbf{A}) \tag{3.4}$$

This gives us another equivalent form of the fundamental law of dynamics

$$\frac{d}{dt}\left\{\frac{m\mathbf{v}}{[1 - (v/c)^2]^{1/2}} + e\mathbf{A}\right\} = - \boldsymbol{\nabla}\,(e\Phi - e\mathbf{v} \cdot \mathbf{A}) \tag{3.5}$$

The foregoing formulas are evidently valid in the case of static fields, but it may be pointed out that they remain applicable in the general case in which Φ and $\mathbf{A}$ are dependent on time. In all cases, these equations must be completed with a term expressing "radiation damping." It is recognized that a particle of charge e, to which acceleration $\dot{\mathbf{v}}$ is imparted, radiates electromagnetic energy. Although velocity remains slight as compared with c, the loss of energy per second is

$$\frac{dW}{dt} = \frac{\mu_0 e^2}{6\pi c}\left(\frac{dv}{dt}\right)^2 \tag{3.6}$$

This loss of energy evidently corresponds to the braking of the particle, expressed by the introduction of a braking force in the equation of motion. The work of this force, per unit of time, is equal to the energy loss by radiation. This is, incidentally, not sufficient for determination of the force, and it is here that we encounter one of the most difficult problems in electromagnetism. Let us limit ourselves to expressing the force in the case in which $v \ll c$

$$F = \frac{\mu_0 e^2}{6\pi c}\,\frac{d^2 v}{dt^2} \tag{3.7}$$

The equation of motion thus completed is a difficult study; even in the simple cases where integration is mathematically possible, the physical consequences are sometimes difficult to elucidate. In fact, this reactive force is always neglected, and Eq. (3.5) simply integrated. On having completed calculation and knowing $\dot{\mathbf{v}}$, it is sometimes useful to compare the value of dW/dt with that of the work of the force due to the electromagnetic fields, in the same time interval. It is only in the case of circular particle accelerators that dW/dt is not negligible. Let us therefore set radiation damping aside and seek to integrate Eq. (3.5), a difficult job, if the problem is to be solved for the general case.

B. Kinetic-Energy Theorem

We multiply the two members of Eq. (3.1) by $\dot{v}$; after some transformations we obtain an integratable combination, which gives us

$$\frac{mc^2}{[1 - (v/c)^2]^{1/2}} + e\Phi = \text{const.} \tag{3.8}$$

A method frequently used by physicists, although a sometimes dangerous convention, is to choose the constant by taking advantage of the fact that the scalar potential is defined to within a constant. The origin of potential is thus chosen at the cathode, where the particles have zero velocity. In this case, we obtain

$$mc^2 \left[\frac{1}{[1 - (v/c)^2]^{1/2}} - 1 \right] = - e\Phi \tag{3.9}$$

By introducing kinetic energy in the first member, and potential energy in the second, total energy is zero if this convention is used. This equation shows that potential energy

$$W_p = - e\Phi \tag{3.10}$$

must be positive, with the foregoing convention. The positively charged particle cannot therefore travel elsewhere than in regions where Φ is negative (and vice versa).

In the case of electrons and protons, W_p expressed in electronvolts, is numerically equal to Φ. It is frequently of interest to compare W_p with the "rest energy" mc^2 of the corpuscle. It will be remembered that mc^2 is 512,000 eV for an electron and 938 MeV for a proton. We therefore pose

$$x = W_p/mc^2 \tag{3.11}$$

which leads us to

$$v = c\,(2x + x^2)^{1/2}/(1 + x) \tag{3.12}$$

Discussion of this formula is simple; if x is much smaller than 1, we have

$$v = (2W_p/m)^{1/2} \tag{3.13}$$

this is the case where Newtonian mechanics can be applied. If, however, x is of the order of 0.2 or 0.3, it is no longer possible to make the foregoing approximation, and general formulas (3.12) must be used. If x is very large, we observe that v remains practically equal to c, thus permitting certain approximations termed "ultrarelativistic."

C. SOME SIMPLE EXAMPLES OF TRAJECTORIES CALCULATION

Rigorous integration of Eq. (3.1) is exceptionally possible. Let us take a few conventional cases

Uniform Electrical Field E. If the particle is released without velocity at the abscissa $x = 0$, it moves parallel to the lines of force according to the time law

$$x = \frac{mc^2}{eE} \left\{ \left[1 + \left(\frac{eEt}{mc} \right)^2 \right]^{1/2} - 1 \right\} \qquad (3.14)$$

If the particle is released with a given velocity, the problem can also be accurately solved; the trajectory is then a catenary curve.

Uniform Magnetic Field. (See Fig. 14.) The Laplace magnetic force is perpendicular to the velocity. It is not able to modify the kinetic energy; thus the value of velocity is constant. Furthermore, as the trajectory is a circular helix with its axis parallel to the lines of force, velocity forms a constant angle θ with it. It can be shown without difficulty that the radius of the cylinder followed by the helix is

$$R = \frac{mv \sin \theta}{eB[1 - (v/c)^2]^{1/2}} \qquad (3.15)$$

whereas the pitch of the helix is

$$p = \frac{2\pi mv \cos \theta}{eB[1 - (v/c)^2]^{1/2}} = 2\pi R \cot \theta \qquad (3.16)$$

The case where $\theta = \pi/2$ is interesting; the path is then a circle, whose radius can be usefully related to the kinetic energy of the particle

$$R = (mc/eB)(2x + x^2)^{1/2} \qquad (3.17)$$

It is recalled that the quantity

$$p = \frac{mv}{[1 - (v/c)^2]^{1/2}} = mc(2x + x^2)^{1/2} \qquad (3.18)$$

is called the momentum and that we therefore have

$$R = p/eB \qquad (3.19)$$

It will be noted that in a uniform magnetic field, a positively charged particle turns in the reverse trigonometric direction (and vice versa) (Fig. 14).

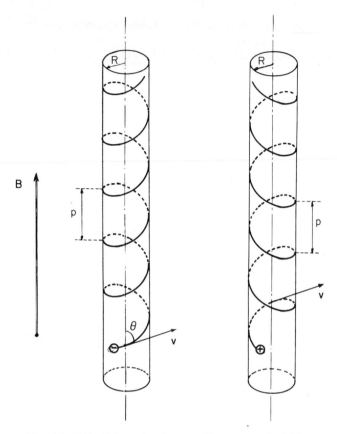

FIG. 14. Helicoidal motion in an uniform magnetic field.

Uniform and Perpendicular Electrical and Magnetic Field.   (See Fig. 15.)
If the problem is studied in a reference system that moves with the constant
velocity, then

$$\mathbf{v}_D = \frac{\mathbf{E} \wedge \mathbf{B}}{B^2} \tag{3.20}$$

which is termed the "drift velocity," we observe that the particle is submitted solely to a magnetic field; in this referential, the particle therefore
follows a circle whose radius can be calculated by using the result of the
foregoing paragraphs. Evidently, what interests us is motion appearing in
the local referential, whereby the Lorentz transformation alows us to pass
from the circular path in the "drifting" referential to the effective path,
in the local referential. Discussion is complicated (Gold, 1954) unless the
quotient E/B is small as compared with c, and the kinetic energy small as

compared with mc^2 (which is fortunately frequent in applications). In this case, the trajectory in the local referential is a "shortened" cycloid, and if the particle is released without velocity, it is even a true cycloid.

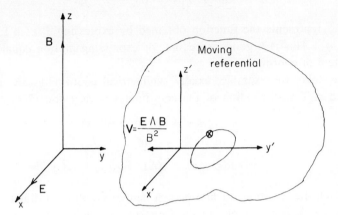

FIG. 15. Particle path in crossed electric and magnetic fields.

D. THE LAGRANGE FORMALISM

Let us return to Eq. (3.5) and project it on Ox: after some calculation, we arrive at

$$\frac{d}{dt}\frac{\partial}{\partial x}\left[-mc^2\left[1-\left(\frac{v}{c}\right)^2\right]^{1/2}+e(\mathbf{v}\cdot\mathbf{A})\right]-\frac{\partial}{\partial x}(-e\Phi+e\mathbf{v}\cdot\mathbf{A})=0 \tag{3.21}$$

which shows that motion equations are identical to the Euler-Lagrange equations applied to the Lagrangian

$$L=-mc^2[1-(v/c)^2]^{1/2}+e(\mathbf{v}\cdot\mathbf{A})-e\Phi \tag{3.22}$$

The equations of movement render "extremal":

$$I=\int_{t_0}^{t_1}L\,dt \tag{3.23}$$

It is known that the Lagrange equations are invariable in relation to punctual transformation of coordinates. We thus have the possibility of writing motion equations in any system of coordinates, q_1, q_2, q_3, if we know:

(a)   the punctual transformation relating x, y, z to q_1, q_2, q_3;

(b)   the expression of the square of the length of velocity, according to the new coordinates.

The motion equations are then

$$\frac{d}{dt}\left(\frac{\partial \mathscr{L}}{\partial \dot{q}_i}\right) - \frac{\partial \mathscr{L}}{\partial q_i} = 0 \tag{3.24}$$

where $\mathscr{L}$ represents the function obtained by expressing L as a function q_1, q_2, q_3. This is a good procedure for expressing motion equations in any system of coordinates.

Let us take, for example, axially symmetrical coordinates around Oz: r, θ, and z. Transformation of the coordinates is easy and gives us

$$\mathscr{L} = - mc^2 \left(1 - \frac{\dot{r}^2 + r^2\,\dot{\theta}^2 + \dot{z}^2}{c^2}\right)^{1/2}$$
$$- e\Phi(r, \theta, z) + e(\dot{r}A_r + r\dot{\theta}A_\theta + \dot{z}A_z) \tag{3.25}$$

from which the equations of motion can be derived without difficulty. The equation for θ, for example, in the frequent case in which $\mathbf{A}$ and Φ are independent of θ, will be as follows:

$$\frac{d}{dt}\left(\frac{mr^2\dot{\theta}}{[1 - (v/c)^2]^{1/2}} + erA_\theta\right) = 0 \tag{3.26}$$

which brings us immediately to the first-order equation termed the "Stormer integral":

$$\frac{mr^2\dot{\theta}}{[1 - (v/c)^2]^{1/2}} + erA_\theta = \text{const} = C \tag{3.27}$$

More generally, a first integral is obtained if one of the variables is not explicitly involved in $\mathscr{L}$. If times does not explicitly appear, we find the integral of the kinetic energy equation (3.9). Let us conclude with an application of the Stormer integral: it frequently occurs that in a region where the field is zero ($A_\theta = 0$), the path is directed towards the axis (it cuts the axis, or else its extension does so). In this case, C is zero, as can be easily ascertained, and $r = 0$ is a solution of the equation. Let us assume, finally, that there is no electrical field. The length of the velocity therefore remains constant. We thus have

$$\dot{r}^2 + r^2\dot{\theta}^2 + \dot{z}^2 = v^2 = \text{const} \tag{3.28}$$

and by replacing $\dot{\theta}$ with its value deduced from the Stormer integral:

$$\dot{r}^2 + \dot{z}^2 = v^2 - \frac{e^2 A_\theta{}^2}{m^2}\left[1 - \left(\frac{v}{c}\right)^2\right] \tag{3.29}$$

The first member is necessarily positive, the second is positive also, and we therefore obtain an interesting equation:

$$\frac{m^2 v^2}{e^2[1 - (v/c)^2]} > A_\theta^2(r, z) \qquad (3.30)$$

This equation determines the regions (r and z) accessible to the particle, certain regions being prohibited. Suitable processing of Eq. (3.29) allows us to determine the regions surrounding the earth into which cosmic radiations of a given energy will not penetrate, because of their repulsion by the earth's magnetic field (Alfven, 1953).

The equations expressing q_1, q_2, and q_3 as a function of t are frequently superfluous. It is sufficient to determine the trajectories and this can be done if the fields are not dependent on time, by taking one of the general coordinates a parameter, for example q_3. We then pose

$$\frac{dq_1}{dq_3} = \frac{\dot{q}_1}{\dot{q}_3} = q_1' \qquad \frac{dq_2}{dq_3} = \frac{\dot{q}_2}{\dot{q}_3} = q_2' \qquad (3.31)$$

The trajectory equations will then be obtained by slight modification of the Lagrange formula, taking into account the theorem of kinetic energy, which makes it possible to express v as the function of the parameters and their derivatives. The expression of dt may be derived from this, so that the time parameter can be removed from $\mathscr{L}$. The Lagrange equations therefore become the trajectory equations.

Let us illustrate this with an example; in the case of coordinates r, θ and z, we have

$$v^2 = \left(\frac{dz}{dt}\right)^2 (1 + r'^2 + r^2\theta'^2) = c^2 \frac{-\frac{2e\Phi}{mc^2} + \left(\frac{e\Phi}{mc^2}\right)^2}{\left(1 - \frac{e\Phi}{mc^2}\right)^2} \qquad (3.32)$$

This equation makes it possible to calculate dt as a function of the parameters their derivatives and dz; so that the integral of the Lagrangian is

$$\int_{z_0}^{z_1} \left\{ mc \left[-\frac{2e\Phi}{mc^2} + \left(\frac{e\Phi}{mc^2}\right)^2 \right]^{1/2} (1 + r'^2 + r^2\theta'^2)^{1/2} \right.$$
$$\left. + eA_z + er'A_r + er\theta'A_\theta \right\} dz \qquad (3.33)$$

The trajectory equations will be the Euler equations at θ and r of the Lagrangian in (3.33). Their solutions will be the functions $r(z)$ and $\theta(z)$. In detailing the foregoing calculations, it will be remembered that in the simple case with which we are concerned, the Lagrangian is defined to within a constant, and that certain simplifications result from this fact.

E. Phase-Space and the Liouville Theorem

We take the Lagrange function again and pose

$$p_i = \partial \mathscr{L} / \partial \dot{q}_i \tag{3.34}$$

thereby introducing three new parameters, the "conjugate momentum." It is generally possible to solve Eq. (3.34) so as to express $\dot{q}_1$, $\dot{q}_2$, $\dot{q}_3$ as functions of q_1, q_2, q_3 and of p_1, p_2, p_3. We thus obtain the H function, namely the Hamiltonian function defined by

$$H(p_i, q_i) = \sum_{i=1}^{3} p_i \cdot \dot{q}_i - \mathscr{L}(q_i, \dot{q}_i) \tag{3.35}$$

We then calculate dH in two different ways and obtain

$$dH = \sum_{i=1}^{3} \left(\frac{\partial H}{\partial q_i} dq_i + \frac{\partial H}{\partial p_i} dp_i \right) \tag{3.36}$$

$$dH = \sum_{i=1}^{3} (\dot{q}_i \, dp_i + p_i \, d\dot{q}_i) - \frac{\partial \mathscr{L}}{\partial q_i} dq_i - \frac{\partial \mathscr{L}}{\partial \dot{q}_i} d\dot{q}_i \tag{3.37}$$

Remembering the definition of p_i and the existence of the Lagrange equations, we may simplify and arrive at

$$\dot{p}_i = - \partial H / \partial q_i \qquad \dot{q}_i = \partial H / \partial p_i \tag{3.38}$$

a canonic form known as the "Hamilton equations." Here again, th se results are valid in the general case where $\mathscr{L}$, and consequently, H are dependent on time.

We are concerned with six equations of the first order that replace three equations of the second order. In the practical case of the calculation of paths, the value of the Hamiltonian equations is essentially in one of their consequences, the Liouville theorem, which we shall establish in closing. What is termed the "phase-space", is a space with six dimensions p_1, p_2, p_3, q_1, q_2, q_3, that is, assuming it to possess Euclidian metrics whose rectangular Cartesian coordinates are represented by the foregoing parameters. The quantity

$$dV = dq_1 \cdot dq_2 \cdot dq_3 \cdot dp_1 \cdot dp_2 \cdot dp_3 \tag{3.39}$$

therefore represents the "volume element" of this space; the Liouville theorem concerns the evolution of this element of volume. We shall demonstrate this in simple case of a single geometrical parameter q. The "phase-space" is then plane, possessing two coordinates p and q (Fig. 16).

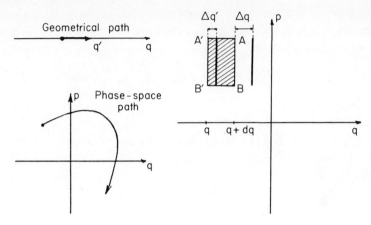

FIG. 16. Phase space and space of configuration in a simple case.

At instant t_0, we draw a small rectangle of area $dp \cdot dq$ and shall study its evolution in time. At the instant $t_0 + \Delta t$, the segment AB has moved by Δq. and if Δt is small, we obtain [refer to Eq. (3.38)]

$$(\Delta q)_{AB} = (\partial H/\partial p)\, \Delta t \qquad (3.40)$$

The opposite segment $A'B'$ has also moved with the difference that the derivative of H must be calculated for the initial position of $A'B'$, which is not that of AB. We finally obtain the width of the element at the end of time Δt:

$$(dq)_f = dq + (\Delta q)_{AB} - (\Delta q)_{A'B'} = dq\left(1 + \frac{\partial^2 H}{\partial q\, \partial p}\, \Delta t\right) \qquad (3.41)$$

The final width of the element is calculated similarly, obtaining

$$(dp)_f = dp\left(1 - \frac{\partial^2 H}{\partial q\, \partial p}\, \Delta t\right) \qquad (3.42)$$

so that final area is

$$(dq \cdot dp)_f = dq \cdot dp\left\{1 - \left(\frac{\partial^2 H}{\partial q\, \partial p}\right)^2 \Delta t^2\right\} \qquad (3.43)$$

At an approximation of the first order at Δt, the area remains unchanged. This is the Liouville theorem, which, in its general form, expresses the fact that the differential element $dp \cdot dq$ of the "phase-space" retains the same value when the points of the border limiting it move according to the laws of dynamics.

It is possible to establish this theorem also, in the general case of six parameters, q_1, q_2, q_3, p_1, p_2, p_3 ; the Liouville theorem is also expressed by an integral invariant:

$$\int\int\int\int\int\int dq_1\, dq_2\, dq_3\, dp_1\, dp_2\, dp_3 = \text{const} \tag{3.44}$$

We shall have also

$$\sum_i \sum_j \int\int\int\int dq_i\, dq_j\, dp_i\, dp_j = \text{const} \qquad (i \neq j) \tag{3.45}$$

and

$$\sum_i \int\int dq_i\, dp_i = \text{const} \tag{3.46}$$

These integrals will be evaluated on four-dimensional or two-dimensional surfaces of phase-space. This last invariant is very interesting; let us give another expression, after an integration with respect to p_i ; we shall have

$$\oint p_i\, dq_i = \text{const} \tag{3.47}$$

The integration path is now, in the geometrical space, a closed curve encircling all the trajectories. It is the "Poincare invariant." Equations (3.34) and (3.27) give

$$\mathbf{p} = \frac{m\mathbf{v}}{[1 - (v/c)^2]^{1/2}} + e\mathbf{A} \tag{3.48}$$

so that the Poincare-invariant is given by

$$\oint \frac{m\mathbf{v} \cdot d\mathbf{s}}{[1 - (v/c)^2]^{1/2}} + e\psi_B = \text{const} \tag{3.49}$$

where ψ_B is the magnetic flux across the closed curve that is used for integration.

In concluding, let us give the expression for the Hamiltonian; we use Equations (3.48), (3.35) and (3.27) to obtain

$$H = c[(\mathbf{p} - e\mathbf{A})^2 + m^2c^2]^{1/2} + e\Phi \tag{3.50}$$

An another expression for H, which is very interesting, is also

$$H = \frac{mc^2}{[1 - (v/c)^2]^{1/2}} + e\Phi \tag{3.51}$$

In the case of fields independent of time, H is equal to the sum of kinetic energy and potential energy; it is constant.

1.1.4. Particle Beams

A. PERTURBATIONS METHOD

Now, we shall study the perturbations method, which is the basis of corpuscular focusing. We shall assume that use of the methods described in the foregoing section, with the aid of powerful electronic computers, has defined the trajectory of a particle—for example, the particle leaving with a velocity v_0, the center of the source, perpendicular to the plane of the latter. In most cases, the other particles of interest are emitted from points on the source near the center, with velocities close to v_0, and following trajectories that are almost normal to the plane of the source. The definition of trajectories of these particles in relation to the trajectory of the "typical particle" and the use of the slight difference in parameters are therefore indicated, for employment of the perturbations method.

We shall completely study a simple example that is frequently encountered. Let us consider a magnetic and electrical focusing system in revolution symmetry around Oz. It is evident that in this system Oz is a possible trajectory. If, therefore, the source is near the axis of the system, and if velocities are at narrow angles to the axis (paraxial conditions), the case outlined above applies. We therefore take the axis Oz as the trajectory of the "typical particle," and establish the equations of motion. It is understood that r and $dr/dz = r'$ are small, and that we shall develop the terms in series as soon as possible.

In developing the electrical potential in series of powers of r, we must remember that the function Φ obeys the Laplace law. We write

$$\Phi(r, z) = \varphi(z) + a(z) \cdot r^2 + b(z) \cdot r^4 + \cdots \tag{4.1}$$

since only the even powers exist, owing to symmetry. On the basis of $\nabla^2 \Phi = 0$, at all orders of approximation, we obtain

$$\Phi(r, z) = \varphi(z) - \frac{1}{4} \varphi''(z) \cdot r^2 + \frac{1}{64} \varphi^{IV}(z) \cdot r^4 + \cdots \tag{4.2}$$

which evidences the importance of the axial potential $\varphi(z)$ and its derivatives. We shall limit development to the terms of the second order. The vector potential may be developed in the same way:

$$A_\theta = \frac{r}{2} B(z) - \frac{r^3}{16} B''(z) + \cdots \tag{4.3}$$

$B(z)$ being the magnetic induction at the axis. We shall now endeavor to find the trajectory taking z as a parameter, as explained in paragraph 1.1.3,D. Equation 3.33 gives us the Lagrangian; if we wish to obtain the first-order equations for r, we must develop r and r' to the second order inclusively. This we shall do; ignoring the terms of no interest (those which are not dependent on the parameters) and which will not appear in the Lagrange equation, we have

$$\mathscr{L} = \frac{1}{2} mc \left[-\frac{2e\varphi}{mc^2} + \left(\frac{e\varphi}{mc^2} \right)^2 \right]^{1/2} (r'^2 + r^2\theta'^2) + \frac{er^2}{2} B \cdot \theta'$$

$$+ \frac{1}{4} \frac{e}{c} r^2 \frac{\varphi'' \left(1 - \frac{e\varphi}{mc^2} \right)}{\left[-\frac{2e\varphi}{mc^2} + \left(\frac{e\varphi}{mc^2} \right)^2 \right]^{1/2}} + \cdots \tag{4.4}$$

The Lagrange formalism produces the path equations. We obtain

$$2mc \left[-\frac{2e\varphi}{mc^2} + \left(\frac{e\varphi}{mc^2} \right)^2 \right]^{1/2} r^2\theta' + \frac{er^2}{2} B = C \tag{4.5}$$

which is simply the Stormer integral; C is determined by initial conditions. We also obtain

$$\frac{d}{dz} \left\{ mc \left[-\frac{2e\varphi}{mc^2} + \left(\frac{e\varphi}{mc^2} \right)^2 \right]^{1/2} r' \right\} - \frac{1}{2} \frac{e}{c} \varphi'' \frac{1 - \frac{e\varphi}{mc^2}}{\left[-\frac{2e\varphi}{mc^2} + \left(\frac{e\varphi}{mc^2} \right)^2 \right]^{1/2}} r$$

$$- mc \left[-\frac{2e\varphi}{mc^2} + \left(\frac{e\varphi}{mc^2} \right)^2 \right]^{1/2} r\theta'^2 - e r B \theta' = 0 \tag{4.6}$$

We must now eliminate θ' between the two equations, in order to obtain the resolving equation, which is linear of the second order and whose solution will give us r. Let us now show quickly the transformations concluding calculation. We write

$$r = u \exp(-j\psi) \tag{4.7}$$

and function $u(z)$ is obtained by solution of the equation

$$\frac{d}{dz}\left\{\left[-\frac{2e\varphi}{mc^2} + \left(\frac{e\varphi}{mc^2}\right)^2\right]^{1/2}\frac{du}{dz}\right\}$$

$$+ \frac{e}{4mc^2\left[-\frac{2e\varphi}{mc^2} + \left(\frac{e\varphi}{mc^2}\right)^2\right]^{1/2}} \cdot \left[\frac{eB^2}{m} - 2\varphi''\left(1 - \frac{e\varphi}{mc^2}\right)\right]u = 0 \tag{4.8}$$

while ψ is obtained by the solution of

$$\frac{d\psi}{dz} = \frac{C\exp(2j\psi)}{u^2\left[-\frac{2e\varphi}{mc^2} + \left(\frac{e\varphi}{mc^2}\right)^2\right]^{1/2}} \tag{4.9}$$

Finally Eq. (4.5) now permits calculation of θ.

Of the numerous particular cases that exist, or are worth mentioning here, concerns the plane trajectories found only in electrical lenses. In this case, $\theta' = 0$ and $C = 0$. If energies are small in relation to mc^2, we obtain

$$r'' + \frac{\varphi'}{2\varphi}r' + \frac{\varphi''}{4\varphi}r = 0 \tag{4.10}$$

which shows that potential ratios only and not absolute potentials are involved. Equations obtained with first-order approximation are evidently linear. It is therefore sufficient to have two solutions in order to obtain the complete solution by linear combination. Furthermore, matrix formalism is widely used, although its study is beyond the scope of the present chapter. The reader is referred to the chapters dealing specially with it. The approximation may subsequently be taken to a higher order, retaining the fourth-power term of r and r' in development of the Lagrange equation, in the case of a system possessing revolution symmetry. If this is done, we obtain nonlinear equations permitting calculation of aberrations. The reader is referred to the chapters dealing with lenses,[1] and especially to Chapter 2.3.

We conclude this section with the fact that the method just discussed may be applied to a particle beam in the vicinity of any given reference path. The Frenet trihedral of the reference path (assumed perfectly known) is taken as referential. A point in space close to this path is determined as follows. From M, the normal plane of the trajectory is lowered to m. The arc of the curve at m is the basic parameter generalizing z in this section. The two other parameters are calculated in the normal plane, following the normal and the binormal (Fig. 17). Calculation is complex, but may be

[1] See Chapters 2.2, 2.3, 2.4, and 2.5.

completed to the various orders of approximation. In this connection, the reader is referred to the fundamental paper by Cotte (1938).

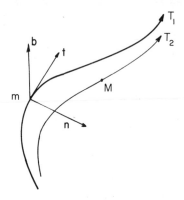

FIG. 17. Determination of a path near a "typical path": T_1 is a typical path; T_2 is another trajectory close to T_1.

B. THE LAGRANGE-HELMHOLTZ INVARIANT

We consider a small object, of area dS_0, perpendicular at the trajectory of "typical particle" (perpendicular to axis Oz, in the case of symmetry around Oz). p_0 is the momentum of typical particle v_0 its velocity, and we examine the phenomena during a very small interval of time dt. All the particles that travel across dS_0 during dt are situated in a small geometrical volume $dS_0 \cdot v_0 \cdot dt$; if the aperture of system is α_0 (suitable diaphragms admit only the particles whose angles of emission are within $+ \alpha_0$ and $- \alpha_0$), the component of $\mathbf{p}$ perpendicular to axis will be $p_0\alpha_0$; thus the small element of momentum space is $p_0^2\alpha_0^2 dp$; dp is an axial small variation of p_0. The volume element of phase-space is

$$dV = p_0^2\alpha_0^2 \, dS_0 \, v_0 \, dt \, dp_0 \qquad (4.11)$$

Some time ago, we examined these particles, during the same interval of time, dt. We now have an area dS, a momentum p, an aperture α, and a velocity v. But the volume element of phase-space is the same, according to the Liouville theorem (Section 1.1.3,E). We shall have

$$p_0^2\alpha_0^2 \, dS_0 \, v_0 \, dp_0 = p^2\alpha^2 \, dS \, v \, dp \qquad (4.12)$$

But we have also, along the trajectory of typical particle,

$$v_0 = dz/dt = \partial H/\partial p \qquad (4.13)$$

according to Hamilton's equation; thus $v \cdot dp$ is equal to the amount of H corresponding to the amount of p. In the case of static fields, energy is conserved, as well as the Hamiltonian. If initially, the particles had energy between W and $W + dW$, after a long time, the energy of the particles would always be in this interval. Consequently,

$$v \, dp = v_0 \, dp_0 \qquad (4.14)$$

and finally

$$p_0{}^2 a_0{}^2 dS_0 = p^2 a^2 dS \qquad (4.15)$$

If we take the expression of p, the momentum of typical particle, in function of φ, and the distribution of potential along this trajectory, we have (according to the kinetic energy theorem)

$$n_0 a_0 \, (dS_0)^{1/2} = na \, (dS)^{1/2} = K \qquad (4.16)$$

where the function n is

$$n = mc \left[-\frac{2e\varphi}{mc^2} + \left(\frac{e\varphi}{mc^2} \right)^2 \right]^{1/2} \qquad (4.17)$$

n is frequently termed, "index of refraction." The relation (4.16) is the Lagrange-Helmholtz theorem; it means that the product formed by multiplying the angular aperture a_0 by the width of beam $dS_0^{1/2}$ by the index is constant along the beam.

C. Space-Charge Complications

The results obtained above show that only beams with $K = 0$ can be made to converge at a point ($dS = 0$). These are called "isogenous beams"; only such beams may become parallel ($a = 0$). The phase-space-domain characteristic of these beams is a curve segment, of zero area. All of the foregoing is based on the assumption that the space charge is left out of consideration. It is worthwhile to end our study by examining the consequences of electromagnetic interaction between particles on beams focus.

Consider an almost-parallel beam of revolution symmetry, of radius R, and carrying current I with acceleration voltage Φ. The particles in the beam carry charge e and are of mass m. The electrical Coulomb force tends to make the beam divergent, whereas the magnetic force tends to make it converge, since two parallel currents in the same direction attract each other. We shall calculate the Newtonian mechanics only, for if we were to attempt to take relativistic effects into account, they would all have to be included, including the fact that the "signals" travel at a velocity c, and therefore take a certain time to cross the beam. We shall therefore content

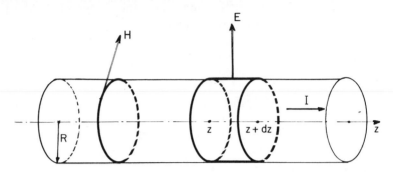

FIG. 18. Fields produced by space charge and electric current of a beam.

ourselves with the Newtonian approximation, which incidentally is very frequently enough. The Gauss theorem (Fig. 18) provides the radial electrical field exerted on the edge of the beam

$$E = \frac{I(m)^{1/2}}{2\,\pi\varepsilon_0(2e\,\Phi)^{1/2}} \frac{1}{R} \tag{4.18}$$

and the Ampere theorem applied to a circle of the same axis as the beam easily provides the magnetic field.

$$B = \frac{\mu_0 I}{2\pi R} \tag{4.19}$$

The force exerted on a particle at the edge of the beam and which tends to move it away from the axis is therefore

$$F = \frac{(me)^{1/2}}{2(2)^{1/2}\pi\varepsilon_0} \frac{I}{\Phi^{1/2}} \frac{1}{R} \tag{4.20}$$

since in Newtonian approximation, the force of magnetic origin is negligible, as can be ascertained by development of the calculation. The radial motion equation is then

$$m\frac{d^2R}{dt^2} = \frac{(me)^{1/2}}{2(2)^{1/2}\pi\varepsilon_0} \frac{I}{\Phi^{1/2}} \frac{1}{R} \tag{4.21}$$

if the beam is not submitted to any other focusing field.

We shall integrate in an interesting case. We assume that the beam of radius R_0 is slightly convergent at abscissa $z = 0$. We have $(dR/dz)_0 = -\alpha_0$; thus without space charge, the beam would focus at the distance R_0/α_0. In fact, the space charge will make it diverge. By integration, (4.21) gives us

$$\frac{m}{2}\left[\left(\frac{dR}{dt}\right)^2 - \left(\frac{dR}{dt}\right)_0^2\right] = \frac{(me)^{1/2}}{2(2)^{1/2}\pi\varepsilon_0}\frac{I}{\Phi^{1/2}}\log\frac{R}{R_0} \qquad (4.22)$$

But by virtue of the kinetic energy theorem, in paraxial approximation

$$\frac{dz}{dt} = \left(\frac{2e\Phi}{m}\right)^{1/2} \qquad (4.23)$$

The effective trajectory will therefore begin at abscissa $z = 0$, with the distance to the axis R_0 and the slope $-\alpha_0$. Then, the slope will diminish and disappear at the abscissa, where the beam will have a minimum radius; it subsequently diverges. Complete calculation is possible but we shall stop at calculation of the radius of the spot which replaces the theoretical punctual focus. At this point, the minimum $dR/dt = 0$. Equation (4.22) then gives us

$$\frac{m}{2}\left(\frac{dR}{dt}\right)_0^2 = e\Phi\alpha_0^2 = -\frac{(me)^{1/2}}{2\pi\varepsilon_0(2)^{1/2}}\frac{I}{\Phi^{1/2}}\log\frac{R}{R_0} \qquad (4.24)$$

The minimum radius of the other spot is thus

$$R_m = R_0 \exp\left[-2\pi\varepsilon_0\left(\frac{2e}{m}\right)^{1/2}\frac{\Phi^{3/2}}{I}\right]\alpha_0^2 \qquad (4.25)$$

Complete calculation shows that other abscissa of the minimum radius spot differs only slightly from that of the theoretical focus. Evidently, all this is of value only if the beam is intense and not very fast, that is to say, if its "perveance" is great, this term designating the quantity

$$P = I/\Phi^{3/2} \qquad (4.26)$$

If the beam has low perveance, R_m is small, and it is generally the aberrations of the focusing systems that determine the minimum spot dimensions. The reader will find the orders of magnitude of the aberrations in the chapters concerning the various types of lenses, especially in Chapter 4.2 of Volume II.

REFERENCES

In the text, we have indicated some examples of calculations; more development on these will be found in the particular references (cited in text):

Alfven, H. (1953). "Cosmical Electrodynamics" Chap. II. Oxford Univ. Press London and New York.
Bertram, S. (1940). *Proc. IRE* **28** 418.
Bertram, S. (1942). *J. Appl. Phys.* **13**, 496.
Cotte, M. (1938). *Ann. Phys. (Paris)* [11] **10**, 333.
Gold, L. (1954). *J. Appl. Phys.* **25**, 683.
Kober, H. (1952). "Dictionary of Conformal Representation." Dover, New York.
Lyle, T. R. (1902). *Phil. Mag.* [6] **3**, 310 .
Pierce, J. R. (1941). *Proc. IRE* **29**, 28.
Regenstreiff, E. (1951). *Ann. Radioélec. Compagn. Franc. Assoc. T.S.F.* **6**, 51, 164, 244, and 299.

BIBLIOGRAPHY

This paper is a general introduction for the calculation of particles beams. We give below some general references to books for more detailed explanations.

1. For calculation of electrostatic or magnetostatic fields:

Buchholz, H. (1957). "Elektrische und magnetische Potentialfelder." Springer, Berlin.
Courant, R. and Hilbert, D. (1962). "Methods of Mathematical Physics." Wiley (Interscience), New York.
Durand, E. (1964). "Electrostatique et Magnétostatique." Masson, Paris.
Kober, H. (1952). "Dictionary of Conformal Representation." Dover, New York.
Landau, L. D., and Lifshitz, E. M. (1962). "The Classical Theory of Fields." Pergamon, Oxford.
Morse, P.M., and Feshbach, H. (1953). "Methods of Theoretical Physics." McGraw-Hill, New York.
Parker, R. J., and Studders, R. S. (1958). "Permanent Magnets and Applications." Wiley, New York.

2. For dynamics of particles:

Corben, H. C. and Stehle, P. (1960). "Classical Mechanics." Wiley, New York.
De Broglie, L. (1950). "Optique électronique et corpusculaire." Hermann, Paris.
Glaser, W. (1952). "Grundlagen des Elektronenoptik." Springer, Berlin.
Goldstein, H. (1959). "Classical Mechanics." Addison-Wesley, Reading, Massachussetts.
Landau, L. D., and Lifshitz, E. M. (1961). "Mechanics." Macmillan (Pergamon), New York.
Pierce, J. R. (1949). "Theory and Design of Electron Beams." Van Nostrand, Princeton, New Jersey.
Sturrock, P.A. (1955). "Static and Dynamic Electron Optics." Cambridge Univ. Press, London and New York.
von Ardenne, M. (1955). "Tabellen für Elektronen und Ionenphysik." VEB Deutscher der Wissenschaften, Berlin.
Whittaker, E. (1952). "A Treatise on the Analytical Dynamics." Cambridge Univ. Press, London and New York.

CHAPTER 1.2

NUMERICAL SOLUTION OF LAPLACE'S AND POISSON'S EQUATIONS AND THE CALCULATION OF ELECTRON TRAJECTORIES AND ELECTRON BEAMS

C. Weber

PHILIPS RESEARCH LABORATORIES N. V. PHILIPS' GLOEILAMPENFABRIEKEN
EINDHOVEN, THE NETHERLANDS

1.2.1. Introduction

We shall deal with the numerical calculation of electron-optical systems using a digital computer. We confine ourselves to electrostatic systems, although the method can be extended to systems with magnetic fields. The examples used concern systems with rotational symmetry, but many formulas are generally valid or need only a slight modification in order to apply them to other than rotationally symmetrical systems.

In Section 1.2.2 we shall describe the numerical solution of Poisson's equation. In that section it is assumed that the space charge is known. The Laplace equation is considered as a special case of Poisson's equation, since with an electronic computer these equations are solved by the same method. In Section 1.2.3 the calculation of electron trajectories in a given potential field is treated. Section 1.2.4 deals with electron beams having thermal velocities. These are nonhomocentric beams, consisting of a great number of electron trajectories. The current density of such a beam cannot be determined in a simple way merely from a number of electron trajectories calculated according to Section 1.2.3. Therefore Section 1.2.4 is devoted to a method of calculating the current density and the space charge of beams of this kind. The method is explained in its application to a high-current electron gun, but it may also be applied to other systems, such as for example electron lenses or a field-free space. For the calculation of the current density and the space charge the potential field must be known. However, for the calculation of the potential field according to Section 1.2.2, the space charge must be known. Therefore we proceed iteratively. First we calculate the potential field with the space charge zero. Then the space charge is calculated in this field. With this space charge the potential is calculated again, then the space charge again, and so on, until a self-consistent solution is obtained. If one knows in advance that the space charge can be neglected, the current density can be calculated in the Laplace field and no iterations are required. Section 1.2.4 also deals with some typical difficulties which arise in the calculation of an electron gun.

1.2.2. The Solution of Laplace's and Poisson's Equations

A. DIFFERENCE EQUATIONS

The numerical solution of a differential equation is usually obtained in a more or less large, but always finite number of points in the region in which the solution is required. Therefore the differential equation must be transformed into a difference equation. Whereas a differential equation gives a relation between the derivatives in a point, the corresponding difference equation gives a relation between the potentials in adjacent points. These adjacent points correspond to points in which the solution is calculated. We can obtain various difference equations, depending upon the number of adjacent points used.

In this section we shall deal with the replacement of the differential equation by a difference equation. We shall derive various difference equations for Poisson's equation with rotational symmetry. However, the method used is very general and can also be applied to derive other difference equations, using other adjacent points and to derive difference equations corresponding to other differential equations. In Sections B–F we shall deal with the solution of a system of difference equations and in Section G we shall say something about the error that is made when we replace a differential equation by a difference equation.

Poisson's equation with rotational symmetry is given by

$$\frac{\partial^2 V}{\partial r^2} + \frac{1}{r}\frac{\partial V}{\partial r} + \frac{\partial^2 V}{\partial z^2} = -\frac{\varrho}{\varepsilon_0} \tag{1}$$

where the z axis is the axis of rotational symmetry, r is the distance to this axis, V is the potential, ϱ is the space charge, which is a given function of r and z, and ε_0 is the dielectric constant of free space. The Laplace equation is considered as a special case of Poisson's equation, since with an electronic computer these equations are solved by the same method.

In order to establish a relation between the potentials of points adjacent to a point P_0, the potential $V(r, z)$ is expanded into a Taylor series around P_0

$$V(r, z) = \sum_{i=0}^{\infty} \sum_{j=0}^{\infty} \frac{V_0^{(i,j)}}{i!\,j!} (r - r_0)^i (z - z_0)^j \tag{2}$$

where (r_0, z_0) are the coordinates of P_0 and

$$V_0^{(i,j)} = \left(\frac{\partial^{i+j} V}{\partial r^i\, \partial z^j} \right)_0$$

Since the potential satisfies (1), there exist relations between the derivatives $V_0^{(i,j)}$. These relations are obtained by differentiating (1) $i - 2$ times with respect to r and j times with respect to z:

$$V^{(i,j)} = -V^{(i-2,j+2)} + \sum_{k=1}^{i-1} \frac{(-1)^{i-k}(i-2)!}{(k-1)!\,r^{i-k}} V^{(k,j)} - \frac{\varrho^{(i-2,j)}}{\varepsilon_0} \tag{3}$$

where $i \geq 2$ and $j \geq 0$. At the left-hand side the potential is differentiated i times with respect to r and at the right-hand side at the most $i - 1$ times. Hence by repeated application of (3) $V^{(i,j)}$ can be expressed in a sum containing only derivatives of the type $V^{(0,m)}$ and $V^{(1,n)}$. This yields the relations

$$V^{(2,0)} = - V^{(0,2)} - \frac{1}{r} V^{(1,0)} - \frac{\varrho}{\varepsilon_0}$$

$$V^{(2,1)} = V^{(0,3)} - \frac{1}{r} V^{(1,1)} - \frac{\varrho^{(0,1)}}{\varepsilon_0} \qquad (4)$$

$$V^{(3,0)} = - V^{(1,2)} + \frac{1}{r} V^{(0,2)} + \frac{2}{r^2} V^{(1,0)} + \frac{\varrho}{\varepsilon_0 r} - \frac{\varrho^{(1,0)}}{\varepsilon_0}$$

and so on.

We now derive a five-point difference equation, which gives a relation between the potentials of P_0 and four neighboring points $P_1, \ldots, P_4$ (Fig. 1). For that purpose we truncate the series (2) after $i + j = 2$ and make use of the relations (4):

$$V(r, z) = V_0 + \left[(r - r_0) - \frac{(r - r_0)^2}{2r_0} \right] V_0^{(1,0)} + (z - z_0) V_0^{(0,1)}$$

$$+ (r - r_0)(z - z_0) V_0^{(1,1)} + [- \frac{1}{2}(r - r_0)^2 + \frac{1}{2}(z - z_0)^2] V_0^{(0,2)}$$

$$- \frac{1}{2}(r - r_0)^2 \frac{\varrho_0}{\varepsilon_0} \qquad (5)$$

In this case we needed only the first of the relations (4), which is the differential equation itself. If a difference equation is derived using more neighboring points, the series (2) must be truncated at a higher order and the other relations must also be used.

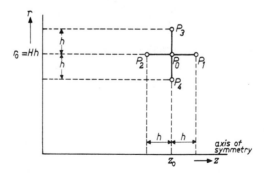

FIG. 1. The situation of the points $P_0, \ldots, P_4$ used for the five-point difference equation in a field with rotational symmetry.

Substitution in (5) of the points $P_1 \ldots, P_4$ and the appropriate potentials $V_1, \ldots, V_4$ yields

$$V_1 = V_0 + hV_0^{(0,1)} + \frac{1}{2} h^2 V_0^{(0,2)}$$

$$V_2 = V_0 - hV_0^{(0,1)} + \frac{1}{2} h^2 V_0^{(0,2)}$$

$$V_3 = V_0 + \left(h - \frac{h^2}{2r_0} \right) V_0^{(1,0)} - \frac{1}{2} h^2 V_0^{(0,2)} - \frac{1}{2} h^2 \frac{\varrho_0}{\varepsilon_0}$$

$$V_4 = V_0 + \left(- h - \frac{h^2}{2r_0} \right) V_0^{(1,0)} - \frac{1}{2} h^2 V_0^{(0,2)} - \frac{1}{2} h^2 \frac{\varrho_0}{\varepsilon_0}$$

(6)

By elimination of $V_0^{(1,0)}$, $V_0^{(0,1)}$ and $V_0^{(0,2)}$ we obtain finally the desired difference equation

$$V_0 = \frac{1}{4} V_1 + \frac{1}{4} V_2 + \left(\frac{1}{4} + \frac{1}{8H} \right) V_3 + \left(\frac{1}{4} - \frac{1}{8H} \right) V_4 + \frac{h^2 \varrho_0}{4\varepsilon_0} \quad (7)$$

where $H = r_0/h$.

At the axis of rotational symmetry $H = 0$ and (7) is not applicable. In this case we must take the limit when $r \rightarrow 0$ of the relations (4):

$$V_{\text{axis}}^{(2,0)} = - \frac{1}{2} V_{\text{axis}}^{(0,2)} - \frac{\varrho_{\text{axis}}}{2\varepsilon_0}$$

$$V_{\text{axis}}^{(2,1)} = - \frac{1}{2} V_{\text{axis}}^{(0,3)} - \frac{\varrho_{\text{axi}}^{(0,1)}}{2\varepsilon_0}$$

(8)

$$V_{\text{axis}}^{(3,0)} = 0$$

and so on.

With (8) we find for the difference equation at the axis (Fig. 2),

$$V_0 = \frac{1}{6} V_1 + \frac{1}{6} V_2 + \frac{2}{3} V_3 + \frac{h^2 \varrho_0}{6\varepsilon_0} \quad (9)$$

Following the same principle other difference equations can also be derived. If the distances of the points $P_1, \ldots, P_4$ to P_0 are arbitrary, $s_1, \ldots, s_4$ (Fig. 3), a difference equation is obtained which is given by Shortley et al. (1947). Also an equation can be derived using eight surrounding points according to Fig. 4 (Durand, 1957). This nine-point difference equation is more accurate than the five-point equation, because now the series (2) can be truncated after $i + j = 4$.

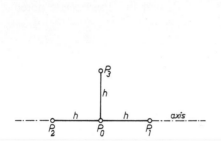

FIG. 2. The points $P_0, \ldots, P_3$ used for the difference equation at the axis of rotational symmetry.

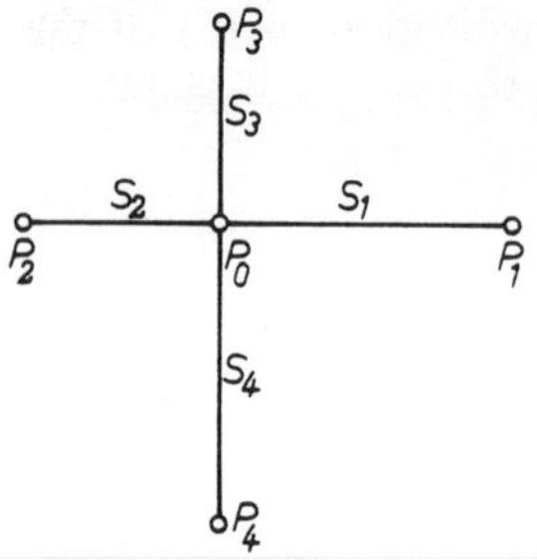

FIG. 3. The points $P_0, \ldots, P_4$ used for the five-point difference equation with unequal distances $s_1, \ldots, s_4$.

We can also derive a nine-point difference equation if the distances $s_1, \ldots, s_8$ are arbitrary (Fig. 5). Again the series (2) is truncated after $i + j = 4$ and we obtain eight expressions that are equivalent to (6). If the points $P_1, \ldots, P_8$ are situated with respect to P_0 as is given in Fig. 5, it turns out that these eight expressions do not include the derivative $V_0^{(1,3)}$. Unfortunately, the elimination of the other seven derivatives leads to a very complicated difference equation. Therefore it is advantageous in this case to perform the elimination with a computer when the numerical values of $s_1, \ldots, s_8$ are specified, instead of substituting the numerical values of $s_1, \ldots, s_8$ in the complicated difference equation.

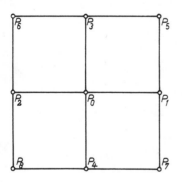

FIG. 4. The points $P_0, \ldots, P_8$ used for the nine-point difference equation.

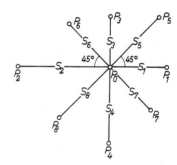

FIG. 5. The points $P_0, \ldots, P_8$ used for the nine-point difference equation with unequal distances $s_1, \ldots, s_8$.

All difference equations given above are obtained by truncating the series (2) and then using the relations (4). However, it is also possible to obtain other difference equations having the same order of accuracy by substituting first (4) into (2) and then truncating the series. After substitution of (4) the coefficient of $V_0^{(1,0)}$ is

$$(r - r_0) - \frac{(r - r_0)^2}{2r_0} + \frac{(r - r_0)^3}{3r_0^2} - \cdots = r_0 \ln \frac{r}{r_0} \tag{10}$$

Similar expressions can be obtained for the coefficients of the other terms. If the space charge does not change much over one mesh length, we obtain instead of (5), in this way

$$V(r, z) = V_0 + r_0 \ln \frac{r}{r_0} \cdot V_0^{(1,0)} + (z - z_0)V_0^{(0,1)} + (r - r_0)(z - z_0) \ln \frac{r}{r_0} \cdot V_0^{(1,1)}$$

$$+ \left[\frac{r_0^2}{2} \ln \frac{r}{r_0} - \frac{(r - r_0)r_0}{2} - \frac{1}{4} (r - r_0)^2 \right] \left(V_0^{(0,2)} + \frac{\varrho_0}{\varepsilon_0} \right)$$

$$+ \frac{1}{2} (z - z_0)^2 V_0^{(0,2)} \tag{11}$$

Some of the higher order terms, as for instance $(r - r_0)^3 V_0^{(1,0)}/3r_0^2$ in (10) are incorporated in (11) but not in (5). Because not all the higher order terms are incorporated in (11), the accuracy of (11) is the same as the accuracy of (5). Instead of (7) we now obtain the difference equation

$$V_0 = \frac{c_1}{c_0} \left(V_1 + V_2 + \frac{h^2 \varrho_0}{\varepsilon_0} \right) + \frac{c_3}{c_0} V_3 + \frac{c_4}{c_0} V_4 \tag{12}$$

where

$$c_0 = \frac{3}{2} \ln \frac{H + 1}{H - 1} + H \ln \frac{H^2}{H^2 - 1}$$

$$c_1 = \frac{1}{4} \ln \frac{H + 1}{H - 1} + \frac{1}{2} H \ln \frac{H^2}{H^2 - 1}$$

$$c_3 = \ln \frac{H}{H - 1}$$

$$c_4 = \ln \frac{H + 1}{H}$$

For large values of H (12) approaches to (7). One can also take linear combinations of (7) and (12). Then the error is also a linear combination of the errors of (7) and (12). One must check that these errors are not multiplied by large numbers, because this can decrease the accuracy considerably. An application of this kind of difference equation has been given by Weber (1963).

52   C. WEBER

B. The Iterative Procedure for the Solution of Poisson's Equation

We shall solve Poisson's equation in a region enclosed by the axis of rotational symmetry and a given boundary (Fig. 6). In order to solve the problem uniquely the potentials on the boundary must be given (the boundary conditions).

Inside the region we draw a network of square meshes. We denote by internal points the mesh points inside the boundary. Mesh points on the axis are also internal points; mesh points on the boundary itself are not. By boundary points we denote the points of intersection of the boundary with the mesh lines (see Fig. 6).

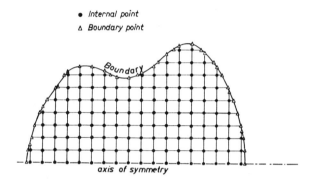

FIG. 6. The region in which the potential is calculated.

At each internal point a difference equation relates the potential of that point to the potentials of neighboring internal and boundary points. For example, we take the difference equation (7) for points that are not surrounded by boundary points, the difference equation (9) for points on the axis, and the difference equation associated with Fig. 3 (Shortley *et al.*, 1947) for internal points near the boundary. If there are N internal points, these difference equations can be considered as N linear equations with N unknowns. The unknowns are the potentials of the internal points. The fixed potentials of the boundary points are considered as constants.

Since N is a large number, it is not possible in practice to solve these equations directly. Therefore we solve the system of equations iteratively. We start with arbitrary potentials at the internal points. Generally, these potentials do not satisfy the system of difference equations. Next we calculate the potential at the first of the N internal points from its neighboring points with the difference equation. The arbitrary initial value at this first internal point is displaced by the calculated value, which satisfies the difference equation. Then the potential of the next internal point is calculated, and so on, until the potentials of all the N internal points have been calculated. Meanwhile, the first point does not satisfy the difference equation any longer, since the potentials of the neighboring points have been changed. Therefore we start again with another cycle, displacing successively the potentials of the N points by the potentials calculated from the difference equations. This cycle is followed by a third cycle, and so on. In the next section we shall prove that this procedure converges; with each cycle the potentials satisfy more closely the difference equations and after a certain number of cycles they satisfy the equations within the required accuracy. The final solution is entirely determined by the fixed boundary potentials and is independent of the arbitrary initial potentials at the internal points.

The displacement of the potentials of the previous iteration cycle by potentials that have just been calculated can be done in two ways: it is called successive displacement if it is done as described above, that is, with the new potentials displacing the potentials of the previous cycle as soon as they are available, so that they are used for the calculation of the subsequent potentials of the same cycle. It is called simultaneous displacement if we wait with the displacement until the end of the cycle, so that all the potentials of a cycle are calculated from potentials of the previous cycle. In an electronic computer simultaneous displacement requires more storage capacity, because we must store the potentials of the previous cycle as well as the potentials of the cycle in question. With successive displacement only one set of potentials must be stored.

We shall first study the simultaneous displacement, which is mathematically simpler and the result of which we shall need in Section D for the study of the successive displacement.

The N internal points are numbered $1, \ldots, n, \ldots, N$. The potential of the nth point is $V(n)$. We shall use the following general notation for the difference equation at the nth point:

$$V(n) = \sum_{k=1}^{N} a_n{}^k V(k) + a_n \tag{13}$$

Since a difference equation relates only potentials of geometrically adjacent points, most $a_n{}^k$ are equal to zero; only the $a_n{}^k$ relating to adjacent points are not equal to zero. Also $a_n{}^n$ is equal to zero. The space charge and the potentials of the boundary points are included in the constant a_n.

The potential of the point n after the mth cycle is denoted by $V^{(m)}(n)$.

With simultaneous displacement we obtain

$$V^{(m)}(n) = \sum_{k=1}^{N} a_n{}^k V^{(m-1)}(k) + a_n \tag{14}$$

If the relaxation procedure converges, there is a final solution $V^{(\infty)}(n)$ that satisfies

$$V^{(\infty)}(n) = \sum_{k=1}^{N} a_n{}^k V^{(\infty)}(k) + a_n \tag{15}$$

After the mth cycle the error $\varepsilon^{(m)}(n)$, appearing because the final solution of the system of difference equations has not yet been obtained, is given by

$$\varepsilon^{(m)}(n) = V^{(m)}(n) - V^{(\infty)}(n) \tag{16}$$

This error must not be confused with other kinds of errors, which are discussed in Section G. Subtraction of (15) from (14) yields the following equations for the error

$$\varepsilon^{(m)}(n) = \sum_{k=1}^{N} a_n{}^k \varepsilon^{(m-1)}(k) \tag{17}$$

By considering the error $\varepsilon^{(m)}(n)$ instead of the potential $V^{(m)}(n)$ we get rid of the term a_n, which contains the space charge and the potentials of the boundary points.

First we shall look for some special solutions of (17), namely, the non-trivial solutions that are multiplied by a constant factor λ after each cycle, λ being independent of m and n. These solutions $e(n)$ have N components ($1 \leq n \leq N$) that satisfy

$$\lambda e(n) = \sum_{k=1}^{N} a_n{}^k e(k) \tag{18}$$

This is a system of N linear homogeneous equations with N unknowns. It has only a nontrivial solution if its determinant is zero. This condition yields an equation of the Nth degree for λ that is usually called the characteristic equation. Each λ yields one or more solutions $e(n)$ that we shall distinguish by the index l, $e_l(n)$ is an eigenfunction or eigenvector of the $N \times N$ matrix $a_n{}^k$ and λ_l is an eigenvalue. It can be shown that there are N linearly independent eigenfunctions if the characteristic equation has N different solutions. If some of the solutions of the characteristic equation are equal to each other, there are two possibilities: Either there are N linearly independent eigenfunctions or there are less than N linearly independent eigenfunctions. In Section C some theorems will be given on eigenvalues and eigenfunctions. A detailed treatment of this subject can be found in the books on matrix theory mentioned at the end of Section C.

Now we return to (17). Before we start with the first cycle, we have an error $\varepsilon^{(0)}(n)$. Since $1 \leq n \leq N$ this error has N components. Suppose that the system has N linearly independent eigenfunctions. Then we can write $\varepsilon^{(0)}(n)$ as a sum of the N linearly independent eigenfunctions:

$$\varepsilon^{(0)}(n) = \sum_{l=1}^{N} \beta_l e_l(n) \tag{19}$$

After one cycle we have according to (17), (18), and (19)

$$\varepsilon^{(1)}(n) = \sum_{k=1}^{N} a_n{}^k \varepsilon^{(0)}(k) = \sum_{k=1}^{N} \sum_{l=1}^{N} \beta_l a_n{}^k e_l(k) = \sum_{l=1}^{N} \beta_l \lambda_l e_l(n)$$

Each eigenfunction is multiplied by a factor equal to its eigenvalue. After m cycles

$$\varepsilon^{(m)}(n) = \sum_{l=1}^{N} \beta_l \lambda_l{}^m e_l(n) \tag{20}$$

If $|\lambda_l| < 1$ this iteration procedure converges to the final solution $\varepsilon^{(\infty)}(n) = 0$. From (20) it can be seen that if the error is resolved in eigenfunctions, with every cycle the eigenfunctions are multiplied by their eigenvalue. This can be compared with an oscillation that is resolved into modes, each mode having its own damping. The rate at which the final state is obtained is determined by the smallest damping. This corresponds to the largest $|\lambda_l|$, which we shall denote by $|\lambda_l|_{\max}$. After a certain number of iterations the term or terms with $|\lambda_l|_{\max}$ will dominate in (20).

The rate of convergence R is usually defined as

$$R = -\ln |\lambda_l|_{\max} \approx 1 - |\lambda_l|_{\max} \tag{21}$$

The approximation is valid if $|\lambda_l|_{\max}$ is nearly one, which is always the case in practice. The number of iterations necessary to decrease the error by a factor $1/e$ is approximately $1/R$.

If N linearly independent eigenfunctions do not exist, a more comprehensive theory, using the Jordan canonical form of a matrix, shows that the iteration procedure still converges, provided that $|\lambda_l| < 1$. The character of the convergence is about the same as that described above.

At the end of Section D some books are mentioned that deal with the theory of the iteration procedure. An example of block diagrams of a program for the solution of Poisson's equation and for the calculation of electron trajectories has been given by Weber (1962–1963).

C. Four Theorems on Eigenvalues and Eigenfunctions

In this section we shall derive four theorems on eigenvalues and eigenfunctions. These theorems yield a mathematical background for the study of the iteration process.

The first theorem concerns the convergence of the iteration process. It mentions conditions that are sufficient in order that the absolute value of every eigenvalue, $| \lambda_l |$, shall be smaller than one. The second theorem yields conditions that are sufficient for the eigenvalues and eigenfunctions to be real. The third theorem concerns the existence of N linearly independent eigenfunctions. The fourth theorem yields sufficient conditions in order that (1) the largest eigenvalue will always be positive; (2) there will only be one eigenfunction belonging to the largest eigenvalue; (3) all the components of this eigenfunction will have the same sign; and (4) this will be the only eigenfunction of which all the components have the same sign.

At a first reading we may take the theorems for granted and continue at Section D. However, it should be noted that there are some cases in which the conditions mentioned in the theorems are not satisfied, although in many cases they are.

It should also be noted that the conditions mentioned in the theorems are sufficient conditions and not necessary conditions.

It will be convenient for later reference to define first some of the properties that our system may have.

Property 1. A system has property 1 if for every n

$$\sum_{k=1}^{N} | a_n{}^k | \leq 1 \tag{22}$$

whereas for at least one value of n the " $<$ " sign applies.

Most difference equations have property 1. The " $<$ " sign applies at points near the boundary, since the boundary potentials are included in the terms a_n.

Property 2. A system has property 2 if the region containing the internal points is connected. This means that we can always connect each internal point to every other internal point with a chain of difference equations, such that each link of this chain connects an internal point $\mathbf{k}$ to an internal point n by means of the difference Eq. (13), with $a_n{}^k \neq 0$.

This property implies that the region is not composed of two entirely separate regions that do not have any connection with each other.

Property 3. A system has property 3 if we can find a set of N positive (nonzero) numbers γ_n such that

$$\gamma_n a_n{}^k = \gamma_k a_k{}^n \tag{23}$$

Example. If the difference equations (7) and (9) are used and if the boundary is composed of parts of mesh lines the numbers γ_n are given by

$$\gamma_n = H \qquad \text{if} \quad H \neq 0$$

$$\gamma_n = 3/16 \qquad \text{if} \quad H = 0$$

The reader may verify that these γ_n satisfy (23).

If we have curved boundaries and if we use the difference equations associated with Fig. 3 (Shortley *et al.*, 1947), generally no γ_n can be found.

Property 4. A system has property 4 if

$$a_n{}^k \geq 0$$

Most difference equations have this property.

Now we shall prove the theorems.

Theorem 1. If a system has the properties 1 and 2, the absolute value of all the eigenvalues is smaller than one, $|\lambda_l| < 1$.

Proof. Suppose that $|\lambda_l| \geq 1$. From (18) we obtain with the triangle inequality

$$|\lambda_l|\,|e_l(n)| \leq \sum_{k=1}^{N} |a_n{}^k|\,|e_l(k)| \tag{24}$$

For at least one value of n, say **n**, $|e_l(n)|$ has its maximum value, which is not equal to zero. We divide (24) by $|e_l(\mathbf{n})|$:

$$|\lambda_l|\,\frac{|e_l(n)|}{|e_l(\mathbf{n})|} \leq \sum_{k=1}^{N} |a_n{}^k|\,\frac{|e_l(k)|}{|e_l(\mathbf{n})|} \tag{25}$$

According to property 1 in at least one point, say the point n_1, the $<$ sign holds in (22). Hence, if $n = n_1$, the right-hand side of (25) is < 1. Since we supposed $|\lambda_l| \geq 1$ this implies that $|e_l(n_1)| < |e_l(\mathbf{n})|$. Now we follow the chain, mentioned in property 2, which connects the points n_1 and **n**.

Consider a link of this chain, which connects the point $\mathbf{k}$ to the point n. Suppose that it is known from the previous link that $|e_l(\mathbf{k})| < |e_l(\mathbf{n})|$. Since $a_n{}^k \neq 0$, it follows from (22) and (25) that also $|e_l(n)| < |e_l(\mathbf{n})|$. At the first point of the chain $|e_l(n_1)| < |e_l(\mathbf{n})|$. Hence we obtain at the last point $\mathbf{n}$ of the chain $|e_l(\mathbf{n})| < |e_l(\mathbf{n})|$. This is obviously incorrect and therefore the assumption $|\lambda_l| \geq 1$ must be incorrect.

Corollary. The system of homogeneous equations

$$\varepsilon^{(\infty)}(n) = \sum_{k=1}^{N} a_n{}^k \varepsilon^{(\infty)}(k)$$

has only the zero solution; otherwise this solution would be an eigenfunction with $\lambda = 1$. This implies that the corresponding nonhomogeneous system (13) and (15) has a unique solution.

Lemma. If a system has property 3 and if $x(i)$ ($1 \leq i \leq N$) are arbitrary complex numbers, the expression

$$\sum_{n=1}^{N} \sum_{k=1}^{N} \gamma_n \, x^*(n) a_n{}^k \, x(k) \qquad (26)$$

is always real. The asterisk denotes the complex conjugate. If we vary the values of $x(i)$ under the auxiliary condition

$$\sum_{k=1}^{N} \gamma_k x^*(k) \, x(k) = 1 \qquad (27)$$

(26) has a stationary value if and only if $x(i)$ is an eigenfunction $e_l(k)$ of $a_n{}^k$, which has been normalized according to (27). The stationary value is equal to the corresponding eigenvalue λ_l.

Proof. We write

$$x(k) = u(k) + i\,v(k) \qquad (28)$$

where $u(k)$ and $v(k)$ are real. Substitution of (28) in (26) and (27) yields, respectively,

$$\sum_{n=1}^{N} \sum_{k=1}^{N} \gamma_n[u(n)u(k) + v(n)v(k)]a_n{}^k \qquad (29)$$

and

$$\sum_{k=1}^{N} \gamma_k[u(k)^2 + v(k)^2] = 1 \qquad (30)$$

The imaginary part of (29) is zero because of (23). It follows from (29) that (26) is always real. The stationary values of (29) under the auxiliary condition (30) are determined with the Langrangian multiplier method. Therefore we must multiply the auxiliary condition (30) by a constant λ, at the moment arbitrary, and subtract it from (29):

$$\sum_{n=1}^{N} \sum_{k=1}^{N} \gamma_n [u(n)u(k) + v(n)v(k)] a_n{}^k - \lambda \left\{ \sum_{k=1}^{N} \gamma_k [u(k)^2 + v(k)^2] - 1 \right\} \quad (31)$$

Because of (30), the added part is zero. We now determine the stationary values of (31), which requires the derivatives of (31) with respect to $u(i)$ and $v(i)$, respectively, to be zero. Using (23) yields

$$\lambda u(i) = \sum_{k=1}^{N} a_i{}^k u(k) \qquad \lambda v(i) = \sum_{k=1}^{N} a_i{}^k v(k) \quad (32)$$

The stationary values of (26) are obtained for $u(i)$, $v(i)$, and λ, which satisfy (32) and (30). The second statement of the lemma follows immediately from (32) and (30). The substitution of a normalized eigenfunction $e_l(i)$ in (26) and the use of (18) yields

$$\sum_{n=1}^{N} \sum_{k=1}^{N} \gamma_n e_l{}^*(n) a_n{}^k e_l(k) = \lambda_l \sum_{n=1}^{N} \gamma_n e_l{}^*(n) e_l(n) = \lambda_l$$

Theorem 2. If a system has property 3, the eigenvalues are real and the eigenfunctions belonging to each eigenvalue can be expressed as a linear combination of one or more real eigenfunctions.

Proof. From the preceding lemma, it is clear that the eigenvalues are real. The eigenfunctions are determined from real linear homogeneous equations and therefore can be expressed as a linear combination of real eigenfunctions.

Theorem 3. If a system has property 3, the matrix $a_n{}^k$ has N linearly independent eigenfunctions.

Proof. We construct the eigenfunctions as follows. Determine the maximum of (26) under the auxiliary condition (27). This maximum exists according to a theorem of Weierstrass, which states that a continuous function defined on a closed and bounded region has a maximum value on at least one point of this region. According to the preceding lemma this must be the largest eigenvalue λ_1 with the corresponding real eigenfunction $e_1(i)$.

Now we determine the maximum value of (26) under the auxiliary conditions (27) and

$$\sum_{k=1}^{N} \gamma_k e_1(k)\, x(k) = 0 \tag{33}$$

According to the theorem of Weierstrass this maximum exists. The substitution of (28) into (33) yields

$$\sum_{k=1}^{N} \gamma_k e_1(k)\, u(k) = 0 \qquad \sum_{k=1}^{N} \gamma_k e_1(k)\, v(k) = 0 \tag{34}$$

Using the Lagrangian multiplier method, we multiply (30) by λ and (34) by v_1 and v_2, respectively. Subtraction from (29) yields (31) with in addition the terms

$$- v_1 \sum_{k=1}^{N} \gamma_k e_1(k) u(k) - v_2 \sum_{k=1}^{N} \gamma_k e_1(k) v(k)$$

Differentiation with respect to $u(i)$ and $v(i)$, respectively, and using (23) yields

$$\tfrac{1}{2} v_1 e_1(i) + \lambda u(i) = \sum_{k=1}^{N} a_i{}^k u(k)$$

$$\tfrac{1}{2} v_2 e_1(i) + \lambda v(i) = \sum_{k=1}^{N} a_i{}^k v(k)$$

$$\tag{35}$$

Multiplication of these expressions by $\gamma_i\, e_1(i)$ and summation over i yields, together with (23) and the auxiliary conditions,

$$\tfrac{1}{2} v_1 = \sum_{i=1}^{N} \sum_{k=1}^{N} \gamma_i e_1(i) a_i{}^k u(k) = \sum_{i=1}^{N} \sum_{k=1}^{N} \gamma_k e_1(i) a_k{}^i u(k)$$

$$= \lambda_1 \sum_{k=1}^{N} \gamma_k e_1(k) u(k) = 0$$

$$\tfrac{1}{2} v_2 = \sum_{i=1}^{N} \sum_{k=1}^{N} \gamma_i e_1(i) a_i{}^k v(k) = \sum_{i=1}^{N} \sum_{k=1}^{N} \gamma_k e_1(i) a_k{}^i v(k)$$

$$= \lambda_1 \sum_{k=1}^{N} \gamma_k e_1(k) v(k) = 0$$

Since $v_1 = v_2 = 0$ it can be seen from (35) that this maximum must be the eigenvalue λ_2 with the corresponding real eigenfunction $e_2(i)$.

Similarly the third eigenfunction is obtained if we determine the maximum value of (26) under the auxiliary conditions (27), (33), and

$$\sum_{k=1}^{N} \gamma_k e_2(k) \, x(k) = 0$$

We proceed in this way until we have N eigenfunctions. Presently we shall prove that these eigenfunctions are linearly independent of each other. Since the region of Weierstrass' theorem has been reduced to zero, from now on no more eigenfunctions can be found which are linearly independent of the others.

In order to prove the linear independency of the eigenfunctions we consider the equations

$$\sum_{i=1}^{N} c_i e_i(k) = 0 \qquad (1 \le k \le N) \tag{36}$$

where the c_i's are the unknowns. Multiplication with $\gamma_k e_j(k)$ and summation over k yields together with the auxiliary conditions

$$\sum_{k=1}^{N} \sum_{i=1}^{N} c_i \gamma_k e_i(k) e_j(k) = \sum_{i=1}^{N} c_i \delta_{ij} = c_j = 0$$

Since the only solution of (36) is the zero solution, the eigenfunctions must be linearly independent.

Theorem 4. If a system has the properties 2, 3, and 4, the largest eigenvalue λ_1 is positive. There is only one linearly independent eigenfunction with eigenvalue λ_1; all the components of this eigenfunction are not equal to zero and have the same sign. This is also the only eigenfunction all of whose components have the same sign.

Proof. If $e_1(k)$ is an eigenfunction corresponding to the largest eigenvalue λ_1 and is normalized according to (27), then $| e_1(k) |$ must be an eigenfunction with eigenvalue λ_1. This can be seen from

$$\sum_{n=1}^{N} \sum_{k=1}^{N} \gamma_n e_1(n) a_n{}^k e_1(k) \le | \sum_{n=1}^{N} \sum_{k=1}^{N} \gamma_n e_1(n) a_n{}^k e_1(k) |$$

$$\le \sum_{n=1}^{N} \sum_{k=1}^{N} \gamma_n | e_1(n) | \, a_n{}^k \, | e_1(k) |$$

Since $| e_1(k) |$ satisfies the normalization (27) and since the left-hand side is equal to λ_1, being the maximum value which this expression can have, the right-hand side must also be equal to λ_1 and $|e_1(k)|$ must be an eigenfunction with eigenvalue λ_1.

Suppose that $|e_1(k)|$ has an element $|e_1(k_1)|$ which is equal to zero. Since $|e_1(k)|$ is an eigenfunction there is at least one other element, say $|e_1(k_2)|$, which is not equal to zero. We connect the point k_2 to the point k_1 with a chain mentioned in property 2. There must be a link of this chain, connecting a point k, with nonzero $|e_1(k)|$, to a point n, with $|e_1(n)| = 0$, such that $a_n{}^k \neq 0$.

According to (18)

$$\lambda_1 |e_1(n)| = \sum_{k=1}^{N} a_n{}^k |e_1(k)| \qquad (37)$$

The left-hand side is zero and the right-hand side is positive. This is a contradiction and therefore the supposition that there exists an element $|e_1(k_1)|$ equal to zero is incorrect, which also implies that no element of $e_1(k)$ is equal to zero. From (37), now with nonzero $|e_1(n)|$, it can be seen that $\lambda_1 > 0$.

Suppose that there is an eigenfunction $x(k)$ with eigenvalue λ_1 that is linearly independent of $|e_1(k)|$. Then the linear combination

$$\frac{x(1)}{|e_1(1)|} |e_1(k)| - x(k)$$

satisfies (18) with $\lambda = \lambda_1$; it has the element zero at the point $k = 1$ and because of the linear independence it has a nonzero element somewhere else. Since we have shown that eigenfunctions with eigenvalue λ_1 have only nonzero elements, this is a contradiction. Thus $x(k)$ must depend linearly upon $|e_1(k)|$. This proves that there is only one linearly independent eigenfunction with eigenvalue λ_1 and that all its components must have the same sign.

Since an eigenfunction $x(k)$ belonging to another eigenvalue satisfies the Eq. (33), all the components of this eigenfunction cannot possibly have the same sign.

Properties of eigenvalues and eigenfunctions can be found in books about matrix theory, for example, those by Zurmühl (1958) and by Bellman (1960). A short survey is given in a paper by Flanders and Shortley (1950).

D. Successive Displacement and Overrelaxation

If we use successive displacement instead of simultaneous displacement, the order in which the points are calculated is significant. Young (1954) introduced an important property concerning the order in which the points are calculated. In literature it is generally known as "property A." The N internal points are divided into I groups: Each group contains a number of points. A point will be indicated by two indices (i, j), where i is the number of the group $(1 \leq i \leq I)$ and j is the number of the point in the group. In general, the groups will contain different numbers of points.

The system of difference equations (13) has property A if we can classify the internal points into groups, such that for the calculation of the potential of each point one needs only the potentials of points of the previous and the following group. In other words, if $a_n{}^k \neq 0$ and n belongs to group number i, k must belong either to group number $i - 1$ or to group number $i + 1$.

The order in which the points are calculated is said to be consistent with this classification into groups if, at the moment that the potential of the point (i, j) is calculated, the required potentials of the group $i - 1$ have already been calculated and the required potentials of the group $i + 1$ have not yet been calculated.

If we calculate successively the points of the groups 1, 2, 3, and so on, this is a consistent order. However, more than one consistent order is often possible, since the definition speaks only of the required potentials for the calculation and not of all the potentials of the groups in question.

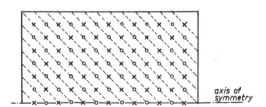

FIG. 7. The classification into groups; the chessboard classification and the diagonal classification.

We shall give an example. We consider the five-point difference equations (7) and (9) in a rectangular region (Fig. 7). We divide the mesh points into two groups, like the black and white fields of a chessboard ($\times$ and $\circ$ in Fig. 7). For the calculation of the points of a group we need only the points of the other group. We get a consistent order if we calculate first the points of the first group and then the points of the second group.

We can also give another classification into groups. Now all the points that lie on the same diagonal (dashed lines in Fig. 7) belong to the same group. The diagonals are numbered from bottom left to top right. For the calculation of a point one needs only potentials of the previous and the following diagonal. A consistent order is obtained if we calculate the points row by row, from the left to the right, starting with the bottom row, followed by the next row, and so on.

If we use the nine-point difference equation, according to Fig. 4, it is not possible to find a classification into groups in accordance with property A. However, we find in practice that fairly good results are obtained if, nevertheless, one applies to these difference equations the results of the following theory.

Suppose the system has property A and is calculated in a consistent order. Instead of the index n we shall use the two indices (i, j). On using successive displacement, (14) changes into

$$V^{(m)}(i,j) = \sum_k \alpha_{i,j}^{i-1,k} V^{(m)}(i-1,k) + \sum_k \alpha_{i,j}^{i+1,k} V^{(m-1)}(i+1,k) + a_{i,j} \quad (38)$$

The rate of convergence of this procedure is very low, even for fast electronic computers. If we consider the potential of a fixed point (i, j) during several cycles, it appears that the potential increases (or decreases) gradually to the final solution. Hence we obtain a faster convergence if we do not use the new value given by the difference equation (38) but use instead a value that we obtain by extrapolation in the direction of the expected change:

$$V^{(m)}(i, j) = V^{(m-1)}(i, j) + \omega \left[\sum_k \alpha_{i,j}^{i-1,k} V^{(m)}(i-1, k) \right.$$
$$\left. + \sum_k \alpha_{i,j}^{i+1,k} V^{(m-1)}(i+1, k) + a_{i,j} - V^{(m-1)}(i, j) \right]$$

where we take $\omega \geq 1$. If $\omega = 1$, Eq. (38) is obtained. The value of ω must be chosen carefully. If we take it too small, the rate of convergence is low. If we take it too large, however, the extrapolation goes too far and the rate of convergence is again low. We shall show later on that there is an optimum value of ω. We can compare this with an oscillation, which is damped undercritically (ω small), overcritically (ω large), or critically (the optimum value of ω). The procedure with $\omega > 1$ is called overrelaxation. If at the same time, successive displacement is used it is called successive overrelaxation (usually abbreviated SOR).

The error satisfies the equation

$$\varepsilon^{(m)}(i, j) = \varepsilon^{(m-1)}(i, j) + \omega \left[\sum_k \alpha_{i,j}^{i-1,k} \varepsilon^{(m)}(i-1, k) \right.$$
$$\left. + \sum_k \alpha_{i,j}^{i+1,k} \varepsilon^{(m-1)}(i+1, k) - \varepsilon^{(m-1)}(i, j) \right] \quad (39)$$

We now look for the eigenfunctions $\mathbf{e}_l(i, j)$ and the eigenvalues $\boldsymbol{\lambda}_l$ of (39). The eigenfunctions and eigenvalues of the system without overrelaxation and with simultaneous displacement (17) are denoted by $e_l(i, j)$ and λ_l, respectively. We shall show that the eigenfunctions $\mathbf{e}_l(i, j)$ of (39) can be expressed in terms of the eigenfunctions $e_l(i, j)$ of (17) as follows:

$$\mathbf{e}_l(i, j) = [\boldsymbol{\lambda}_l^{1/2}]^i e_l(i, j) \tag{40}$$

We substitute (40) into (39), on the assumption that (40) is an eigenfunction with eigenvalue $\boldsymbol{\lambda}_l$

$$\boldsymbol{\lambda}_l[\boldsymbol{\lambda}_l^{1/2}]^i e_l(i, j) = [\boldsymbol{\lambda}_l^{1/2}]^i e_l(i, j) + \omega \left(\sum_k a_{i,j}^{i-1,k} \boldsymbol{\lambda}_l [\boldsymbol{\lambda}_l^{1/2}]^{i-1} e_l(i-1, k) \right.$$
$$\left. + \sum_k a_{i,j}^{i+1,k} [\boldsymbol{\lambda}_l^{1/2}]^{i+1} e_l(i+1, k) - [\boldsymbol{\lambda}_l^{1/2}]^i e_l(i, j) \right) \tag{41}$$

Since $e_l(i, j)$ is an eigenfunction of (17) the first two terms between the large parentheses may be replaced by

$$\lambda_l [\boldsymbol{\lambda}_l^{1/2}]^{i+1} e_l(i, j)$$

Now division of (41) by $[\boldsymbol{\lambda}_l^{1/2}]^i e_l(i, j)$ yields

$$\boldsymbol{\lambda}_l = 1 + \omega \, \lambda_l \boldsymbol{\lambda}_l^{1/2} - \omega \tag{42}$$

This relation is independent of (i, j). Hence if $\boldsymbol{\lambda}_l$ satisfies (42) the expression (40) represents eigenfunctions of (39). From (42) we derive

$$\boldsymbol{\lambda}_l^{1/2} = \tfrac{1}{2} \, \omega \, \lambda_l \pm \tfrac{1}{2}[\omega^2 \lambda_l^2 - 4(\omega - 1)]^{1/2}$$
$$\boldsymbol{\lambda}_l = \tfrac{1}{2} \, \omega^2 \lambda_l^2 - \omega + 1 \pm \tfrac{1}{2} \, \omega \, \lambda_l [\omega^2 \lambda_l^2 - 4(\omega - 1)]^{1/2} \tag{43}$$

If the system has property A, one can easily show that if (17) has the eigenvalue λ_l with the eigenfunction $e_l(i, j)$ it must also have the eigenvalue $- \lambda_l$ with the eigenfunction $(-1)^i e_l(i, j)$. We can also show that such a pair λ_l, $- \lambda_l$ gives only two substantially different eigenfunctions and eigenvalues of (39). The $\pm$ sign in (43) does not give us more different eigenvalues and eigenfunctions.

By means of (43) the eigenvalues $\boldsymbol{\lambda}_l$ are expressed in λ_l and ω. It is remarkable that the $\boldsymbol{\lambda}_l$ do not depend upon the chosen subdivision into groups or on the chosen order. It is only required that a division into groups shall exist in accordance with property A and that the order shall be consistent. The eigenfunctions (40) do depend upon the chosen subdivision into groups. If the number of groups I is large the factor in (40) containing the square root is small for points with a large i. In this context it is found that it may be advantageous in some cases to use the chessboard division with two groups instead of the diagonal division with many groups.

We can prove that the eigenfunctions (40) are linearly independent if the functions $e_l(i, j)$ are linearly independent. If $\lambda_l = 0$ the expressions (40) are not eigenfunctions, because then all the components are zero. From (42) it can be seen that $\lambda_l = 0$ implies $\omega = 1$. It can be shown that there are usually not N linearly independent eigenfunctions if $\omega = 1$. Since in practice the optimum value of ω is used and not $\omega = 1$, we shall not go into details about this exceptional case.

Now we shall investigate the way in which $|\lambda_l|$ depends upon ω in the region $1 < \omega < 2$. Suppose that the λ_i's are real (cf. Theorem 2, Section C). The order of the λ_i's is taken such that

$$\lambda_1 \geq \lambda_2 \geq \cdots \geq \lambda_l \geq \lambda_{l+1} \geq \cdots$$

First we consider an arbitrary pair λ_l, $-\lambda_l$. In view of the chosen order, $-\lambda_l$ must be equal to λ_{N-l+1}. Without loss of generality we can take l such that $l < N - l + 1$. It follows from (43) that this pair yields two different real and positive values λ_l if $\omega < \omega_l$ where ω_l is given by

$$\omega_l = 2/[1 + (1 - \lambda_l^2)^{1/2}] \tag{44}$$

There are two complex values of λ_l if $\omega > \omega_l$. These values are complex conjugate and their absolute value is $\omega - 1$, independent of λ_l. In Fig. 8 the solid curve shows $|\lambda_l|$ as a function of ω. If $\omega < \omega_l$ the curve has two

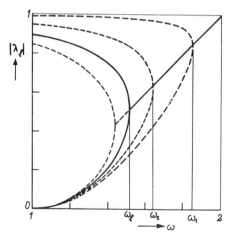

Fig. 8. $|\lambda_l|$ as a function of the overrelaxation parameter ω.

branches. The upper branch will be denoted by λ_l, the lower branch by λ_{N-l+1}. If $\omega > \omega_l$ both values of $|\lambda_l|$ are equal to $\omega - 1$. The dashed curves show $|\lambda_l|$ for other pairs λ_l, $-\lambda_l$.

From Fig. 8 it can be seen that at a fixed value of ω, the largest value of $|\lambda_l|$ is determined by the largest value of λ_l, namely λ_1. The fastest convergence is obtained for that value of ω of which the largest $|\lambda_l|$ is as small as possible. This ω is given by

$$\omega_1 = 2/[1 + (1 - \lambda_1^2)^{1/2}] \approx 2 - 2[2(1 - \lambda_1)]^{1/2} \qquad (45)$$

The approximation is valid if λ_1 is nearly one, which is always the case in practice. For this value of ω all the $|\lambda_l|$ have the same value, namely,

$$|\lambda_l|_{\omega_1} = \omega_1 - 1 \approx 1 - 2[2(1 - \lambda_1)]^{1/2} \qquad (46)$$

The rate of convergence **R** is given by a formula equivalent to (21)

$$\mathbf{R} \approx 2[2(1 - \lambda_1)]^{1/2} \qquad (47)$$

Since λ_1 is independent of the space charge and the boundary potentials, ω_1 and **R** are also independent of these quantities. However, they do depend on the mesh length and on the shape of the boundary. In Section F an example will be given.

If a system has the properties 2, 3, and 4 in accordance with Theorem 4, the largest eigenvalue has only one linearly independent eigenfunction $e_1(i, j)$ and all the components of this function have the same sign and are not equal to zero. If $\omega < \omega_1$ it follows from (43) that the eigenvalue which has the largest absolute value, λ_1, is real and positive. There exists only one corresponding eigenfunction, given by (40), whose components are real, all of which have the same sign and are not equal to zero.

A further investigation shows that convergence is also obtained in the region $0 < \omega < 1$. However, this region is not interesting, since the convergence in this region is slower than the convergence at $\omega = \omega_1$.

The theory of successive overrelaxation (SOR) was developed between 1950 and 1960, when high-speed electronic computers became available. Before that time the calculation was performed with desk machines, using a kind of unsystematic overrelaxation that depended on the insight of the person doing the calculations (Southwell, 1946). Outstanding papers concerning the development of SOR were those by Frankel (1950) and Young (1954). Numerous papers, mostly highly mathematical, have appeared on this subject. A survey of the mathematical theory can be found in the books by Forsythe and Wasow (1960) and by Varga (1962), which also contain extensive reference lists. Further developments were the successive line overrelaxation (SLOR) and the alternating direction implicit methods (ADI). They can be found in the books just mentioned. The ADI methods were also treated in detail by Birkhoff et al. (1962). On the one hand, these methods yield a faster convergence; but on the other, they require more computing time per cycle and their programming is more complicated.

E. SOME PRACTICAL QUANTITIES, CHARACTERIZING THE CONVERGENCE OF THE ITERATION PROCESS

In practice the final solution is not known; thus the error (16) cannot be traced during the iteration process. In order to ascertain whether the iteration has been carried out far enough, the difference is usually calculated between the potentials of two successive cycles:

$$\Delta V^{(m)}(n) = V^{(m-1)}(n) - V^{(m)}(n) \tag{48}$$

This difference is not stored, but during each cycle the following quantities are determined from it:

$$| \Delta V^{(m)} |_{\max} = \max_n | \Delta V^{(m)}(n) |$$

$$\Sigma \Delta V^{(m)} = \sum_{n=1}^{N} \Delta V^{(m)}(n) \tag{49}$$

$$\Sigma | \Delta V^{(m)} | = \sum_{n=1}^{N} | \Delta V^{(m)}(n) |$$

These are the quantities that must yield information about the progress of the iteration process. If the calculations are carried out with an infinite number of decimal places they approach to zero. Since in practice the calculations are carried out with a finite number of decimal places, they approach to a small finite value.

The behavior of the quantities (49) is quite different in the regions $\omega < \omega_1$ and $\omega > \omega_1$. We shall investigate first the region $\omega < \omega_1$. We have seen that λ_1 (the eigenvalue with the largest absolute value) is real and positive in this region. After a certain number of iterations this eigenvalue will dominate over the other eigenvalues in the expression which is equivalent to (20). Since λ_1 has only one eigenfunction $e_1(n)$ the error $\varepsilon^{(m)}(n)$ can be approximated by

$$\varepsilon^{(m)}(n) \approx \beta_1 \lambda_1{}^m e_1(n) \tag{50}$$

From this expression it can be seen that with every cycle the error will decrease by a factor λ_1. From (16), (48), and (50) we obtain

$$\Delta V^{(m)}(n) = \beta_1 \lambda_1^m[(1/\lambda_1) - 1]e_1(n) = [(1/\lambda_1) - 1]\varepsilon^{(m)}(n) \qquad (51)$$

This shows that $\Delta V^{(m)}(n)$ also decreases with every cycle by a factor λ_1. Figure 9 shows how the potential $V^{(m)}(n)$ approaches to its limit $V^{(\infty)}(n)$.

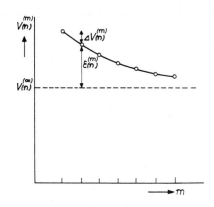

FIG. 9. The convergence to the final solution in the region $\omega < \omega_1$.

If $|e_1(n)|$ has its largest component at the point $\mathbf{n}$, we find from (49) and (51)

$$|\Delta V^{(m)}|_{\max} = |\beta_1| \lambda_1^m[(1/\lambda_1) - 1]| |e_1(\mathbf{n})| = [(1/\lambda_1) - 1]| |\varepsilon^{(m)}(\mathbf{n})| \quad (52)$$

Hence $|\Delta V^{(m)}|_{\max}$ also decreases with every cycle by a factor λ_1. The relation between $|\Delta V^{(m)}|_{\max}$ and the largest error $|\varepsilon^{(m)}(\mathbf{n})|$ is given by (52). Since λ_1 is usually about one, the error is much larger than $|\Delta V^{(m)}|_{\max}$. The quantity $|\Delta V^{(m)}|_{\max}$ itself gives a too favorable impression of the error.

We have shown that all the components of $e_1(n)$ have the same sign. Hence the absolute value of $\Sigma \Delta V^{(m)}$ is nearly the same as $\Sigma |\Delta V^{(m)}|$. With every cycle both quantities decrease by a factor λ_1.

We obtain a quite different pattern if $\omega > \omega_1$. Now all the eigenvalues λ_l are complex and they have the same absolute value, namely $\omega - 1$, so that we must take them all into account and the approximation (50) cannot be made. It follows from (43) that the eigenvalues always occur in pairs, which are complex conjugate. We denote the argument of λ_l by ψ_l. Instead of writing the real error as a sum of complex eigenfunctions, we can also write it as a sum of the real functions

$$[|\,\lambda_l\,|^{1/2}]^i\, e_l(i, j) \cos \tfrac{1}{2}\, i\, \psi_l \qquad (53)$$

$$[|\,\lambda_l\,|^{1/2}]^i\, e_l(i, j) \sin \tfrac{1}{2}\, i\, \psi_l \qquad (54)$$

Here i is the number of the group, not the imaginary unit. If we multiply (54) by the imaginary unit and add it to [or subtract it from] (53), we obtain the eigenfunctions (40). It can be seen that after m cycles the functions (53) and (54) change into

$$[|\,\lambda_l\,|^{1/2}]^{2m+i}\, e_l(i, j) \cos \tfrac{1}{2}\, (2\, m + i)\, \psi_l$$

$$[|\,\lambda_l\,|^{1/2}]^{2m+i}\, e_l(i, j) \sin \tfrac{1}{2}\, (2\, m + i)\, \psi_l$$

If the error is written as a sum of these functions many different arguments ψ_l occur. However, all the $|\,\lambda_l\,|$ have the same value, namely, $\omega - 1$. The cosines and the sines cause a very irregular convergence, but the error remains between given limits (Fig. 10). Sometimes we even find an increase of the error instead of a decrease. Only if we consider a large number of cycles can convergence be established with an average decrease of a factor $\omega - 1$ per cycle.

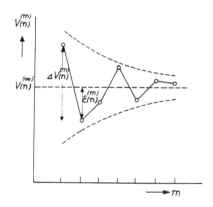

FIG. 10. The convergence to the final solution in the region $\omega > \omega_1$.

The sign of $\Delta V^{(m)}(n)$ changes frequently if m or n are changed. Hence, in contrast to the case $\omega < \omega_1$, the sign of $\Sigma \Delta V^{(m)}$ also changes if a range of values m is considered and the absolute value of $\Sigma \Delta V^{(m)}$ is usually appreciably smaller than $\Sigma \mid \Delta V^{(m)} \mid$. From Fig. 10 it can be seen that $\Delta V^{(m)}(n)$ has about the same order of magnitude as $\varepsilon^{(m)}(n)$. Hence in this case

$|\Delta V^{(m)}|_{\max}$ can give directly an impression of the order of magnitude of the error.

If during the calculation we change the ω from $\omega > \omega_1$ to $\omega < \omega_1$ we find that $|\Delta V^{(m)}|_{\max}$ and $\Sigma |\Delta V^{(m)}|$ decrease rapidly. However, it can be seen from Figs. 9 and 10 that this decrease does not imply a sudden decrease of the error.

The difference between $\omega < \omega_1$ and $\omega > \omega_1$ changes gradually in a small region near $\omega = \omega_1$. If ω is only slightly smaller than ω_1, λ_1 is only slightly larger than the absolute value of the other eigenvalues, which are complex. Hence the approximation (50) is not valid and the irregularity due to the complex eigenvalues can already be noticed.

F. Determination of ω_1

In order to determine ω_1 we must know the largest eigenvalue λ_1 of (18). For some particular boundaries we can determine λ_1. For that purpose we suppose that the eigenfunction that corresponds to λ_1 does not change rapidly over one mesh length. Then the system of difference equations (18), which is built up from the difference equations (7) and (9), can be transformed back to the differential equation:

$$\frac{\partial^2 V}{\partial r^2} + \frac{1}{r} \frac{\partial V}{\partial r} + \frac{\partial^2 V}{\partial z^2} = 4 \frac{\lambda - 1}{h^2} V \tag{55}$$

Eigenfunctions are nonzero solutions, that satisfy this equation with boundary condition zero. By separation of the variables, Eq. (55) can be solved

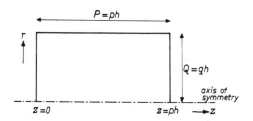

FIG. 11. The region for which ω_1 is calculated.

for some particular boundary configurations. With the boundary of Fig. 11 we find

$$\lambda_1 = 1 - \frac{\pi^2}{4p^2} - \frac{x_1^2}{4q^2} \tag{56}$$

where x_1 is the first zero point of the Bessel function $J_0(x)$. The corresponding eigenfunction $e_1(r, z)$ is given by

$$e_1(r, z) = J_0\left(\frac{x_1 r}{qh}\right) \sin \frac{\pi z}{ph}$$

It can be seen that this eigenfunction has the same sign in the whole region in question. If p and q are large it changes slowly over one mesh length, hence the replacement of the difference equation by the differential equation is justified. The differential equation (55) is not valid for rapidly changing eigenfunctions, as for example, the one corresponding to $-\lambda_1$.

If $P = ph$ and $Q = qh$ and if p and q are large, we find for expressions (56), (45), and (46), approximately,

$$\lambda_1 = 1 - \frac{1}{2} h^2 \left[\frac{\pi^2}{2P^2} + \frac{x_1^2}{2Q^2}\right]$$

$$\omega_1 \approx 2 - 2h \left[\frac{\pi^2}{2P^2} + \frac{x_1^2}{2Q^2}\right]^{1/2}$$

$$|\lambda_l|_{\omega_1} \approx 1 - 2h \left[\frac{\pi^2}{2P^2} + \frac{x_1^2}{2Q^2}\right]^{1/2}$$

With these expressions we find for the rate of convergence (21) with simultaneous displacement and $\omega = 1$, approximately,

$$R \approx \frac{1}{2} h^2 \left(\frac{\pi^2}{2P^2} + \frac{x_1^2}{2Q^2}\right)$$

The number of iterations necessary to decrease the error by a factor $1/e$ is inversely proportional to h^2. On the other hand, if we use successive overrelaxation with $\omega = \omega_1$ the rate of convergence (47) is approximately given by

$$\mathbf{R} \approx 2h \left[\frac{\pi^2}{2P^2} + \frac{x_1^2}{2Q^2}\right]^{1/2}$$

Now the number of iterations necessary to decrease the error by a factor $1/e$ is inversely proportional to h, which is particularly important if h is small. For instance, if $p = 50$ and $q = 25$, $\mathbf{R}$ is 49 times larger than R, which is a considerable saving of time.

The number of points that must be calculated is inversely proportional to h^2. Hence the required time is inversely proportional to h^4 if $\omega = 1$ and inversely proportional to h^3 if $\omega = \omega_1$.

Another approach to the problem of determining ω_1 has been given by

Carré (1961). We take an ω smaller than ω_1 and carry out a number of iterations. Now we can determine an approximate value of λ_1 by taking the ratio of $\Sigma \, | \, \Delta V^{(m)} \, |$ at two successive cycles. Because $\Sigma \, | \, \Delta V^{(m)} \, |$ is the sum of all the values $\Delta V^{(m)}(n)$, it is more appropriate for the determination of λ_1 than $| \, \Delta V^{(m)} \, |_{\max}$. From λ_1 we can calculate λ_1 (42) and from λ_1 we can calculate ω_1 (45).

The determination of λ_1 depends upon the precision of the approximation (50). Hence it is advantageous if the difference between λ_1 and the next eigenvalue λ_2 is as large as possible. From Fig. 8 it can be seen that this is the case for $\omega = \omega_2$, where ω_2 is determined from (44) with $\lambda_l = \lambda_2$. Since in practice λ_2 is unknown, ω_2 cannot be determined from (44). Therefore Carré uses the estimation

$$\omega_2 = \omega_1 - \tfrac{1}{4}(2 - \omega_1) \tag{57}$$

In order to determine ω_1 one proceeds as follows.

A number of iterations is carried out with a low value of ω, which is certainly below ω_1 (for example, $\omega = 1.4$). The approximate values of λ_1 and ω_1 are then found, after which ω_2 is determined with (57). A number of iterations is once more carried out and a better approximation of ω_1 is obtained. Again (57) is applied and a number of iterations is carried out with the new value of ω_2. This procedure is repeated until the change of ω_1 is smaller than the tolerance $0.05 \, (2 - \omega_1)$. At that moment ω_1 has been determined accurate enough and all the subsequent iterations are carried out with $\omega = \omega_1$.

G. TRUNCATION AND ROUND-OFF ERRORS

We shall first deal with the truncation error, which is the error made by approximating the differential equation (1) by a system of difference equations.

The difference equation (7) has been derived by truncating the series (2) after $i + j = 2$. We can investigate the order of magnitude of the truncation error by truncating the series (2) after $i + j = 4$. Then the following terms must be added to the right-hand side of (7):

$$- \frac{h^4}{24 r_0} V_0^{(3,0)} - \frac{1}{48} h^4 V_0^{(4,0)} - \frac{1}{48} h^4 V_0^{(0,4)} \tag{58}$$

In this expression no use has been made of (4) to replace the higher order derivatives. If r_0 remains constant, (58) is proportional to h^4 when $h \to 0$. On the other hand, if $H = r_0/h$ remains constant, (58) is also proportional to h^4 when $h \to 0$, because now $r_0 = Hh \to 0$ and we can use the approximation

$$V_0^{(3,0)} \approx r_0 V_{\text{axis}}^{(4,0)}$$

The truncation error of the difference equation (9) is given by

$$-\tfrac{1}{36} h^4 V_{\text{axis}}^{(4,0)} - \tfrac{1}{72} h^4 V_{\text{axis}}^{(0,4)} \tag{59}$$

This expression is also proportional to h^4.

The expressions (58) and (59) are the errors made if V_0 is calculated with the difference equation from correct values $V_1, \ldots, V_4$. However, we not only have one difference equation but a system of difference equations. The potentials $V_1, \ldots, V_4$ also have errors, because they are also calculated from difference equations. Only the boundary values are known exactly. In practice, the truncation error of the difference equation itself has the same sign in large regions. If $V_1, \ldots, V_4$ have a certain error, the potential V_0 (which is calculated from $V_1, \ldots, V_4$ with the difference equation) will have the same error anyhow, and besides it will have a truncation error. Hence the error of the system of difference equations (7) and (9) will be larger than the truncation error of one difference equation. It is found that the error of the system is of the order of h^2, whereas the truncation error of one difference equation is of the order of h^4.

We can illustrate this with a one-dimensional problem. Consider the equation

$$d^2y/dx^2 = 0 \tag{60}$$

We take the boundaries at $x = 0$ and $x = 1$. The dependent variable y is zero at the boundaries. We make a division into meshes with meshwidth h. The difference equation corresponding to (60) is

$$y(x) = \tfrac{1}{2} y(x - h) + \tfrac{1}{2} y(x + h)$$

In this particular case the truncation error of the difference equation is zero and we obtain the exact solution $y = 0$.

Now we introduce deliberately an error of the order of h^4 in the difference equation:

$$y(x) = \tfrac{1}{2} y(x - h) + \tfrac{1}{2} y(x + h) + ch^4$$

where c is a constant. The exact solution of this system of difference equations is

$$y(x) = cx(1 - x)h^2$$

From this solution it can be seen that the error of the system of difference equations is of the order of h^2 although the error of one difference equation is only of the order of h^4.

In some cases one can give a theoretical estimate of the order of magnitude of the error. The methods used are usually based on the work of Gerschgorin (1930). A survey can be found in the book by Forsythe and Wasow (1960).

Richardson (1910) pointed out that if we calculate the potential with a fine meshwidth h_1 and with a coarse meshwidth h_2, a more accurate solution can be obtained by extrapolation. We write the potential at the point (r, z) calculated with the meshwidth h as $V(r, z; h)$. Suppose that the error of the system of difference equations is of the order of h^a when $h \to 0$, then

$$V(r, z; h) - V(r, z; 0) = a(r, z; h)h^a \qquad (61)$$

where $a(r, z; 0) \neq 0$. Suppose that $a(r, z; h)$ can be expanded into a power series with respect to h, then

$$V(r, z; h_1) - V(r, z; 0) = a(r, z; 0)h_1^a + \text{higher order terms}$$

$$V(r, z; h_2) - V(r, z; 0) = a(r, z; 0)h_2^a + \text{higher order terms}$$

Elimination of $a(r, z; 0)$ yields

$$V(r, z; 0) = V(r, z; h_1) + \frac{h_1^a}{h_2^a - h_1^a}[V(r, z; h_1) - V(r, z; h_2)]$$

$$+ \text{higher order terms} \qquad (62)$$

Sometimes this gives a remarkable increase of accuracy. However, we must be very careful that (62) is in fact valid.

In the first place, we must take care that the solution does not have derivatives that are infinitely large at some point. This is the case, for instance, if the boundary has interior angles that are larger than 180°. We must also be careful if the boundaries do not coincide with the mesh lines (curved boundaries). It may happen, particularly when the difference equations near this boundary are less accurate than the difference equations at the interior points, that $a(r, z; h)$ of (61) cannot be expanded into a power series (Wasow, 1955).

We can easily be in error about the value of a to be used with the nine-point difference equations of Fig. 4 (Durand, 1957). The difference equations themselves have an error of the order of h^6. The truncation errors of the five-point difference equations (7) and (9) lose a factor h^2 if we pass from the difference equation itself to the system of difference equations. Hence it seems reasonable to expect the same for the nine-point difference equation. This yields $\alpha = 4$. However, if we calculate a given problem in practice with different meshwidths, we often find $\alpha = 2$. In these cases the boundary has interior angles larger than $180°$, or there are irregularities in the boundary potentials, so that the potential field has an infinitely large derivative at some point.

In practice the calculations are carried out with a finite number of decimal places. This causes a round-off error. If we always round off to the same side, say to the next lower value, this yields a local error that has the same sign everywhere. In the same way as with the truncation error, this gives a much larger error in the final solution. In order to avoid this, we must round off the numbers carefully, so that on an average one half of them are increased and one half of them decreased. Now the sign of the local error changes frequently and this causes a much smaller round-off error in the final result. If these precautions are taken the round-off error is negligible with respect to the truncation error in most practical cases. A further treatment of the round-off error can be found in the book by Forsythe and Wasow (1960).

1.2.3. The Calculation of Electron Trajectories

A. The Equations of Motion of an Electron

In this section we shall deal with the equations of motion of an electron. They are very suitable for the calculation of electron trajectories. These differential equations are generally valid and are not restricted to paraxial rays. In this connection Section B deals with the numerical solution of ordinary differential equations and the numerical evaluation of integrals. The method described herein can be used to solve the just-mentioned equations of motion and most other ordinary differential equations and integrals that occur in electron optics.

Consider a system having rotational symmetry. Suppose that Poisson's equation (1) has been solved in the r–z plane at the mesh points of a square grid. If we calculate an electron trajectory with an arbitrary initial velocity, the trajectory does not show any rotational symmetry. Therefore it is advantageous to calculate the trajectory in the Cartesian coordinate system x, y, z instead of in a cylindrical coordinate system. The z axis of the Cartesian coordinate system will coincide with the axis of symmetry. The trajectory equations in the Cartesian coordinate system are

$$\ddot{x} = \frac{e}{m}\frac{\partial V}{\partial x} = \frac{e}{m}\frac{x}{r}\frac{\partial V}{\partial r}$$

$$\ddot{y} = \frac{e}{m}\frac{\partial V}{\partial y} = \frac{e}{m}\frac{y}{r}\frac{\partial V}{\partial r} \qquad (63)$$

$$\ddot{z} = \frac{e}{m}\frac{\partial V}{\partial z}$$

where $r = [x^2 + y^2]^{1/2}$, $-e$ is the electron charge and m is the electron mass. Each dot denotes differentiation with respect to time. The right-hand sides of these equations are functions of x, y, and z.

In order to solve Eqs. (63) we must be able to calculate the derivatives of the potential at an arbitrary point (r, z) from the potentials at the mesh points of the r–z plane. To that end, we determine the mesh point (r_0, z_0), which is nearest to the point (r, z). The derivatives of the potential at the point (r, z) are calculated from the Taylor expansion around the mesh point (r_0, z_0):

$$\frac{\partial V}{\partial r} = \left(\frac{\partial V}{\partial r}\right)_0 + \left(\frac{\partial^2 V}{\partial r^2}\right)_0 (r - r_0) + \left(\frac{\partial^2 V}{\partial r \partial z}\right)_0 (z - z_0)$$

$$\frac{\partial V}{\partial z} = \left(\frac{\partial V}{\partial z}\right)_0 + \left(\frac{\partial^2 V}{\partial r \partial z}\right)_0 (r - r_0) + \left(\frac{\partial^2 V}{\partial z^2}\right)_0 (z - z_0)$$

The derivatives at the point (r_0, z_0) can be calculated from the potentials of neighboring mesh points (Fig. 12):

$$\left(\frac{\partial V}{\partial r}\right)_0 = \frac{V_3 - V_4}{2h} \qquad\qquad \left(\frac{\partial V}{\partial z}\right)_0 = \frac{V_1 - V_2}{2h}$$

$$\left(\frac{\partial^2 V}{\partial r^2}\right)_0 = \frac{V_3 + V_4 - 2V_0}{h^2} \qquad\qquad \left(\frac{\partial^2 V}{\partial z^2}\right)_0 = \frac{V_1 + V_2 - 2V_0}{h^2}$$

$$\left(\frac{\partial^2 V}{\partial r \partial z}\right)_0 = \frac{V_5 - V_6 - V_7 + V_8}{4h^2}$$

If r is small the first derivative of the potential with respect to r is also small, so that in (63) we must divide two small numbers. This division is avoided if we use the approximation

$$\frac{1}{r}\frac{\partial V}{\partial r} \approx \left(\frac{\partial^2 V}{\partial r^2}\right)_{axis}$$

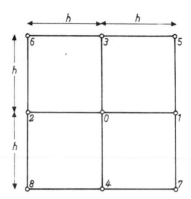

Fig. 12. The nine points used for the calculation of the derivatives of V.

B. The Numerical Solution of Ordinary Differential Equations and Integrals

Most digital computers have subroutines available for solving ordinary differential equations and integrals. They may be used to solve Eqs. (63), the paraxial ray equation, third-order aberration integrals, and so on. Often these subroutines change the step length during the calculation so that the truncation error can be neglected, whereas the number of required steps is kept as small as possible.

We shall deal with a third-order Runge-Kutta method to solve a system of n second-order ordinary differential equations without first derivative. The method is adapted to this particular kind of differential equations and therefore it is usually faster than general subroutines that can also be used for other kinds of differential equations.

The equations to be solved are written as

$$y_i'' = f_i(x; y_1, \ldots, y_n) \tag{64}$$

where $1 \leq i \leq n$, $y_1, \ldots, y_n$ are the dependent variables, x is the independent variable and a prime denotes differentiation with respect to x. The functions f_i are given functions of the $n + 1$ arguments. They are also written as $f_i[x; y_j]$. Examples are: (1) the equations of motion of an electron (63), where x, y, and z are the dependent variables and t is the independent variable; (2) the equation of a paraxial electron trajectory if the time is chosen as independent variable; and (3) Picht's equation for a paraxial electron trajectory, where z is the independent variable.

In order to solve Eqs. (64) at an initial point $x = x_0$ the values of $y_i(x_0)$ and $y_i'(x_0)$ must be given. If δ is a small interval we shall calculate numerically at $x = x_0 + \delta$ the values of $y_i(x_0 + \delta)$ and $y_i'(x_0 + \delta)$. Other steps can be calculated successively by repeating the same procedure. First, we shall describe the performance of the calculation. To begin with, we must evaluate successively the $2n$ auxiliary quantities

$$k_i^{(0)} = \delta f_i[x_0 \; ; y_j(x_0)]$$
$$k_i^{(1)} = \delta f_i[x_0 + \tfrac{2}{3}\,\delta; y_j(x_0) + \tfrac{2}{3}\,y_j'(x_0)\,\delta + \tfrac{2}{9}\,k_j^{(0)}\delta]$$

(65)

The values of $y_i(x_0 + \delta)$ and $y_i'(x_0 + \delta)$ are then calculated with the equations

$$y_i(x_0 + \delta) = y_i(x_0) + \delta[y_i'(x_0) + \tfrac{1}{4}\,k_i^{(0)} + \tfrac{1}{4}\,k_i^{(1)}]$$
$$y_i'(x_0 + \delta) = y_i'(x_0) + \tfrac{1}{4}\,k_i^{(0)} + \tfrac{3}{4}\,k_i^{(1)}$$

(66)

The terms neglected are of the order of δ^4.

For the proof of these formulas we expand $y_i(x_0 + \delta)$ and $y_i'(x_0 + \delta)$ into a Taylor series:

$$y_i(x_0 + \delta) = y_i(x_0) + y_i'(x_0)\delta + \tfrac{1}{2}\,y_i''(x_0)\delta^2 + \tfrac{1}{6}\,y_i'''(x_0)\delta^3 + \cdots$$
$$y_i'(x_0 + \delta) = y_i'(x_0) + y_i''(x_0)\delta + \tfrac{1}{2}\,y_i'''(x_0)\delta^2 + \tfrac{1}{6}\,y_i^{IV}(x_0)\delta^3 + \cdots$$

(67)

We shall show that the difference between (66) and (67) is of the order of δ^4. Therefore we must expand the auxiliary quantities (65) into a Taylor series. In order to avoid lengthy formulas we shall also use the notation y_0 for the independent variable x, so that $y_0 = x$, $y_0' = 1$ and $y_0'' = 0$. We introduce the function f_0, which is always identical to zero. Consequently (64) is valid for $i = 0$ and auxiliary functions $k_0^{(0)}$ and $k_0^{(1)}$ are given by (65). These two functions are identical to zero. We shall use the abbreviations

$$f_{i,0} = f_i[x_0; y_j(x_0)]$$

$$f_{i,j} = \left(\frac{\partial f_i}{\partial y_j}\right)_{x=x_0}$$

$$f_{i,jk} = \left(\frac{\partial^2 f_i}{\partial y_j\,\partial y_k}\right)_{x=x_0}$$

The series expansion of (65) up to the third order of δ is

$$k_i^{(0)} = \delta f_{i,0}$$

$$k_i^{(1)} = \delta f_{i,0} + \delta^2 \sum_{j=0}^{n} f_{i,j} \left\{ \frac{2}{3} y_j'(x_0) + \frac{2}{9} k_j^{(0)} \right\}$$

$$+ \frac{1}{2} \delta^3 \sum_{j=0}^{n} \sum_{k=0}^{n} f_{i,jk} \frac{4}{9} y_j'(x_0) y_k'(x_0)$$

$$= \delta f_{i,0} + \frac{2}{3} \delta^2 \sum_{j=0}^{n} f_{i,j} y_j'(x_0)$$

$$+ \frac{2}{9} \delta^3 \left\{ \sum_{j=0}^{n} f_{i,j} f_{j,0} + \sum_{j=0}^{n} \sum_{k=0}^{n} f_{i,jk} y_j'(x_0) y_k'(x_0) \right\}$$

(68)

Differentiation of Eq. (64) yields

$$y_i''(x_0) = f_{i,0}$$

$$y_i'''(x_0) = \sum_{j=0}^{n} f_{i,j} y_j'(x_0)$$

(69)

$$y_i^{IV}(x_0) = \sum_{j=0}^{n} \sum_{k=0}^{n} f_{i,jk} y_j'(x_0) y_k'(x_0) + \sum_{j=0}^{n} f_{i,j} y_j''(x_0)$$

$$= \sum_{j=0}^{n} \sum_{k=0}^{n} f_{i,jk} y_j'(x_0) y_k'(x_0) + \sum_{j=0}^{n} f_{i,j} f_{j,0}$$

Now (68) can be written as

$$k_i^{(0)} = \delta y_i''(x_0)$$

$$k_i^{(1)} = \delta y_i''(x_0) + \tfrac{2}{3} \delta^2 y_i'''(x_0) + \tfrac{2}{9} \delta^3 y_i^{IV}(x_0)$$

(70)

Substitution of (70) into (66) shows that the difference between (66) and (67) is of the order of δ^4.

For the determination of the step length δ it is important to know the magnitude of the truncation error. In practice, it is not possible to calculate this error exactly. However, we can get an impression about its magnitude if we determine the third-order terms of (67), namely, $\tfrac{1}{6} y_i'''(x_0) \delta^3$ and $\tfrac{1}{6} y_i^{IV}(x_0) \delta^3$. Zonneveld (1964) derived formulas for these terms. Therefore we must first evaluate n auxiliary quantities

$$k_i^{(2)} = \delta f_i[x_0 + \delta; y_j(x_0 + \delta)]$$

(71)

We expand (71) in a Taylor series. Using (67) and (69), we obtain

$$k_i^{\{2\}} = \delta f_{i,0} + \delta^2 \sum_{j=0}^{n} f_{i,j}\{y_j'(x_0) + \tfrac{1}{2}\delta y_j''(x_0)\} + \tfrac{1}{2}\delta^3 \sum_{j=0}^{n}\sum_{k=0}^{n} f_{i,jk} y_j'(x_0) y_k'(x_0)$$

$$= \delta y_i''(x_0) + \delta^2 y_i'''(x_0) + \tfrac{1}{2}\delta^3 y_i^{IV}(x_0)$$

Now it can be seen that the third-order terms of (67) are approximated by

$$\varepsilon_i = \delta[-\tfrac{1}{4}k_i^{(0)} + \tfrac{1}{4}k_i^{(1)}] \approx \tfrac{1}{6}\delta^3 y_i'''(x_0)$$

$$\varepsilon_i' = \tfrac{1}{2}k_i^{(0)} - \tfrac{3}{2}k_i^{(1)} + k_i^{(2)} \approx \tfrac{1}{6}\delta^3 y_i^{IV}(x_0)$$

The terms neglected in these approximations are of the order of δ^4.

Suppose that tolerances τ_i and τ_i' are given for ε_i and ε_i', respectively. These tolerances will depend upon the required accuracy for the electron trajectory. The step length must be determined such that the truncation error is not too large:

$$\varepsilon_i \leq \tau_i \quad \text{and} \quad \varepsilon_i' \leq \tau_i' \quad (1 \leq i \leq n) \tag{72}$$

On the other hand, the number of required steps must be as small as possible. This is achieved if the equality sign holds for at least one of the inequalities (72). The step length that satisfies these requirements is called the optimum step length $\boldsymbol{\delta}$. In order to determine $\boldsymbol{\delta}$ we calculate first a step with the arbitrary step length δ and we calculate also ε_i and ε_i'. Since ε_i and ε_i' are approximately proportional to δ^3, the optimum step length $\boldsymbol{\delta}$ is given by the smallest of the quantities

$$\delta \left(\frac{\tau_i}{\varepsilon_i}\right)^{1/3} \quad \text{and} \quad \delta \left(\frac{\tau_i'}{\varepsilon_i'}\right)^{1/3} \quad (1 \leq i \leq n)$$

If $\delta > \boldsymbol{\delta}$, the step δ must be rejected and another step must be calculated with step length $\boldsymbol{\delta}$. If $\delta \leq \boldsymbol{\delta}$ the step δ may be accepted and we may calculate a next step. In order to determine the step length of the next step we can best use a linear extrapolation from two previous optimum step lengths $\boldsymbol{\delta}_1$ and $\boldsymbol{\delta}_2$ at the points x_1 and x_2, respectively. The approximated optimum step length at x_3 is then given by

$$\boldsymbol{\delta}_3 = \boldsymbol{\delta}_2 + (\boldsymbol{\delta}_2 - \boldsymbol{\delta}_1)\frac{x_3 - x_2}{x_2 - x_1}$$

We should use a somewhat smaller step length than the optimum value $\boldsymbol{\delta}$, since the value of $\boldsymbol{\delta}$ is determined only approximately and a small error in the value of $\boldsymbol{\delta}$ may cause a rejection of the step; whereas, if a small safety margin had been used, the step would probably have been accepted.

Often the evaluation of the functions f_i is very time consuming, so that the time required for the solution of the differential equations is determined mainly by the number of times that the functions f_i must be evaluated. The quantities $k_i^{(2)}$ are equal to the $k_i^{(0)}$ of the next step. Therefore using this third-order Runge-Kutta method we must evaluate two times per step the functions f_i, at least if the step is accepted, which is nearly always the case, since we applied a linear extrapolation of the step length and a safety margin. A corresponding fourth-order Runge-Kutta method would require four times per step the evaluation of the functions f_i. This drawback is usually not compensated by the larger step length that may be used.

A survey of Runge-Kutta expressions with formulas for the last term of the series expansion has been given by Zonneveld (1964). He derives formulas for first-order differential equations and for second-order differential equations with or without first derivatives at various orders of accuracy. We shall deal only with one other case, namely, the numerical evaluation of integrals such as the third-order aberration integrals. The integral is written as

$$y(x) = \int_{x_0}^{x} f(\xi)d\xi$$

It can be transformed to a differential equation of the first order:

$$y'(x) = f(x)$$

Here f depends only on x. The fourth-order Runge-Kutta formulas for this equation are

$$k^{(0)} = \delta f(x_0)$$
$$k^{(1)} = \delta f(x_0 + \tfrac{1}{2}\delta)$$
$$k^{(2)} = \delta f(x_0 + \delta)$$
$$y(x_0 + \delta) = y(x_0) + \tfrac{1}{6}[k^{(0)} + 4k^{(1)} + k^{(2)}]$$

This is the well-known Simpson formula. The last term that has been taken into account is of the order of δ^4. Its value is given by

$$\varepsilon = -\tfrac{2}{3}k^{(0)} + 4k^{(1)} + 2k^{(2)} - \tfrac{16}{3}k^{(3)} \approx \tfrac{1}{24}\delta^4 y^{IV}(x_0)$$

where

$$k^{(3)} = \delta f(x_0 + \tfrac{3}{4}\delta)$$

The proof is similar to the proof described in the foregoing. The step length is also determined similarly.

1.2.4. The Calculation of Electron Beams Taking into Account Thermal Velocities

A. Curvilinear Paraxial Electron Trajectories

In this section we shall deal with the calculation of the current density and the space charge of an electron beam with thermal velocities. It is difficult to determine the current density and the space charge of such a beam merely from a number of electron trajectories, calculated according to Section 1.2.3. Therefore we shall deal with another way of calculating these quantities. The method will be explained in its application to a high-current electron gun, because with this gun several aspects of the calculation can be shown. The theory is also applicable, however, to other kinds of beams. For example, the current density may be calculated in low-current electron guns, where the space charge is negligible, but where the thermal velocities must be taken into account. We may also calculate the current density of electron beams in lenses if thermal velocities and space charge must be considered. In a field-free region or in an accelerating field the defocussing of electron beams due to space charge and thermal velocities can also be determined.

Before we deal with the calculation of electron beams in Section C, we shall first derive some required properties of curvilinear paraxial rays (Section A) and of Gaussian beamlets (Section B).

We assume an electrostatic electron optical system, having rotational symmetry. The axis of symmetry coincides with the z axis of the Cartesian coordinate system x, y, z. Consider an electron that moves in the x–z plane (y velocity $= 0$). Because of the rotational symmetry the forces in the y direction are zero and the electron will not leave the x–z plane. The trajectory of this special electron will be called the central trajectory.

We define a coordinate system ξ, y, ζ with respect to this central trajectory (Fig. 13). The coordinate ζ is taken along the central trajectory, ξ is taken perpendicular to it in the x–z plane and y can be used as the third coordinate. The radius of curvature of the central trajectory, P, is a function of ζ. It is taken positive if the curvature is in the positive ξ direction.

Now we shall describe the motion of another arbitrary electron in the coordinate system ξ, y, ζ. First we determine the distance ds between the points Q_1, with coordinates ξ, y, ζ, and Q_2 with coordinates $\xi, y, \zeta + d\zeta$ (Fig. 14):

$$ds = \frac{P - \xi}{P} \, d\zeta$$

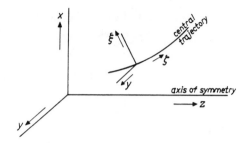

FIG. 13. The curved coordinate system ξ, y, ζ.

The velocity in the ζ direction is given by

$$v_\zeta = \frac{ds}{dt} = \left(1 - \frac{\xi}{P}\right)\dot{\zeta} \tag{73}$$

Because we deal with a curved coordinate system the ζ coordinate is compressed (or expanded) if $\xi \neq 0$ and the velocity in the ζ direction is not given by $\dot{\zeta}$. The velocities in the ξ and the y directions are as usual equal to $\dot{\xi}$ and $\dot{y}$.

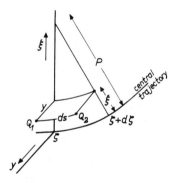

FIG. 14. The compression of ζ caused by the curvature of the central trajectory.

The centrifugal force in the coordinate system ξ, y, ζ is given by $- mv_\zeta^2/(P - \xi)$. This yields the following equations of motion in this system

$$\ddot{\xi} = \eta \frac{\partial V}{\partial \xi} - \frac{v_\zeta^2}{P - \xi}$$

$$\ddot{y} = \eta \frac{\partial V}{\partial y} \tag{74}$$

where $\eta = e/m$.

The energy equation is

$$\dot{\xi}^2 + \dot{y}^2 + v_\zeta{}^2 - 2\eta V - 2\eta\varepsilon = 0 \tag{75}$$

where ε is the velocity (in volts) of the electron under consideration when the potential is zero. The additive constant of the potential has been chosen such that the potential is zero, where the velocity of the electron at the central trajectory is zero.

Now we shall consider the trajectories of electrons near the central trajectory with velocities that do not differ much from the velocity of the central trajectory. These trajectories are usually called curvilinear paraxial electron trajectories. We shall neglect the terms with $\dot{\xi}^2$, y^2, $\dot{\xi}^2$, $\dot{y}^2$, and ε. The potential and its derivatives are approximated by a series expansion with respect to ξ and y:

$$V(\xi, y, \zeta) = \varphi(\zeta) + \varphi_\xi(\zeta)\xi$$

$$\frac{\partial V(\xi, y, \zeta)}{\partial \xi} = \varphi_\xi(\zeta) + \varphi_{\xi\xi}(\zeta)\xi \tag{76}$$

$$\frac{\partial V(\xi, y, \zeta)}{\partial y} = \varphi_{yy}(\zeta)y$$

where

$$\varphi(\zeta) = \varphi = V(0, 0, \zeta)$$

$$\varphi_\xi(\zeta) = \varphi_\xi = \left(\frac{\partial V}{\partial \xi}\right)_{\xi=y=0}$$

$$\varphi_{\xi\xi}(\zeta) = \varphi_{\xi\xi} = \left(\frac{\partial^2 V}{\partial \xi^2}\right)_{\xi=y=0}$$

$$\varphi_{yy}(\zeta) = \varphi_{yy} = \left(\frac{\partial^2 V}{\partial y^2}\right)_{\xi=y=0}$$

First derivatives with respect to y and second derivatives with respect to ξ and y are zero because of the rotational symmetry. They are omitted in (76).

Introduction of the paraxial approximations in the energy equation (75) yields

$$v_\zeta{}^2 = 2\eta[\varphi + \varphi_\xi\xi] \tag{77}$$

We use this expression to eliminate $v_\zeta{}^2$ from (74). We assume that ξ is small compared to the radius of curvature of the central trajectory, so that ξ^2/P^2

can be neglected. With the approximations (76) we obtain from (74)

$$\ddot{\xi} = \eta \varphi_\xi + \eta \varphi_{\xi\xi}\xi - \frac{2\eta\varphi}{P} - \frac{2\eta\varphi_\xi}{P}\xi - \frac{2\eta\varphi}{P^2}\xi$$

$$\ddot{y} = \eta \varphi_{yy} y$$

(78)

Since the central trajectory is an electron ray $\xi = y = \ddot{\xi} = \ddot{y} = 0$ must be a solution of (78). This yields

$$P = \frac{2\varphi}{\varphi_\xi}$$

(79)

With (79) we reduce (78) to

$$\ddot{\xi} = \eta \left[\varphi_{\xi\xi} - \frac{3}{2} \frac{\varphi_\xi^2}{\varphi} \right] \xi$$

$$\ddot{y} = \eta \varphi_{yy} y$$

(80)

In these equations the derivatives are with respect to time and φ is a function of ζ. In order to solve these equations we must use the relation between ζ and time. This relation is given by the energy equation (77) together with (73). After substitution of this relation in (80), the paraxial approximations can be made. This implies that for this purpose we may approximate the relation between ζ and time by

$$\dot{\zeta}^2 = 2\eta\varphi$$

Integration yields

$$t - t_0 = \int_0^\zeta \frac{d\zeta'}{[2\eta\varphi(\zeta')]^{1/2}}$$

where t_0 is the time at which the electron is at $\zeta = 0$.

Equations (80) are the curvilinear paraxial ray equations. They are similar to those derived by Wendt (1942–1943) and Sturrock (1952–1953). In our derivation we did not use the Laplace equation, and therefore Eqs. (80) are also valid for an electron optical system with space charge.

Since Eqs. (80) are homogeneous and linear, the general solution of each equation can be written as a linear combination of two special solutions. For the first equation we take the special solutions $u(t - t_0)$ and $v(t - t_0)$; for the second equation, $\mathbf{u}(t - t_0)$ and $\mathbf{v}(t - t_0)$. These special solutions are chosen such that they satisfy the initial conditions

$$u(0) = \mathbf{u}(0) = 0 \qquad \dot{u}(0) = \dot{\mathbf{u}}(0) = 1$$

$$v(0) = \mathbf{v}(0) = 1 \qquad \dot{v}(0) = \dot{\mathbf{v}}(0) = 0$$

(81)

Now the general solution of (80) can be written as

$$\xi = \dot{\xi}_0 u(t - t_0) + \xi_0 v(t - t_0)$$
$$y = \dot{y}_0 u(t - t_0) + y_0 v(t - t_0)$$

$$(82)$$

Henceforth the index zero will refer to quantities at $\zeta = 0$.

The functions u and v have to satisfy the first of Eqs. (80):

$$\ddot{u} = \eta \left[\varphi_{\xi\xi} - \frac{3}{2} \frac{\varphi_\xi^2}{\varphi} \right] u \qquad (83)$$

$$\ddot{v} = \eta \left[\varphi_{\xi\xi} - \frac{3}{2} \frac{\varphi_\xi^2}{\varphi} \right] v \qquad (84)$$

We multiply the first equation by v and the second equation by u. Subtraction of the equations obtained yields

$$\ddot{u}v - \ddot{v}u = d/dt(\dot{u}v - u\dot{v}) = 0$$

Integration and application of the initial conditions (81) gives us the Wronskian

$$\dot{u}v - u\dot{v} = \text{const} = 1 \qquad (85)$$

In a similar way we obtain

$$\dot{\mathbf{u}}\mathbf{v} - \mathbf{u}\dot{\mathbf{v}} = 1 \qquad (86)$$

Differentiation of (82) with respect to time yields

$$\dot{\xi} = \dot{\xi}_0 \dot{u} + \xi_0 \dot{v}$$
$$\dot{y} = \dot{y}_0 \dot{u} + y_0 \dot{v}$$

$$(87)$$

From (82) and (87) we obtain the reverse relations, which we shall need later on.

$$\xi_0 = \xi \dot{u} - \dot{\xi} u \qquad \dot{\xi}_0 = - \xi \dot{v} + \dot{\xi} v$$
$$y_0 = y\dot{\mathbf{u}} - \dot{y}\mathbf{u} \qquad \dot{y}_0 = - y\dot{\mathbf{v}} + \dot{y}\mathbf{v}$$

$$(88)$$

Near the cathode the derivation of the paraxial ray equations (80) is not valid. For example, near the cathode all the terms of (75) are of the same order of magnitude, so that the approximation (77) cannot be made. Nevertheless, it can be shown (Francken and Dorrestein, 1951) that the paraxial ray equations are also valid near the cathode, provided that near the cathode the longitudinal forces are large compared to the transverse forces and, as before, the transverse velocities are not too large. This is usually the case if the radius of curvature of the cathode is large compared to the width of the paraxial region. However, the longitudinal forces are not large compared to the transverse forces in a cathode ray tube near cutoff or, in some cases, near the edge of the beam (Hasker and Groendijk, 1962).

B. GAUSSIAN BEAMLETS

In this section we shall deal again with a given potential field having rotational symmetry. We consider a central trajectory, with paraxial rays around it (see Section A).

Assume a flat cathode at $\zeta = 0$ and suppose that the current density at this cathode is given by

$$\frac{i_b}{\pi R_0^2} \exp - \left\{ \frac{\xi_0^2 + y_0^2}{R_0^2} \right\} \tag{89}$$

The index zero will refer to quantities at $\zeta = 0$ and i_b is the total current, which is obtained by integration of (89) with respect to ξ and y. Since the electrons move in the positive ζ direction, the current i_b is negative. The current density (89) is a Gaussian function around the central trajectory, with a $1/e$ value R_0. How this current density is obtained physically is not important at the moment. We also assume, at the moment, that the space charge of the emitted electrons does not disturb the given rotationally symmetrical field.

The electrons are emitted thermally from the cathode. We shall only consider the transverse velocity spread and neglect the chromatic aberration caused by the longitudinal velocity spread. The electrons leave the cathode with longitudinal velocity zero. The potential of the cathode is taken equal to zero. Then at the cathode the electron-density function in the phase space is given by

$$\frac{2(-i_b)}{e\pi^2 \theta_0^2 R_0^2} \exp - \left\{ \frac{\xi_0^2 + y_0^2}{R_0^2} + \frac{\dot{\xi}_0^2 + \dot{y}_0^2}{\theta_0^2} \right\} \delta(v_{\zeta 0}^2) \tag{90}$$

where $\theta_0^2 = 2kT/m$, k is Boltzmann's constant, T is the cathode temperature, and δ is the Dirac delta function. Multiplication of (90) by $- ev_{\zeta 0}$ and integration with respect to the velocity coordinates yields the current density (89).

We take R_0 small and assume that all the electrons emitted according to (90) are paraxial with respect to the central trajectory. Of course there are electrons with large ξ_0, y_0, $\dot{\xi}_0$, or $\dot{y}_0$, which have nonparaxial trajectories, but we suppose that their number is relatively small. Thus we make no appreciable error by considering these few electrons as paraxial. The electrons constitute a thin beamlet around the central trajectory. Following Dorrestein (1950) we shall call it a Gaussian beamlet.

According to Liouville's theorem the electron-density function in the phase space is constant along an electron trajectory. Hence if we want to know the electron density at the point ξ, y, ζ, $\dot{\xi}$, $\dot{y}$, v_ζ of the phase space, we must trace back to the cathode ($\zeta = 0$). The coordinates at $\zeta = 0$ are given by (88) and by the energy relation

$$v_{\zeta 0}^2 = v_\zeta{}^2 - 2\eta V(\xi, y, \zeta) \tag{91}$$

In this energy relation paraxial approximations have been made concerning the velocities, but, for convenience later on, the potential has not been approximated.

Substitution of (88) and (91) in (90) yields the desired electron-density function. If we arrange the terms in the exponent and apply (85) and (86), we obtain the electron-density function

$$\frac{2(-i_b)}{e\pi^2\theta_0{}^2R_0{}^2}\exp-\left\{\frac{\xi^2}{R^2}+\frac{y^2}{\mathbf{R}^2}+\frac{(\dot{\xi}-\mu\xi)^2}{\theta^2}+\frac{(\dot{y}-\boldsymbol{\mu}y)^2}{\boldsymbol{\theta}^2}\right\}$$

$$\times\,\delta\{v_\zeta{}^2-2\eta V(\xi, y, \zeta)\} \tag{92}$$

where

$$R^2 = \theta_0{}^2u^2 + R_0{}^2v^2 \qquad \mathbf{R}^2 = \theta_0{}^2\mathbf{u}^2 + R_0{}^2\mathbf{v}^2$$

$$\frac{1}{\theta^2} = \frac{u^2}{R_0{}^2}+\frac{v^2}{\theta_0{}^2} \qquad \frac{1}{\boldsymbol{\theta}^2}=\frac{\mathbf{u}^2}{R_0{}^2}+\frac{\mathbf{v}^2}{\theta_0{}^2} \tag{93}$$

$$\frac{\mu}{\theta^2}=\frac{u\dot{u}}{R_0{}^2}+\frac{v\dot{v}}{\theta_0{}^2} \qquad \frac{\boldsymbol{\mu}}{\boldsymbol{\theta}^2}=\frac{\mathbf{u}\dot{\mathbf{u}}}{R_0{}^2}+\frac{\mathbf{v}\dot{\mathbf{v}}}{\theta_0{}^2}$$

From these expressions we obtain the relations

$$R\theta = R_0\theta_0 \qquad\qquad \mathbf{R}\boldsymbol{\theta} = R_0\theta_0 \tag{94}$$

$$R\dot{R} = \theta_0{}^2u\dot{u} + R_0{}^2v\dot{v} \qquad \mathbf{R}\dot{\mathbf{R}} = \theta_0{}^2\mathbf{u}\dot{\mathbf{u}} + R_0{}^2\mathbf{v}\dot{\mathbf{v}} \tag{95}$$

$$R\ddot{R} = \theta_0{}^2u\ddot{u} + R_0{}^2v\ddot{v} + \frac{\theta_0{}^2R_0{}^2}{R^2} \qquad \mathbf{R}\ddot{\mathbf{R}} = \theta_0{}^2\mathbf{u}\ddot{\mathbf{u}} + R_0{}^2\mathbf{v}\ddot{\mathbf{v}} + \frac{\theta_0{}^2R_0{}^2}{\mathbf{R}^2} \tag{96}$$

$$\mu = \frac{\dot{R}}{R} \qquad\qquad \boldsymbol{\mu} = \frac{\dot{\mathbf{R}}}{\mathbf{R}} \tag{97}$$

An investigation of (92) shows that, as expected, the only possible ζ velocity is $[2\eta V(\xi, y, \zeta)]^{1/2}$. The current-density distribution is obtained if we multiply (92) by $-ev_\zeta$ and integrate over the velocities. On using (94) this yields

$$\frac{i_b}{\pi R \mathbf{R}} \exp -\left\{ \frac{\xi^2}{R^2} + \frac{y^2}{\mathbf{R}^2} \right\}$$

The current-density distribution is Gaussian, as at the cathode. However, now the $1/e$ values are different in the ξ and in the y direction. From (92) it can be seen that the transverse velocity spread is also Gaussian. The magnitude of the spread in the ξ and y directions is determined by θ and $\boldsymbol{\theta}$, respectively. If $\theta = \boldsymbol{\theta} = 0$ there is no spread in the transverse direction and the only possible transverse velocity is $\mu\xi$, $\boldsymbol{\mu}y$. With (97) it can be seen that in this case we have a homocentric beam with a Gaussian current density distribution. If θ and $\boldsymbol{\theta}$ are not equal to zero, the spread of the transverse velocities is centered around the rays of the corresponding homocentric beam with $\theta = \boldsymbol{\theta} = 0$. The relation (94) states that the product of the "radius" of the beamlet and the "average transverse velocity spread" is independent of ζ. This invariance is well known in electron optics. Among others it can be found in slightly different forms in papers by Langmuir (1937), Herrmann (1958), and Hasker and Groendijk (1962).

It is possible to derive differential equations for R and $\mathbf{R}$. We multiply (83) by $\theta_0{}^2 u$ and (84) by $R_0{}^2 v$. After adding the equations and using (93) and (96), we obtain the equation for R:

$$\ddot{R} - \eta \left[\varphi_{\xi\xi} - \frac{3}{2} \frac{\varphi_\xi{}^2}{\varphi} \right] R - \frac{\theta_0{}^2 R_0{}^2}{R^3} = 0 \tag{98}$$

Similarly, we obtain

$$\ddot{\mathbf{R}} - \eta \varphi_{yy} \mathbf{R} - \frac{\theta_0{}^2 R_0{}^2}{\mathbf{R}^3} = 0 \tag{99}$$

If R_0 and θ_0 are given, we can calculate R and $\mathbf{R}$ from (98) and (99). The initial conditions are obtained from (93), (95), and (81):

$$R_{\zeta=0} = R_0 \qquad \mathbf{R}_{\zeta=0} = R_0$$

$$\dot{R}_{\zeta=0} = 0 \qquad \dot{\mathbf{R}}_{\zeta=0} = 0$$

Equations (98) and (99) are the usual paraxial ray equations with an additional term, which represents the thermal spread. The numerical calculation can be performed with the method described in Section 1.2.3,B. These equations can also be used to study the defocussing of paraxial beams by space charge and by thermal transverse velocities (Weber, 1964).

The space charge of the beamlet is obtained if we multiply (92) by $-e$ and integrate it with respect to the velocities. With (94) this yields

$$\frac{i_b}{\pi R\mathbf{R}[2\eta V(\xi, y, \zeta)]^{1/2}} \exp -\left\{\frac{\xi^2}{R^2} + \frac{y^2}{\mathbf{R}^2}\right\} \qquad (100)$$

The ξ coordinate can be expressed in the x coordinate (Fig. 15):

$$x = r_c + \xi/\cos \chi \qquad (101)$$

where r_c is the distance of the central trajectory to the axis and χ is the angle between the central trajectory and the axis, $\tan \chi = dr_c/dz$. The relation (101) is valid on the assumption that the radius of curvature is large compared with ξ. Using (101), we can write the space charge (100) as

$$\frac{i_b}{\pi R\mathbf{R}(2\eta V)^{1/2}} \exp -\left\{\frac{(x - r_c)^2 \cos^2 \chi}{R^2} + \frac{y^2}{\mathbf{R}^2}\right\}$$

The $1/e$ value in the x direction is a factor $1/\cos \chi$ larger than in the ξ direction because we have an oblique cross section through the beamlet.

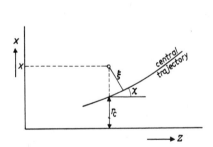

FIG. 15. The relation between x and ξ.

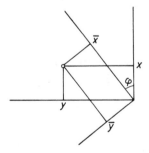

FIG. 16. The rotated coordinate system $\bar{x}$, $\bar{y}$.

Until now we have considered a beamlet around a central trajectory in the x–z plane. Since the system has rotational symmetry, we can rotate the central trajectory and the beamlet over an angle φ. The relation between the rotated coordinate system $\bar{x}$, $\bar{y}$ (Fig. 16) and the fixed system x, y, is given by

$$\bar{x} = x \cos \varphi + y \sin \varphi \qquad \bar{y} = -x \sin \varphi + y \cos \varphi$$

In the fixed coordinate system the space charge at the point $(x, 0)$ due to the rotated beamlet is

$$\frac{i_b}{\pi R \mathbf{R}(2\eta V)^{1/2}} \exp - \left\{ \frac{(x \cos \varphi - r_c)^2 \cos^2 \chi}{R^2} + \frac{x^2 \sin^2 \varphi}{\mathbf{R}^2} \right\}$$

Suppose that we have M beamlets, numbered 1, 2, ..., m, ..., M. Each beamlet is obtained by rotation over an angle $\varphi = 2\pi m/M$, so that the beamlets are situated at equal distances in angular direction. The total space charge of these beamlets at the point $(x, 0)$ of the fixed coordinate system is

$$\frac{i_a}{\pi M R \mathbf{R}(2\eta V)^{1/2}}$$

$$\times \sum_{m=1}^{M} \exp - \left\{ \frac{(x \cos 2\pi m/M - r_c)^2 \cos^2 \chi}{R^2} + \frac{x^2 \sin^2 2\pi m/M}{\mathbf{R}^2} \right\} \quad (102)$$

where $i_a = M i_b$ is the total current of the M beamlets.

If we take M infinitely large, the sum changes into an integral. Because the distance between the beamlets in angular direction is now infinitely small, the resulting space charge will have rotational symmetry and we obtain an "annular beamlet." Because of the rotational symmetry, x can be replaced by r. With $2\pi m/M = \vartheta$ and $2\pi/M = d\vartheta$, we obtain from (102) the space charge of the annular beamlet

$$\varrho_a(r, z) = \frac{i_a}{2\pi^2 R \mathbf{R}(2\eta V)^{1/2}} \int_0^{2\pi} \exp - \left\{ \frac{(r \cos \vartheta - r_c)^2 \cos^2 \chi}{R^2} + \frac{r^2 \sin^2 \vartheta}{\mathbf{R}^2} \right\} d\vartheta$$
$$(103)$$

The current density in the z direction of an annular beamlet is given by the same expression, multiplied by $(2\eta V)^{1/2} \cos \chi$.

At a flat cathode $R = \mathbf{R} = R_0$ and $\cos \chi = 1$. Now (103) can be integrated and we obtain for the current density at the cathode of an annular beamlet

$$j_a(r_0) = \frac{i_a}{\pi R_0^2} I_0 \left\{ \frac{2 r_{c0} r_0}{R_0^2} \right\} \exp - \left\{ \frac{r_0^2 + r_{c0}^2}{R_0^2} \right\} \quad (104)$$

where I_0 is the modified Bessel function of the first kind and of zero order.

C. CALCULATION OF ELECTRON BEAMS

We shall describe now the calculation of the electron beam in an electron gun. We take into account the transverse thermal velocity spread and neglect the longitudinal thermal velocity spread. The separate parts of the calculation will be treated first and finally a synthesis will be given. The subject of Sections B and C was treated by Weber (1965).

First, the total beam is divided into a number of annular beamlets, each having its own central trajectory. These central trajectories leave the cathode with zero initial velocity and at regular distances in the r direction, $r_{co} = n\Delta r$ where Δr is the radial spacing between the central trajectories and the integer $n = 0, 1, 2, \ldots$. The properties of a single annular beamlet belonging to a given central trajectory have been described in the previous section.

The current of an annular beamlet, i_a, is different for the different beamlets and will be denoted by $i_a(r_{co})$. These currents are chosen such that the sum of the cathode current densities of all the annular beamlets is equal to the actual current density $j(r_0)$. At the moment we shall consider $j(r_0)$ as a given function of r_0; later on we shall show how it is calculated from the potential field. Of course, it is mathematically not possible to write a given function $j(r_0)$ as a sum of a finite number of functions (104); but in practice a useful approximation can be obtained.

The width of an annular beamlet, which is determined by R_0, must be taken of the same order of magnitude as the spacing between the central trajectories Δr. If we take R_0 too small, the current will be concentrated near the central trajectories, and between the central trajectories the current density will be low. If we take R_0 too large, changes of $j(r_0)$ as a function of r_0 cannot be represented adequately. A suitable choice is $R_0 = \Delta r$.

In practice it is appropriate to take the current of an annular beamlet equal to

$$i_a(r_{co}) = 2\pi \int_{r_{co}-\frac{1}{2}\Delta r}^{r_{co}+\frac{1}{2}\Delta r} j(r)r\,dr$$

With this choice a reasonable approximation of $j(r_0)$ is obtained and the total current of the actual beam is equal to the sum of the currents of the annular beamlets.

The current density at the cathode is determined by the condition that the field strength at the cathode shall be zero. The condition causes an infinitely large space charge near the cathode. This singularity yields difficulties with the numerical calculations in this region. Therefore, we divide the region of the electrode system into two parts and calculate these parts separately, using different numerical methods. The first part, region I, is the region between the cathode and a plane at a distance $d = 1\frac{1}{2}$ mesh lengths from the cathode. The second part, region II, is the region from a plane at a distance of one mesh length from the cathode to the end of the electrode system.

The region II, which has no singularities, is calculated with the numerical methods described in the previous sections. The two regions overlap each other over one half mesh length. In this overlapping region the potential and the field strength of both regions must be equal to each other. How these regions are fitted to each other will be shown later on.

First we shall deal with the calculation of region I. This region is very thin and therefore in the radial direction the physical quantities change slowly, provided that the distances concerned are comparable to the thickness of region I. Therefore the z dependence of the potential may be described by the formula of the space-charge limited diode

$$V(r, z) = \left(\frac{9}{4\varepsilon_0 (2\eta)^{1/2}} \right)^{2/3} [-j(r)]^{2/3} z^{4/3} \tag{105}$$

where the cathode is assumed to be at $z = 0$.

The derivative with respect to z is

$$\frac{\partial V(r, z)}{\partial z} = \left[\frac{1}{\varepsilon_0} \left(\frac{6}{\eta} \right)^{1/2} \right]^{2/3} [-j(r)]^{2/3} z^{1/3} \tag{106}$$

Assume that this derivative is known as a function of r at the distance d from the cathode (d is the thickness of region I). Then with (106) the function $j(r)$ can be calculated as

$$-j(r) = \varepsilon_0 \left(\frac{\eta}{6} \right)^{1/2} \frac{E(r)^{3/2}}{d^{1/2}} \tag{107}$$

where

$$E(r) = \left[\frac{\partial V(r, z)}{\partial z} \right]_{z=d} \tag{108}$$

Substitution of (106) in (105) yields the potential distribution in region I expressed in $E(r)$.

$$V(r, z) = \frac{3}{4} E(r) \frac{z^{4/3}}{d^{1/3}} \tag{109}$$

With this potential distribution the equations of motion of the central trajectory are

$$\ddot{r}_c = \eta \frac{\partial V}{\partial r} = \frac{3}{4} \eta \left(\frac{dE(r)}{dr} \right)_{r=r_c} \frac{z_c^{4/3}}{d^{1/3}}$$

$$\ddot{z}_c = \eta \frac{\partial V}{\partial z} = \eta \frac{E(r_c)}{d^{1/3}} z_c^{1/3} \tag{110}$$

The initial conditions at $t = t_0$ are $r_c = r_{c0}$ and $z_c = \dot{r}_c = \dot{z}_c = 0$. In order to solve (110) the function $E(r_c)$ is expanded into a power series of $r_c - r_{c0}$. We expand $r_c - r_{c0}$ and z_c into power series of $t - t_0$. After substituting these series in (110) and equating equal powers of $t - t_0$, we obtain for the leading terms of the time-dependent series

$$r_c - r_{c0} = \frac{\eta^3}{1440} \frac{E(r_{c0})^2}{d} \left(\frac{dE(r)}{dr} \right)_{r=r_{c0}} (t - t_0)^6$$

$$z_c = \left(\frac{\eta}{6} \right)^{3/2} \frac{E(r_{c0})^{3/2}}{d^{1/2}} (t - t_0)^3$$

(111)

In a similar way we can solve the equations for R (98) and $\mathbf{R}$ (99). Using the potential (109) and the value of z_c given by (111), we obtain for the first terms of the series

$$R = \mathbf{R} = R_0 + \frac{\theta_0^2}{2R_0} (t - t_0)^2 - \frac{\theta_0^4}{8R_0^3} (t - t_0)^4$$

(112)

A difference between R and $\mathbf{R}$ is found in the higher order terms.

We have solved region I and we have now reached the stage wherein we can give a survey of the calculation of the whole system. We start with the solution of the Laplace equation for the whole system (region I + region II), using the method described in Section 1.2.2. From this Laplace field we calculate $E(r)$ (108). With (107) the current density at the cathode, $j(r_c)$ can be determined.

Now we calculate in region I the first annular beamlet. An electron on the central trajectory of this beamlet arrives at the plane $z = d$ at the time t_d. This time is calculated from the second equation of (111):

$$t_d - t_0 = \left[\frac{6d}{\eta E(r_{c0})} \right]^{1/2}$$

At this plane the value of r_0, R, and $\mathbf{R}$ can be determined from (111) and (112). We also calculate the time derivatives of these quantities at the time t_d. Thus we obtain the initial conditions for the calculation in region II and we can perform the calculation in that region. At the mesh points of region II we determine the space charge of the annular beamlet by means of (103). The other annular beamlets are calculated in a similar way. The total space charge at the mesh points is obtained by addition of the space charges of the annular beamlets.

Using (109) we calculate the potential at one mesh length from the cathode. The Laplace potentials are replaced by these new potentials.

Then the Poisson equation is solved in region II, where the potentials at one-mesh-length distance from the cathode are used as fixed boundary potentials.

From the Poisson field we can again determine $E(r)$ with (108) and in a similar way we calculate the next cycle (these cycles must not be confused with the cycles of the potential calculation of Section 1.2.2). We proceed until a stationary situation is reached. It can be seen that in this stationary situation, the potential and the field strength in the overlapping part of the regions I and II fit one another.

If we calculate a gun in the way described above, we find that the final result is approached like a damped oscillation. An example is given in Fig. 17.[1] The full line in this figure shows the total beam current of the gun in

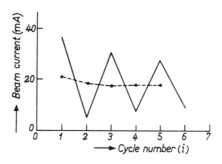

FIG. 17. The total current of the gun of the $2\frac{1}{2}$-mm "Philips" reflex klystron DX 237 at the various cycles of the calculation without underrelaxation (solid line) and with underrelaxation (dashed line).

a $2\frac{1}{2}$-mm "Philips" reflex klystron DX 237 at the various cycles of the calculation. It can be seen that the convergence is very slow. A faster convergence is obtained if we use underrelaxation (Kirstein and Hornsbey, 1964). The oscillations are due essentially to the way in which the current density is calculated. Oscillations of the cathode current density cause similar oscillations of the space charge in the whole region. A too large value of $E(r)$ causes a current density, which is too high (107). Because of the large space charge caused by this high current density at the next cycle, $E(r)$ will be too low. The low value of $E(r)$ causes in turn too large a value of $E(r)$ in the next cycle, and so on. This oscillation can be damped if we do not use the $E(r)$ calculated from (108), but use instead a function somewhere

[1] Figures 17–19 were calculated at the Philips Computing Center by Mr. Th. P. M. de Grefte and Mr. H. B. Nota.

between the previous $E(r)$ and the $E(r)$ calculated from (108):

$$E^{(i)}(r) = E^{(i-1)}(r) + a \left[\left(\frac{\partial V(r, z)}{\partial r} \right)_{z=d} - E^{(i-1)}(r) \right] \qquad (113)$$

where $E^{(i)}(r)$ is the function used at the ith cycle and a is the underrelaxation parameter, which is smaller than one. In practice a suitable value of a is 0.6 or 0.7. Also at the first cycle the current, which is calculated from the Laplace field, will be too large. Underrelaxation at the first cycle is obtained if we take $E^{(0)}(r) = 0$ and then use (113) to calculate the function $E^{(1)}(r)$ that is applied in the first cycle. As shown in Fig. 17 (dashed line), with the underrelaxation the oscillation is damped and convergence is much faster. The underrelaxation is equivalent to the overrelaxation used in the solution of Poisson's equation in Section 1.2.2.

Figure 18 shows the gun of the $2\frac{1}{2}$-mm "Philips" reflex klystron DX 237. It consists of a cathode (0V), a beam-forming electrode ($-$ 10 V; radius of the aperture, 290 μ) and an anode (2500 V; radius of the aperture, 75 μ).

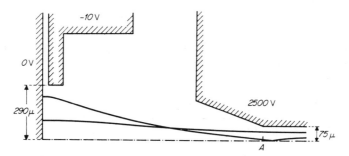

FIG. 18. The gun of the $2\frac{1}{2}$-mm "Philips" reflex klystron DX 237, with two central trajectories.

Two central trajectories, which were calculated in the space-charge field, are drawn. The inner trajectory does not cross the axis, but the outer trajectory does, because of spherical aberration. Figure 19 shows the calculated current density at the cathode and at the anode (at a plane through the point A of Fig. 18). The dashed line shows the cathode current density on another scale. Since the gun is only approximately a Pierce gun, the cathode current density is not rectangular. The calculated cathode current (18 mA) is in full agreement with the measured current. From the anode current density it can be seen that the electrons fill almost entirely the anode aperture, while only a small part hits the anode. This is also in agreement with measurements. The variation of the current density over the cross

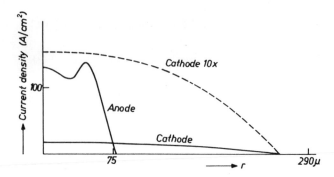

FIG. 19. The current density at the cathode and at the anode. The dashed line presents the cathode current density on a scale that is enlarged 10 times.

section can be explained by a caustic surface, which is caused by lens errors. This caustic surface is also found in other cross sections. Because of the thermal velocities the current density at the caustic surface is not infinitely large.

REFERENCES

Bellman, R. (1960). "Introduction to Matrix Analysis." McGraw-Hill, New York.

Birkhoff, G., Varga, R.S., and Young, D. (1962). *Advan. Computers* **3**, 189.

Carré, B. A. (1961). *Computer J.* **4**, 73.

Dorrestein, R. (1950). *Philips Res. Rept.* **5**, 116.

Durand, E. (1957). *Compt. Rend.* **244**, 2355.

Flanders, D. A., and Shortley, G. (1950). *J. Appl. Phys.* **21**, 1326.

Forsythe, G. E., and Wasow, W. R. (1960). "Finite-difference Methods for Partial Differential Equations." Wiley, New York.

Francken, J. C., and Dorrestein, R. (1951). *Philips Res. Rept.* **6**, 323.

Frankel, S.P. (1950). *Math. Tables Aids Computation* **4**, 65.

Gerschgorin, S. (1930). *Z. Angew. Math. Mech.* **10**, 373.

Hasker, J., and Groendijk, H. (1962). *Philips Res. Rept.* **17**, 401.

Herrmann, G. (1958). *J. Appl. Phys.* **29**, 127.

Kirstein, P. T., and Hornsbey, J. S. (1964). *IEEE Trans. Electron Devices* **11**, 196.

Langmuir, D. B. (1937). *Proc. IRE* **25**, 977.

Richardson, L. F. (1910). *Phil. Trans. Roy. Soc.* **A210**, 307.

Shortley, G., Weller, R., Darby, P., and Gamble, E. H. (1947). *J. Appl. Phys.* **18**, 116.

Southwell, R. V. (1946). "Relaxation Methods in Theoretical Physics." Oxford Univ. Press (Clarendon), London and New York.

Sturrock, P.A. (1952–1953). *Phil. Trans. Roy. Soc.* **A245**, 155.

Varga, R. S. (1962). "Matrix Iterative Analysis." Prentice-Hall, Englewood Cliffs, New Jersey.

Wasow, W. (1955). *Z. Angew. Math. Phys.* **6**, 81.

Weber, C. (1962–1963). *Philips Tech. Rev.* **24**, 130.

Weber, C. (1963). *Proc. IEEE* **51**, 252.

Weber, C. (1964). *Proc. IEEE* **52**, 996.

Weber, C. (1965). *In* "Microwave Tubes, Proceedings of the 5th International Congress, Paris, 1964," p. 47. Academic Press, New York.

Wendt, G. (1942–1943). *Z. Physik* **120**, 710.

Young, D. M. (1954). *Trans. Am. Math. Soc.* **76**, 92.

Zonneveld, J. A. (1964). "Automatic Numerical Integration." Mathematisch Centrum, Amsterdam.

Zurmühl, R. (1958). "Matrizen." Springer, Berlin.

CHAPTER 1.3

ANALOGICAL METHODS FOR RESOLVING
LAPLACE'S AND POISSON'S EQUATIONS

Jan C. Francken

UNIVERSITY OF GRONINGEN
THE NETHERLANDS

Introduction

In physics we use analog methods to determine quantities for which direct measurements or calculations are difficult to perform. In general, the analog is a model in which a physical quantity can be described with the same differential equation as the quantity to be determined. In some cases, the analogy is quite obvious. One example is the electrolytic tank that is used as an analog of electrostatic fields. In other cases, there is no visual resemblance between the analog and the original problem. One example is found in elementary physics in which the electrical oscillator circuit is used as the analog of a mechanical pendulum.

In this chapter we shall discuss analog methods for the numerical solution of the differential equations of static electric and magnetic fields that are used in electron optics and particle dynamics. Although at the present time the numerical computation of such fields is greatly facilitated by the development of fast digital computers with large memories, we still use analog methods in those cases in which large computers are not readily available. As a matter of fact, analogs can be considered to be (analog) computers themselves. A combination with other (small) analog or digital computers can lead to results comparable to those that can be obtained with very large digital computers. 'Many small problems can be solved more quickly (and less costly) with the aid of an analog rather than with a digital computer. From a didactic point of view, solving problems with an analog computer can be much more revealing than solving problems with a digital computer. Furthermore, systems with only a few planes of symmetry (such as *quadrupole lenses*) can hardly be computed numerically, even with extremely fast and large digital computers.

In such cases, one of the analogs to be treated in this chapter, that is, the *electrolytic plotting tank*, offers interesting possibilities. With the aid of this analog, the *Laplace equation* can be solved and, with certain limitations, also the *Poisson equation*. Kirchhoff (1845) was the first person to publish the principle of this analog whereas Kennely and Whiting (1906) were the first to use an electrolyte. An extensive and thorough survey of the mathematical foundations as well as the application to numerous physical problems has been given by Malavard (1956a,b,c); this survey contains an extensive bibliography.

Section 1.3.1 is devoted to the principles and practical realization of the electrolytic tank analog and its application to the plotting of electrostatic and magnetic fields.

In Section 1.3.2 we discuss the *resistance network analog*. Contrary to the electrolytic tank analog, the resistance network is not a direct analog for solving a partial differential equation. Instead, it provides a solution of an equation of finite differences that corresponds to such a differential equation. An analog of this kind has been proposed already by Gershgorin in 1929 (published in Russian), and, independently, by Hogan (1943). However, credit has to be given to Liebmann (1949, 1950a) for showing the method to be superior to the electrolytic tank in regards to accuracy.

In this Section 1.3.2 we shall describe networks for solving the following: (1) Laplace and Poisson equations for both plane and rotationally symmetrical electrostatic fields; (2) the Laplace equation for the magnetic scalar potential in plane and rotational symmetrical fields; (3) the equation for the vector potential in plane fields and for the magnetic flux in rotationally symmetrical fields. A more exhaustive treatise, including other analogs and applications in various fields of physics, has been written by Karplus (1958).

The evaluation of field plots and their use for determining electron or ion trajectories is outside the scope of this chapter. The same holds for automatic and semiautomatic ray-tracing methods that can be applied together with either the resistance network or electrolytic tank analog. For these, we refer the reader to a survey article by Liebmann (1950c) for the earlier literature and to the bibliography (Verster 1960–1961; de Beer *et al.*, 1961–1962; van Duzer *et al.*, 1963) for some recent papers.

1.3.1. The Electrolytic Plotting Tank

A. ELECTROSTATIC FIELDS WITHOUT SPACE CHARGE

1. *General Theory of the Tank*

a. *The Laplace equation.* In a system consisting of conducting electrodes submerged in an electrolyte, the current density is determined by the relation:

$$\mathbf{j} = - \sigma \operatorname{grad} \varphi \tag{1.1}$$

where $\mathbf{j}$ is the current density vector, σ the conductivity, and φ the potential.

In the absence of current sources, the current density is divergence-free. Thus

$$\operatorname{div} \mathbf{j} = 0 \tag{1.2}$$

From (1.1) and (1.2), we get

$$\sigma \, \Delta \varphi + \operatorname{grad} \sigma \cdot \operatorname{grad} \varphi = 0 \tag{1.3}$$

Therefore, in a homogeneous electrolyte in which the conductivity is a constant in space, Laplace's equation is fulfilled. Thus

$$\Delta \varphi = 0 \tag{1.4}$$

The potential field in the electrolyte, therefore, is analogous with a field in a dielectric medium, provided the boundary conditions are similar. In view of the linearity of Laplace's equation (1.4), this means that both the geometrical dimensions and the potential can be scaled to convenient values. We can measure the potential in the electrolyte with the aid of a probe, connected to a voltmeter with a high internal resistance in order to minimize field disturbance. It may be noted that a direct measurement is not possible for the actual fields (for example, *in vacuo*) because of the high resistance of the dielectric.

b. *Boundary conditions.* The solution of (1.4) is fully determined if, on a number of surfaces, one of which completely encloses the field region, either of the following quantities is given

(a)   The potential φ;
(b)   The normal component of the potential gradient $\partial \varphi / \partial n$;
(c)   A functional relation between φ and $\partial \varphi / \partial n$.

The solution is also determined in those cases in which in some parts of the boundaries, one of these conditions is fulfilled, and, in other parts, another. In many practical problems case (a) arises where boundaries are formed by equipotential electrodes. In the tank metallic electrodes, fed with the appropriate potential, are used to establish this type of boundary condition.

As for case (b), we can distinguish between two possibilities:

(b1) $$\partial \varphi / \partial n = 0 \qquad (1.5)$$

From (1.1) it follows that in this case in the tank the normal component of the current density vanishes:

$$j_n = 0 \qquad (1.6)$$

This condition is obviously fulfilled at the surface of *insulating boundaries.*

Equation (1.5) is also the condition for a *plane of symmetry* of the potential field (mirror plane). This can be made clear in the following way.

Consider the field between two equal charges. Obviously, the midplane between these charges is a plane of symmetry (Fig. 1.1). For symmetry reasons, the normal component of the field strength has to be zero in this plane; that is, Eq. (1.5) holds. Since the potential in either half of the field is uniquely determined by one charge, the boundary conditions in the midplane and the (zero) potential at infinity, a field-half can be generated by a single charge and an insulating plane, positioned in place of the midplane of the original charges and extending into infinity. At this boundary, again, Eq. (1.5) is valid. We can use similar reasoning with other field configurations.

It will be clear, therefore, that all insulating boundaries constitute mirror planes. In particular, the surface of the electrolyte acts as a plane of symmetry. This fact can be used by making the surface a plane of symmetry

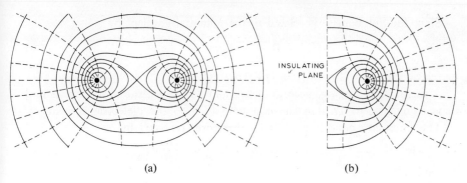

FIG. 1.1. (a) The electrostatic field of two equal charges; ———— equipotentials; – – – fieldlines. (b) The electrostatic field of one charge opposite a flat insulating plane. This field is the same as one half of the field in (a).

of the field to be plotted. Measurements in this plane can easily be made, since the probe need not be submerged in the electrolyte.

In most applications of the electrolytic tank measurements are confined to such a plane of symmetry. In many cases (see Sections 1.3.1,A, 3a and 3b) measurements in one plane suffice; in other cases, measurements in two such planes are required to be able to compute the complete field.

The latter category accommodates fields with two mutually perpendicular planes of symmetry, such as those which occur with quadrupole lenses. These types of fields, because of the large number of cells required, are difficult to evaluate with the aid of numerical methods (computers, resistance networks); for that reason, the electrolytic plotting tank is still an extremely useful device.

(b2)
$$\frac{\partial \varphi}{\partial n} = -\frac{q_{\text{eff}}}{\varepsilon} \qquad (1.7)$$

This is the case when the field region is bounded by a charged insulator, carrying an "effective" surface-charge density q_{eff}.[1] An example is the potential distribution in the belt region of a van de Graaff generator.

The analog of (1.7) for the tank can be found from the following considerations. At a charged, insulating surface, the normal component of the flux density **D** that is due to the charge is given by

$$D_n = q_{\text{eff}} \qquad (1.8)$$

[1] The relation between q_{eff} and the total charge density q depends on the boundary configuration. For a thin insulator in a plane of symmetry $q_{\text{eff}} = \frac{1}{2}q$.

Since the flux density is related to the potential gradient by

$$\mathbf{D} = - \varepsilon \operatorname{grad} \varphi \tag{1.9}$$

relation (1.7) follows from (1.8) and (1.9).

In the tank analog, according to (1.1) and (1.9), the current density $\mathbf{j}$ is equivalent to $\mathbf{D}$, and the conductivity σ to the dielectric constant ε. The boundary condition (1.7) is, therefore, replaced by

$$\partial \varphi / \partial n = - j_n / \sigma \tag{1.10}$$

with

$$j_n = (\sigma / \varepsilon)\, q_{\mathrm{eff}} \tag{1.11}$$

This condition can be realized in the tank by injecting currents from the boundary. For this purpose, the (insulating) boundary is supplied with thin and narrow metallic strips, fed through (variable) resistors in order to establish the required current-density distribution (Miroux, 1958).

2. Practical Realization

a. *General principle of measurement.* Field plotting with the electrolytic tank involves the measurement of the potential of a probe with respect to the potentials of the electrodes in the tank. To this purpose, electrodes with the highest and lowest potential, respectively, and a precision decade potentiometer, are fed in parallel from a voltage source. A null reading instrument, such as an electronic voltmeter or an oscilloscope, is connected between the potentiometer moving contact and the probe (Fig. 2.1.). If intermediate voltages are required for other electrodes, these can be supplied by a voltage divider.

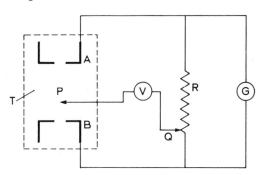

FIG. 2.1. Basic circuit for field-plotting with the electrolytic tank. T, tank; A, B, electrodes; P, probe; R, potentiometer with moving arm Q; G, generator; and V, null-reading voltmeter.

In general, more accurate results can be obtained if a problem is split into two-voltage problems. A specific solution is then constructed by using the superposition principle for linear, homogeneous differential equations. Where it is required to measure the field strength directly, two (Sander and Yates, 1953) or even four probes at a fixed distance are used (Verster, 1963) and the potential differences between these probes have to be measured.

b. *Polarization.* The bridge circuit cannot be fed from a direct voltage source, since the electrolyte would decompose by electrolysis. However, the use of alternating current does not solve the whole problem, since now *polarization effects* upset the simple linear relation (1.1). The electrodes -electrolyte system does not behave like an ohmic resistance; but, as a result of polarization, a reactive element is added to the resistance. The impedance of the model can, approximately, be simulated by a series connection of a resistance R_e, representing the bulk electrolyte, terminated at both ends by an impedance Z_p, consisting of a resistance R_p and a capacitance C_p. (Fig. 2.2.)

This simulation can be made clear as follows. Consider a simple cell (Fig. 2.3), consisting of two flat, parallel electrodes in a cylindrical tube,

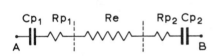

FIG. 2.2. Series connection representing the tank impedance. R_e, resistance of bulk electrolyte; R_{p_1}, R_{p_2} (ohmic) "interface resistance" between electrode and electrolyte; C_{p_1} and C_{p_2}, capacitances representing polarization effects; and A and B, electrodes.

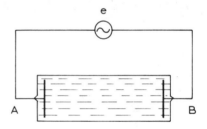

FIG. 2.3. Simple test cell for showing effect of polarization.

filled with an electrolyte. When an (alternating) voltage is applied between the electrodes, an ionic current sets in and ions give up their charges at the electrodes. The atoms formed in this way (for example, H and O atoms from H^+ and OH^- ions) give rise to the formation of double layers causing different work functions at the electrodes. Initially, the resulting *electromotive force* (emf) of this element *in statu nascendi*, can be assumed to be proportional to the electrode coverage, and therefore, to the amount of transported charge:

$$e_p = \lambda \int_0^t j \, dt = (\lambda/A) \int_0^t i \, dt \tag{2.1}$$

where e_p is the instantaneous polarization potential; λ a constant; A the electrode area; and i the total current; the current density being uniform over the electrode area A in this case. By feeding the electrodes from an external source e, we have

$$i(R_e + R_p) = e - (\lambda/A) \int_0^t i \, dt \qquad (2.2)$$

In this equation R_e represents the resistance of the bulk electrolyte and R_p the "interface" resistance between electrolyte and electrodes.

Equation (2.2) is equivalent to that for a series connected RC circuit with

$$R = R_e + R_p \quad \text{and} \quad C = A/\lambda \qquad (2.3)$$

Since polarization occurs at both electrodes, the circuit of Fig. 2.2 is a better analog.

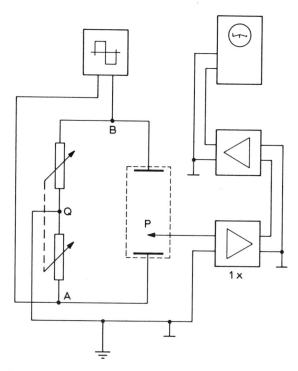

FIG. 2.4. Block diagram of measuring circuit, using square-wave excitation. The bridge is fed with a floating square-wave generator. The bridge voltage $V_P - V_Q$ (V_Q connected to ground) is fed to a cathode follower, that is amplified and displayed on an oscilloscope.

Polarization effects can be minimized by using suitable combinations of electrolyte and electrode materials and a high frequency of the driving voltage (see following subsections, c and d). However, because of the effect of stray capacitances, the null setting becomes less sensitive with increasing frequency. For this reason, a Wagner ground and adjustable capacitors have been used to obtain satisfactory balancing (Kennedy and Kent, 1956).

c. *Use of square-wave excitation.* This method was introduced by Sander and Yates (1953) and has been used by several others (Miroux, 1958; Verster 1960–1961), as well as by the author. It appears to be the most sensitive and accurate method and, therefore, its properties will be discussed in some detail. A block diagram of the circuit is given in Fig. 2.4. The bridge circuit is fed by a floating square-wave generator. The moving arm of the potentiometer is connected to ground, and the potential difference (PD) between the probe and ground [$(V_P - V_Q)$ in Fig. 2.5] is mini-

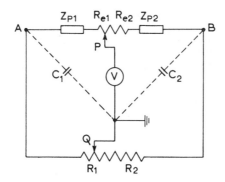

FIG. 2.5. Equivalent circuit of the bridge of Fig. 2.4. R_1 and R_2, resistances of the potentiometer arms; and C_1 and C_2, parasitic capacitances of the electrodes toward ground (for the remaining symbols, see Fig. 2.2).

mized. This PD can be obtained as the difference between two PD's [$(V_P - V_A)$ and $(V_Q - V_A)$ in Fig. 2.5]. The waveform of the latter PD may be as depicted in Fig. 2.6b, showing "negative" spikes with exponential decay at each change of sign of the input wave $(V_B - V_A)$ (Fig. 2.6a). Since, at any time, the sum of the PD's $(V_Q - V_A)$ and $(V_B - V_Q)$ must equal the amplitude a of the input voltage $(V_B - V_A)$, as depicted in Fig. 2.6a, the waveform $V_B - V_Q$ shows "positive" spikes (Fig. 2.6c). With another setting of the bridge, the waveforms $(V_Q - V_A)$ and $(V_B - V_Q)$ may appear reversed.

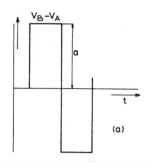

(a)

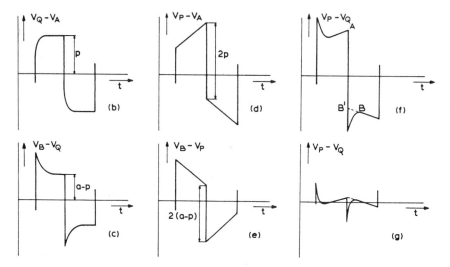

FIG. 2.6. Waveforms appearing between various points of the bridge circuit of Fig. 2.4. (a) Input square wave; (b and c) Square wave distorted by spikes with fast decay, resulting from parasitic capacitances (C_1 and C_2 in Fig. 2.5); (d and e) Effect of polarization at the electrode surfaces showing slow decay; (f and g) Wave shape resulting from the combined effects of parasitic capacitances and polarization.

The waveform in Fig. 2.6b can be explained as follows. Suppose the bridge setting to be such that

$$C_1/C_2 > R_2/R_1 \quad \text{(see Fig. 2.5)}$$

In this case, the PD ($V_Q - V_A$), as determined by the instantaneous displacement current through C_1 and C_2 at voltage reversal, is smaller than that, as determined by the ratio of resistances $R_1:R_2$. The excess charge can only be supplied or removed through the resistances R_1 and R_2, involving exponential transient currents.

Turning our attention to the probe, the PD $(V_P - V_A)$ will vary with time as depicted in Fig. 2.6d (or 2.6e). This wave shape results from the fact that, as a result of polarization at the electrode surfaces, a small PD that increases with time after each reversal appears between the electrode and the bulk electrolyte. This PD results from the combined effect of the interface resistance R_p and the increasing polarization emf e_p, as discussed in the previous paragraph.

The (nearly) vertical parts of the curves are mainly caused by the resistance of the electrolyte R_e and, to a much smaller extent, to R_p. The slope of the plateau results from the time-dependent emf e_p [see Eq. (2.1)].

The combined effect of e_p and R_p can be represented by the "interface impedances" Z_{p1} and Z_{p2} (see Fig. 2.5). The potential difference across Z_p, for relatively small time intervals, can be expressed by

$$v_p = [(\delta/\sigma) + \lambda t]\, j = (\delta + \lambda \sigma t)E \tag{2.4}$$

where v_p is the instantaneous polarization PD; E the field strength; δ, λ are constants; σ is the conductivity; and t the time subsequent to a change of current.

Careful measurements made by Sander and Yates (1953) showed δ to be independent of the electrolyte concentration. The parameter λ proved to be strongly dependent on the state of the electrode surface. From the wave shapes $(V_Q - V_A)$ and $(V_P - V_A)$ (Figs. 2.6b, d) it follows that a typical off-balance wave shape $(V_P - V_Q)$ is represented by Fig. 2.6f. Here the height AB' is a measure for the true PD $(V_P - V_Q)$, that is, if the effects of the stray capacitances C_1 and C_2 and the polarization capacitances C_{p1} and C_{p2} would be eliminated. The bridge is balanced by minimizing this PD (Fig. 2.6g). It should be noted, however, that in this case a residual error is caused by the interface resistance, represented by δ/σ in (2.4). The effect of this error is an apparent displacement of the electrodes by an amount δ.

It is possible to correct the signal $(V_P - V_Q)$ for the effects of stray capacitance and polarization (Miroux, 1958). To this purpose, from the square wave signal V_{AB} (Fig. 2.4) a sawtooth signal is derived by an integrating circuit, as well as a "spike" by a differentiating circuit. The time constant, amplitude, and phase of each signal can be adjusted. These correction signals are amplified and then added to the amplified signal V_{PQ} before the latter is fed to the oscilloscope.

Polarization effects at the *probe* can be made negligibly small by using a null-indicating instrument with a high impedance. In this case j in Eq.

(2.4) can be made sufficiently small. The amplitude a should be chosen small enough and the frequency f high enough in order to prevent excessive polarization at the electrodes; that is, the linear relation (2.4) should hold. However, an upper limit to the frequency f is imposed by the condition that the period of the square wave has to be long compared to the decay time of the "spikes" in Figs. 2.6b and c, in order not to upset the linearity of the greater part of the plateau. Typical values are

$$a = 6 \text{ V} \qquad f = 1000 \text{ cps}$$

d. *Choice of materials.* The material generally used for the probe is platinum wire about 0.2 mm in diameter. It is chemically inactive with weak electrolytes and has a small polarization PD. Although several kinds of *electrolytes* have been used in the past, *deionized water* seems to give the best results, because of its small conductivity. The current density j in (2.4) is very small in this case (Verster, 1963). However, in many cases ordinary *tap water* yields satisfactory results. The choice of the *electrode material* should be based on small values of δ and $\lambda\sigma$ (or λ/ϱ) in Eq. (2.4). In Table I values are given for electrode materials, generally considered to be suitable (Einstein, 1951; Sander and Yates, 1953; Verster, 1963). Although the best combination is obtained by using silver coated electrodes with deionized water, the other combinations also give satisfactory results, provided the time t is small enough to allow for the linear approximation (2.4).

TABLE I

VALUES OF δ AND λ/ϱ FOR VARIOUS ELECTRODE MATERIALS WITH DEIONIZED WATER AND TAPWATER [a]

	Stainless steel		Graphited brass		Silvercoated brass	
	δ (meters)	λ/ϱ (meters/s)	δ (meters)	λ/ϱ (meters/s)	δ (meters)	λ/ϱ (meters/s)
Deionized water ($\varrho = 10^4$ Ωmeters)	$16 \cdot 10^{-6}$	0.03	$17 \cdot 10^{-6}$	0.02	$12 \cdot 10^{-6}$	0.02
Tapwater ($\varrho = 27$ Ωmeters)	$2.3 \cdot 10^{-6}$	0.11	$8 \cdot 10^{-6}$	0.18	$22 \cdot 10^{-6}$	0.04

[a] These values are taken from measurements made in the Department of Technical Physics at the University of Groningen. The method of measurements is similar to that described by Sander and Yates. (1953. p. 174).

As for the *tank material*, an insulating, chemically inert, coating of the tank walls (such as polyvinylchloride) is recommended. Although the base material could be any metal (including sheet iron), it is recommended not to use metallic tank walls, in order to avoid large capacitances from the electrodes towards ground. Such capacitances would increase C_1 and C_2 (Fig. 2.5).

e. *Accuracy.* Inaccuracies may result for two reasons. First of all, experimental errors arise due to *inaccurate measurements*. The main source of error, in this case, may be the geometrical precision of the electrodes and positioning of the probe (Einstein, 1951). If accurate results are required, utmost care should be taken in machining and positioning the electrodes and determining the probe position.

Furthermore, the electrolyte should have uniform and isotropic conductivity. Notably, the surface should be kept clean. In cases in which tap water can be used, a slowly running water supply has definite advantages.

Polarization effects can be kept small, as shown in the previous subsections. The residual displacement error should be kept small with respect to the dimensions of the model.

The *meniscus* at the electrodes can cause a comparatively large error, which is not confined to regions close to the electrodes, but extends a considerable distance from them (Einstein, 1951). This meniscus can be eliminated by adjusting the electrolyte level, such that it coincides with the top of the electrodes. These have to be machined and adjusted with extreme care, in this case.

On taking all possible precautions, a precision of 0.2%, relative to the total voltage, is claimed by Einstein (1951). More generally, accuracies of the order of 1% have been reported.[2]

A second type of inaccuracy arises from the fact that the analogy between the actual problem and that represented by the model is not exact. The main source of error—in this case, the proximity of tank walls—has been discussed extensively by Kennedy and Kent (1956).

Insulating walls represent mirror planes, conducting walls are equipotentials (refer to Section 1.3.1,A,1). Generally, it is necessary to keep "open" parts of the model well away from the walls. A useful "rule of thumb" is to make the distance from a gap between electrodes to the tank wall equal to at least twice the gap width.

---

[2] A summary of the results reported up to 1950 can be found in a survey article by Liebmann (1950c).

Other imperfections in the analogy can arise from the finite conductivity of metallic electrodes (corrosion, bad connections) or from surface conduction of insulating boundaries.

3. Special Shapes of the Electrolytic Tank

a. *Two-dimensional problems.* These problems can be solved in a shallow tank with an insulating bottom. In practice, a ruled glass bottom is a great aid in the location of the electrodes (Kennedy and Kent, 1956). However, for most problems of this type, the resistance network (see Section 1.3.2) is a more accurate analog, whereas such problems can also be solved with relative ease with the aid of digital computers (cf. Chapter 1.2).

However, for problems involving *curved boundaries* the tank offers advantages. It may be noted that in cases in which a moderate accuracy suffices, the use of *resistance paper,*[3] with painted, conducting electrodes is a more rapid and easy method (reviewed by Liebmann, 1953). Recently, the use of thin metallic films deposited on a glass substrate has been reported (Peters, 1965).

b. *Axially symmetrical system.* Simple electrodes for such systems can be used if the electrolytic tank is wedge shaped (Bowman-Manifold and

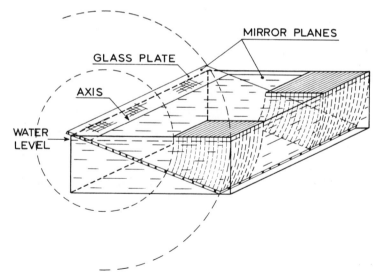

FIG. 3.1. Wedge tank with model of a two-electrode electrostatic aperture lens.

---

[3] A good type of graphited paper is manufactured by Felix Heinrich Schoeller, G.m.b.H. (Düren, Germany). It has a resistance of 450 ohms per square.

Nicoll, 1938). An inclined, insulating bottom acts as a mirrorplane in the same way as the surface of the electrolyte (Fig. 3.1). The line of intersection of these planes is the axis of rotational symmetry. If the angle of the wedge is sufficiently small, the sector-shaped electrodes can have straight sides rather than curved ones, simplifying their manufacture.

A serious drawback of the wedge tank is the effect of capillary forces. These give rise to meniscuses forming at the axis and at the electrodes (Fig. 3.2). For this reason, measurements near both the axis and the elec-

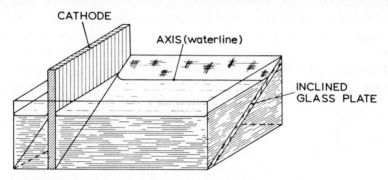

FIG. 3.2. Schematic illustration of capillary effects in a wedge tank.

trodes (for example, a cathode) tend to be very inaccurate. Various techniques have been used to minimize the meniscus effect (Kennedy and Kent, 1956). Van Duzer et al. (1963) could eliminate the meniscus at the axis of symmetry by adding a wetting agent (Du Pont Alkanol D.W., 0.314%) to tap water and selecting the correct slope of the tank bottom (10°). The wedge tank offers definite advantages for problems involving space charge (see Section 1.3.1,B). Otherwise, a greater precision in the vicinity of the axis can be obtained by making the wedge angle θ equal to π, that is, by using a deep tank and semicircular electrodes.

The electrolytic tank has advantages over the resistance network for the solution of axially symmetrical problems (see Section 1.3.2) in cases where curved electrodes are used.

B. ELECTROSTATIC FIELDS WITH SPACE CHARGE

4. Principle of Experimental Procedures

The electrolytic tank can be adapted to solve potential problems in which the field is determined by the Poisson equation:

$$\Delta\varphi = \operatorname{div}\operatorname{grad}\varphi = -\,\varrho/\varepsilon_0 \tag{4.1}$$

where ϱ represents the (volume) charge density *in vacuo*. If the space charge is caused by an *electron stream* with density j_c, then Eq. (4.1) can be written as

$$\text{div grad } \varphi = C j_c/(\varphi)^{1/2} \tag{4.2}$$

C is a constant, depending on units and φ is the potential measured relative to the cathode.

  a. *Shallow tank with variable depth.* A shallow tank with an insulating floor can be used to solve *two-dimensional problems* (Section 1.3.1,A,3), the potential φ being independent of the vertical coordinate z. If the depth h of such a tank is made a *slowly* changing function of x and y, the potential will still be practically independent of z. Therefore, denoting the electrolyte surface by $z = 0$, we have

$$\varphi(x, y, z) \simeq \varphi(x, y, 0) = \varphi(x, y) \tag{4.3}$$

Moreover, the vertical component $\mathbf{j}_z$ of the current density will be small compared with the horizontal component $\mathbf{j}_{xy}$. Application of the continuity condition for the current into a small prism with height h and sides dx and dy leads to

$$\text{div } \mathbf{J}(x, y) = 0 \tag{4.4}$$

where the "two-dimensional current density" $\mathbf{J}(x, y)$ can be approximated by

$$\mathbf{J}(x, y) \simeq h(x, y) \, \mathbf{j}_{x\,y}(x, y, 0) \tag{4.5}$$

The horizontal component of the current density at the surface is related to the potential by

$$\mathbf{j}_{xy}(x, y, 0) = - \sigma \text{ grad } \varphi (x, y) \tag{4.6}$$

making use of (4.3). From (4.4)–(4.6) it follows that

$$\Delta\varphi(x, y) + \frac{1}{h(x, y)} \{\text{grad } h(x, y)\} \cdot \text{grad } \varphi(x, y) = 0 \tag{4.7}$$

This equation is formally equivalent to (4.2) if the function $h(x, y)$ satisfies the relation

$$\text{grad } \varphi \cdot \text{grad}\{\ln h(x, y)\} = - C j_c/(\varphi)^{1/2} \tag{4.8}$$

Musson-Genon (1947) used this equation by (iteratively) shaping the bottom of a shallow tank. The method can also be adapted to a wedge tank to solve problems with *rotational symmetry*.

The procedure, being iterative, is laborious and the solution is approximate as a result of the simplifying assumption made for the independence of the current density from z.

b. *Current injection.* A model corresponding directly with Poisson's equation [(4.1)] is obtained in the tank if a current is injected with density j':

$$\Delta\varphi = -j'/\sigma \qquad (4.9)$$

Since it is impossible to inject current uniformly into the tank, discrete current injectors have to be used. This method has been investigated by Hollway (1955) and Loukochkov (1956) and further developed for use with the wedgetank by van Duzer and Brewer (1959).

5. *Practical Realization of the Current-Injection Method*

a. *Two-dimensional fields.* Integrating both Eqs. (4.1) and (4.9) over a small volume element δV yields the relation

$$\frac{\delta j'}{\delta q} = -k_v \frac{\sigma}{\varepsilon_0} \qquad (5.1)$$

where k_v is the scale factor. This equation determines the current $\delta j'$ to be injected into a volume element δV if the amount of charge δq in the corresponding element is known. Hollway (1955) has realized the charge simulation by current injection with the aid of submerged, vertical line sources, protruding through the insulating tank bottom. Such line sources represent charges in vertical cylinders of radius r_1, if two conditions are fulfilled. The depth of the submersion must be equal to r_1 and the depth of the bottom of the source has to be made equal to one half the depth of the tank (Fig. 5.1a, $y_2 = 0.5y_3$). This means that the current lead has to be partly insulated, as shown in Fig. 5.1.

The first requirement ensures that the maximum of the field gradient, measured at the surface of the electrolyte, is at a distance r_1 from the center of the source (Fig. 5.1b). The second requirement proves to yield the best match for the distribution of the gradient near the maximum, again measured at the liquid surface.

The field of a long cylinder of charge of uniform density ϱ and radius r_1 is given by

$$\frac{d\varphi}{dr} = -r \frac{\varrho}{2\varepsilon_0} \qquad \text{for} \quad 0 < r < r_1 \qquad (5.2)$$

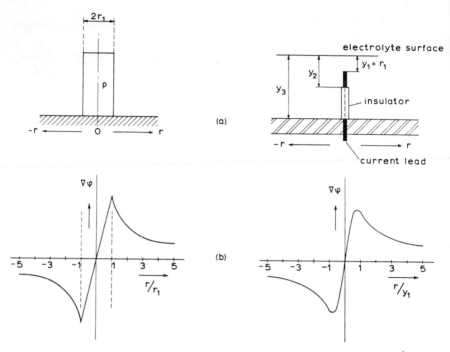

FIG. 5.1. Simulation of a charged cylinder by a submersed line source. (Courtesy: *Hollway*, 1954.) (a) Cylinder (left) with radius r_1 and volume charge density ϱ can be simulated by a thin current source, extending from a distance y_1 to a distance y_2 from the electrolyte surface. This line source is represented by a wire probe, that is insulated between y_2 and the bottom of the tank (y_3) and is fed through that bottom. (b) Field gradient distribution as a function of the distance from the center of a charged cylinder (left) and a line source (right).

and

$$\frac{d\varphi}{dr} = -\frac{1}{r}\frac{r_1^2\varrho}{2\varepsilon_0} \quad \text{for} \quad r > r_1 \qquad (5.3)$$

At a great distance from the source, the current density in the tank, will be nearly independent of depth. Therefore, if the source carries a current i

$$\frac{\partial\varphi}{\partial r} = -\frac{1}{r}\frac{i}{2\pi\sigma y_3} \qquad (5.4)$$

The relation between i and ϱ, taking scalefactors into account, follows from (5.3) and (5.4). From approximate calculations and measurements Hollway (1955) arrived at the optimum values mentioned for the heights y_1 and y_2 (Fig. 5.1a). For small values of y_1 $(y_1 < 0.2y_3)$, the crest in the gradient distribution becomes flatter. A better approximation can be obtained by using several, equally spaced sources extending from the bottom of the tank up to a distance r_1 from the water surface (Hollway, 1955).

A cloud of charge (usually an electron or ion beam) is simulated by dividing the region of interest into cylindrical, partly overlapping cells such that the total area of the circles equals the area of the space-charge region. However, in close proximity to a conducting electrode (for example, a cathode), the gradient at the liquid surface is slightly reduced. This effect can be partly corrected for by raising the injection currents accordingly (Hollway, 1955).

b. *Axially symmetrical systems.* The method of space-charge simulation by current injection can.be used here by placing current leads in the bottom of a wedge tank (see Section 1.3.1,A, 3b). When the charge cloud (beam of charged particles) itself has axial symmetry and is nearly cylindrical, the charge may be represented by point sources placed along the axis of symmetry, that is, the "waterline" in Fig. 3.2, p. 115 (Hollway, 1955).

A more generally applicable method has been developed by van Duzer and Brewer (1959).

In the deep parts of the wedge tank wire probes, as described in the previous paragraph, are used. In the shallow parts of the tank when the water depth is comparable to the radius r_1 of the desired cell in the sloping bottom, distributed sources of high resistivity are provided. Near the axis, these sources are uniformly distributed in order to obtain a uniform divergence of current at the liquid surface. In somewhat deeper zones the sources are made smaller to achieve the same result. Patterns and source types as used by van Duzer and Brewer (1959) are shown in Figs. 5.2 and 5.3.

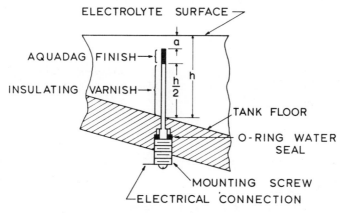

FIG. 5.2. Wire-probe current source for simulation of uniform space-charge density in deep regions of a wedge tank. (Courtesy: *van Duzer and Brewer*, 1959).

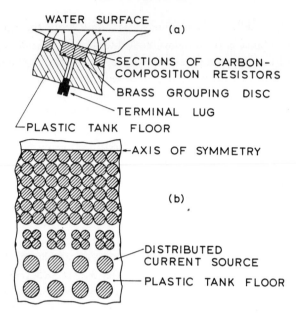

FIG. 5.3. Space-charge-simulation buttons used as distributed sources in the shallow-water region of the tank. (a) Cross section of one type of current source; (b) Top view of button pattern. (Courtesy: *van Duzer and Brewer*, 1959).

C. MAGNETIC FIELDS

6. Simulation of the Scalar Magnetic Potential

a. *Definition of the scalar potential.* Magnetic fields are characterized by the field strength **H** and the magnetic induction **B**. These quantities are related by the equation

$$\mathbf{B} = \mu\mathbf{H} = \mu_r\mu_0\mathbf{H} \tag{6.1}$$

where μ is the permeability, μ_r the relative permeability, and $\mu_0 = 4\pi \cdot 10^{-7}$ $V \cdot sec/A \cdot m$. For stationary fields, Maxwell's equations for **H** and **B** can be simplified to

$$\mathrm{rot}\,\mathbf{H} = \mathbf{j}_w \tag{6.2}$$

and

$$\mathrm{div}\,\mathbf{B} = 0 \tag{6.3}$$

where $\mathbf{j}_w$ is the current density in the wires producing the field. For regions outside current carriers, Eq. (6.2) reduces to

$$\mathrm{rot}\,\mathbf{H} = 0 \tag{6.2'}$$

This equation is fulfilled if **H** can be deduced from a *scalar potential*:

$$\mathbf{H} = - \operatorname{grad} \psi \tag{6.4}$$

From (6.3), (6.1), and (6.4)

$$\mu \operatorname{div} \operatorname{grad} \psi + \operatorname{grad} \mu \cdot \operatorname{grad} \psi = 0 \tag{6.5}$$

For regions with constant μ, we have

$$\operatorname{grad} \mu = 0$$

and, therefore, (6.5) reduces to the Laplace equation:

$$\Delta \psi = 0 \tag{6.6}$$

In cases where the magnetic fields are generated by current carrying wires, the function ψ is a multiple-valued function of the coordinates. Its value depends on the number of times the path of integration of **H** from (6.4) encircles the current carriers. The potential ψ can, in this case, be made a single-valued function by introducing a cut, which prevents the path of integration from enclosing the current. The scalar magnetic potential ψ is then defined in a singly connected region, bounded by a surface where (6.2') is valid.

b. *Boundary conditions.* Equipotential boundaries are formed by high permeability pole-pieces ($\mu = \infty$), since in this case it follows from (6.1) to (6.3) that the tangential component of **H**, $\mathbf{H}_{\mathrm{tang}} = 0$. On the other hand, along any **H**-line, the normal derivative of ψ vanishes. Therefore, any known **H**-line can be used as a boundary, with the condition:

$$H_n = \partial \psi / \partial n = 0 \tag{6.7}$$

c. *Analogy in the tank.* Since the Laplace equation is solved in the electrolytic tank, the scalar potential can be simulated if the correct boundary conditions are established. The simulation analogy is shown in Table II.

In Fig. 6.1 an example is given for the simulation of a solenoidal magnet with high permeability pole-pieces and cover (van Duzer *et al.*, 1963). In this case, the solenoid surface is represented by an insulating boundary, the pole pieces and cover by conducting electrodes. The "cut" is made in the plane of symmetry of the field. It is formed by two conductors, back-to-back, with a voltage difference equivalent with the number of ampere turns NI of the coil. Thus, the electric field lines are parallel to the axis

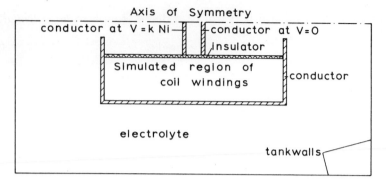

FIG. 6.1. Simulation of a solenoidal magnet with high-permeability polepieces and cover (Courtesy: *van Duzer, et al.*, 1963).

TABLE II

SIMULATION ANALOGY OF THE SCALAR MAGNETIC POTENTIAL

Magnetic field	Electrolytic tank
Scalar potential, ψ	Electrolytic potential, φ
Field intensity, $\mathbf{H}$	Field strength, $\mathbf{E}$
Flux density, $\mathbf{B}$	Current density, $\mathbf{j}$
Permeability, μ	Conductivity, σ

here, in as much as the induction lines are in the actual magnetic field. Apart from disturbing effects due to the tank walls and because the actual permeability is not infinite, the boundary conditions themselves are also approximations. Indeed, the solenoid surface near the coil ends does not necessarily coincide with $\mathbf{H}$-lines; but for relatively long magnets the analogy is quite satisfactory.

Saturated pole-pieces or other regions with relatively low permeability μ cannot be represented by conducting electrodes. Instead, such regions can be simulated by using electrolytes with a higher conductivity than that of the bulk electrolyte. From Table II it can be deduced that the following relation should be fulfilled:

$$\frac{\sigma'}{\sigma} = \frac{\mu}{\mu_0} = \mu_r$$

In this equation σ' and σ are the conductivity of the particular region and that of the bulk electrolyte, respectively, the latter representing vacuum ($\mu = \mu_0$). Since, obviously, the two regions with different electrolytes should be separated by impermeable walls, on one hand, but have to be electrically connected, on the other, an approximate solution can be found by using thin, insulating separation walls pierced with a great number of conducting pins (Fig. 6.2).

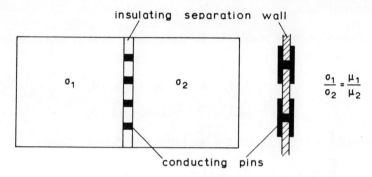

FIG. 6.2. Boundary simulation between regions with different permeability, μ_1 and μ_2.

7. Simulation of the Magnetic Vector Potential

a. *Definition of the vector potential.* For regions containing current carriers, instead of the scalar potential ψ, a vector quantity **A** can be defined by

$$\mathbf{B} = \text{rot } \mathbf{A} \tag{7.1}$$

If $\mathbf{A}(x, y, z)$ exists, Eq. (6.3) is automatically fulfilled. The quantity **A** is called the *magnetic vector potential.* Since it is not completely determined by (7.1), it is customary to let **A** also satisfy the condition

$$\text{div } \mathbf{A} = 0 \tag{7.2}$$

From (6.1), (6.2), (7.1), and (7.2) we have

$$\varDelta\mathbf{A} + \mu\{(\text{rot } \mathbf{A}) \times \text{grad } (1/\mu)\} + \mu\mathbf{j}_w = 0 \tag{7.3}$$

in which, generally, $\varDelta\mathbf{A}$ is defined by

$$\varDelta\mathbf{A} = \text{grad div } \mathbf{A} - \text{rot rot } \mathbf{A} \tag{7.4}$$

In Cartesian coordinates

$$\varDelta\mathbf{A} = (\nabla \quad \nabla)\mathbf{A} = \nabla^2\mathbf{A} \tag{7.4'}$$

For regions with constant μ, (7.3) reduces to

$$\Delta\mathbf{A} + \mu\mathbf{j}_w = 0 \qquad (7.5)$$

This equation can be solved with the electrolytic tank in cases where the magnetic potential has one component only, perpendicular to the plane of the flux density vector $\mathbf{B}$. This is the case for two-dimensional and axially symmetrical fields.

b. *Two-dimensional problems.* In these cases, there is one component only of both $\mathbf{A}$ and $\mathbf{j}_w$:

$$A_x = A_y = 0 \qquad \mathbf{A} = A(x, y)\mathbf{i}_z \qquad (7.6)$$

$$(j_w)_x = (j_w)_y = 0 \qquad \mathbf{j}_w = j_w(x, y)\mathbf{i}_z \qquad (7.7)$$

From (7.5)–(7.7) it follows that A satisfies the *Poisson equation*:

$$\Delta A(x, y) = -\mu j_w(x, y) \qquad (7.8)$$

Since, in this case Eq. (7.1) reduces to

$$\mathbf{B} = \nabla \times \mathbf{A} = \mathbf{i}_1 \frac{\partial A}{\partial y} - \mathbf{i}_2 \frac{\partial A}{\partial x} = -\mathbf{i}_3 \times (\nabla A) \qquad (7.9)$$

it follows that the $\mathbf{B}$ *lines are equipotentials for the vector potential,* since they are perpendicular to grad $\mathbf{A}$. From this and from what has been said in Section 1.3.1,C, 6a, it appears that in two-dimensional fields the determination of the scalar potential and the vector potential A are *conjugate problems*: Field lines in the one problem are equipotentials in the other and vice versa.

Furthermore, denoting the distance between equipotentials for $A = A_n$ and $A = A_n + dA$ by ds, it follows from Eq. (7.9) that

$$B \equiv |\mathbf{B}| = dA/ds$$

Since the magnetic flux between two $\mathbf{B}$ lines in plane fields is given by

$$dM = B \, ds$$

it follows that

$$dM = dA$$

This means that the flux between successive equipotentials is equal to the increment $\delta A = A_{n+1} - A_n$. The total flux between pole pieces, therefore, can easily be calculated if plots are made with a constant increment δA.

By setting up the tank analog for Eq. (7.8), using current injection (current density $\mathbf{j}'$), as described in Subsections 4b and 5a, the following analogy exists between the quantities in the actual field and in the tank (Table III).

TABLE III

SIMULATION ANALOGY OF THE MAGNETIC VECTOR POTENTIAL IN PLANE FIELDS

Magnetic field	Electrolytic tank
Magnetic potential, $A_z = A(x, y)$	Electric potential, $\varphi(x, y)$
Flux density, $\mathbf{B}$	$\mathbf{i_3} \times (\nabla\varphi) = -\,\mathbf{i_3} \times \dfrac{\mathbf{j}(x, y)^a}{\sigma}$
Current density,[b]	Injected current density,[c]
$\mathbf{j}_w = \mathbf{j}_w(x, y)\mathbf{i}_z$	$\mathbf{j}' = j'(x, y)\mathbf{i}_z$
Permeability, μ	Specific resistance, $\varrho_w = 1/\sigma$

[a] $\mathbf{j}(x, y)$ is the current density in the electrolyte.
[b] $\mathbf{j}_w(x, y)$ is the current density in wires.
[c] $\mathbf{j}'(x, y)$ is the injected current density.

Concerning the boundary conditions, curves with constant A—that is, (known) $\mathbf{B}$ lines—are simulated by high-conductivity electrodes. If the net current flow into the simulated part of the field differs from zero, the excess current will then flow to such equipotential electrodes (Peierls, 1946). Pole pieces with high permeability ($\mu = \infty$) are represented by insulating boundaries ($\varrho_w = \infty$).

In Fig. 7.1, for the case of two concentric, long cylinders, the tank analogs are compared for measuring either $\psi(x, y)$ or $A(x, y)$ directly. From this example and, more generally, from the discussions in this subsection and Subsection 6, it will be clear that for two-dimensional problems, the electrolytic tank can be set up, either as a *direct analog* or as a *conjugate analog*. In the first case the potential in the tank corresponds with the required function. In Fig. 7.1a this is the scalar potential distribution; in Fig. 7.1b, the vector potential distribution. However, these functions can also be found from the conjugate analog by measuring the field gradient. Streamlines, namely, lines of maximum gradient, in this case correspond to the required function. For example, in Fig. 7.1a equipotentials for $A(x, y)$ can be found in this way and, in Fig. 7.1b equipotentials for $\psi(x, y)$.

The gradients can be determined, either by graphical or numerical methods from the potential distributions or, for example, by using two closely

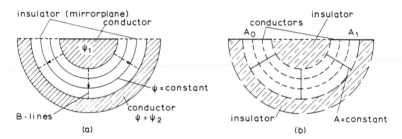

FIG. 7.1 Two ways for setting up a tank analog for determining two-dimensional magnetic fields, illustrated on the case of two concentric, long cylinders. (a) Direct measurement of the scalar magnetic potential, $\psi(x, y)$. (b) Direct measurement of the value of the magnetic vector potential, $A(x, y)$.

spaced probes, rotatable around an axis through one of these. The incremental direction as well as the magnitude of an equipotential can then be simultaneously determined.

c. *Axially symmetrical fields.* As a result of the symmetry, both $\mathbf{A}(r, z)$ and $\mathbf{j}_w(r, z)$ are in the tangential direction, both quantities being independent of θ

$$\mathbf{A}(r, z) = A(r, z)\mathbf{i}_\theta \tag{7.10}$$

$$\mathbf{j}_w(r, z) = j_w(r, z)\mathbf{i}_\theta \tag{7.11}$$

Equation (7.5) in this case, using cylindrical coordinates, reduces to

$$\frac{\partial^2 A}{\partial z^2} + \frac{\partial}{\partial r}\left(\frac{\partial A}{\partial r} + \frac{A}{r}\right) = -\mu j_w(r, z) \tag{7.12}$$

The boundary conditions for this equation along **B**-lines or high-permeability pole-pieces are not as simple as in the two-dimensional case. Therefore, Eq. (7.12) is not very suitable for solving with an electrolytic tank (nor with a resistance network analog). This will be clear from the following argument.

From Eq. (7.1) for axially symmetrical fields, the following relations can be derived. Thus

$$B_z = \frac{1}{r} \frac{\partial}{\partial r} (rA) \qquad (7.1')$$

$$B_r = - \frac{\partial A}{\partial z} \qquad (7.2')$$

The normal and the tangent to a boundary can be determined by the unit vectors. Thus

$$\mathbf{n}_I = \quad \mathbf{i}_z \cos \alpha + \mathbf{i}_r \sin \alpha \qquad (7.3')$$

$$\mathbf{t}_I = - \mathbf{i}_z \sin \alpha + \mathbf{i}_r \cos \alpha \qquad (7.4')$$

where α is the angle between the normal to the boundary and the positive z-axis. Therefore, the normal and tangential components of grad A, using Eqs. (7.1') and (7.2'), are given by

$$(\text{grad } A)_n = \frac{\partial A}{\partial z} \cos \alpha + \frac{\partial A}{\partial r} \sin \alpha = - B_r \cos \alpha + \left(B_z - \frac{A}{r} \right) \sin \alpha \ (7.5')$$

$$(\text{grad } A)_t = - \frac{\partial A}{\partial z} \sin \alpha + \frac{\partial A}{\partial r} \cos \alpha = B_r \sin \alpha + \left(B_z - \frac{A}{r} \right) \cos \alpha \ (7.6')$$

Along $\mathbf{B}$ lines, $B_n = 0$. Therefore,

$$\mathbf{B} \cdot \mathbf{n}_I = B_z \cos \alpha + B_r \sin \alpha = 0 \qquad (7.7')$$

From (7.6') and (7.7'):

$$(\text{grad } A)_t = - A/r \cos \alpha \qquad (7.8')$$

At high permeability pole pieces, $B_t = 0$. Therefore,

$$\mathbf{B} \cdot \mathbf{t}_I = - B_z \sin \alpha + B_r \cos \alpha = 0 \qquad (7.9')$$

From (7.5') and (7.9') we get

$$(\text{grad } A)_n = - A/r \sin \alpha \qquad (7.10')$$

The boundary condition (7.10') can be established with relative ease in a resistance-network, but the condition (7.8') is much more complicated.

A method for solving Eq. (7.12) is, to introduce the magnetic flux through a disk with radius r_0 centered on the axis of symmetry. Thus

$$M(r_0 , z) = \iint_S \mathbf{B} \cdot d\boldsymbol{\sigma} = 2\pi \int_0^{r_0} r B_z \, dr \qquad (7.13)$$

Substitute Eq. (7.1′) in (7.13). This leads to

$$M(r_0, z) = 2\pi r_0 A(r_0, z) \tag{7.14}$$

From (7.12) and (7.14) the partial differential equation for the magnetic flux can be found. Thus

$$\frac{\partial^2 M}{\partial z^2} - \frac{1}{r}\frac{\partial M}{\partial r} + \frac{\partial^2 M}{\partial r^2} = -2\pi\mu r j_w(r, z) \tag{7.15}$$

The boundary conditions in this case are the same as those for $A(x, y)$ in two-dimensional fields. Along **B**-lines

$$(\operatorname{grad} M)_t = 0 \tag{7.16}$$

At a high-permeability pole piece

$$(\operatorname{grad} M)_n = 0 \tag{7.17}$$

These conditions can be derived as follows.

From Eqs. (7.14), (7.1′), and (7.2′) we find

$$\frac{\partial M}{\partial z} = 2\pi r \frac{\partial A}{\partial z} = -2\pi r B_r \tag{7.11′}$$

$$\frac{\partial M}{\partial r} = 2\pi \frac{\partial}{\partial r}(r A) = 2\pi r B_z \tag{7.12′}$$

From Eqs. (7.3′) and (7.4′) we obtain

$$(\operatorname{grad} M)_n = \mathbf{n}_I \cdot \operatorname{grad} M = \frac{\partial M}{\partial z}\cos\alpha + \frac{\partial M}{\partial r}\sin\alpha$$
$$= 2\pi r(-B_r\cos\alpha + B_z\sin\alpha) = -2\pi r B_t \tag{7.13′}$$

using (7.9′), (7.11′), and (7.12′). Similarly,

$$(\operatorname{grad} M)_t = \mathbf{t}_I \cdot \operatorname{grad} M = -\frac{\partial M}{\partial z}\sin\alpha + \frac{\partial M}{\partial r}\cos\alpha$$
$$= 2\pi r(B_r\sin\alpha + B_z\cos\alpha) = 2\pi r B_n \tag{7.14′}$$

Along **B**-lines Eq. (7.7′) holds, yielding (7.16) from (7.14′). At a high-permeability pole piece, Eqs. (7.9′) and (7.13′) lead to (7.17).

Equation (7.15) differs from the Poisson equation only by the sign of the second term. This fact has been used by Peierls and Skyrme (1949) to obtain an approximate solution by superimposing the results obtained with a wedge tank on that obtained in a tank for two-dimensional problems. With the aid of these tanks, solutions M_1 and M_2 are obtained satisfying the equations:

$$\frac{\partial^2 M_1}{\partial r^2} + \frac{1}{r}\frac{\partial M_1}{\partial r} + \frac{\partial^2 M_1}{\partial z^2} = -2\pi\mu r j_w(r, z) \qquad (7.18)$$

$$\frac{\partial^2 M_2}{\partial r^2} + \frac{\partial^2 M_2}{\partial z^2} = -2\pi\mu r j_w(r, z) \qquad (7.19)$$

An approximate solution of (7.15) for regions near the axis of symmetry is (Peierls and Skryme, 1949):

$$M(r, z) = 2M_2(r, z) - M_1(r, z) \qquad (7.20)$$

However, a better precision can generally be obtained by using a special resistance network for solving Eq. (7.18) directly (see the following section).

1.3.2. The Resistance Network

A. Theory of the Resistance Network for Solving the Laplace and the Poisson Equations

1. *Finite Difference Operators*

The left-hand side of the Laplace or Poisson equation is, in rectangular coordinates

$$\frac{\partial^2 \varphi}{\partial x^2} + \frac{\partial^2 \varphi}{\partial y^2} + \frac{\partial^2 \varphi}{\partial z^2} \equiv \varDelta\varphi(x, y, z) \qquad (1.1)$$

The differential operator $\varDelta$ (or ∇^2) is referred to as the *Laplacian operator*. The effect of the operator on the function $\varphi(x, y, z)$ will generally depend on the point in space at which it is applied (that is, $\varDelta\varphi$ is again a function of x, y, z). If, however, $\varphi(x, y, z)$ is a solution of the Laplace equation, the effect is zero in every point, namely,

$$\varDelta\varphi = 0 \qquad (1.2)$$

From the differential operator $\varDelta$ a *finite difference operator L* will be derived that, operating on the same function at the same point, produces almost the same effect as $\varDelta$ itself, namely,

$$L\varphi \simeq \Delta\varphi \tag{1.3}$$

Consider an arbitrary point $P_0(x_0, y_0, z_0)$ and the three pairs of points P_1 and P_2, P_3 and P_4, P_5 and P_6, lying at distances $-a$ and a from P_0 in the x, y and z directions, respectively [Fig. (1.1)].

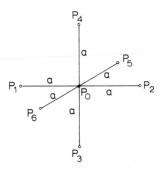

FIG. 1.1. Location of mesh points used in deriving the finite difference equation for equal meshes.

By using the Taylor theorem, the values of φ in these points can be expressed in series in terms of the values of $\varphi(x_0, y_0, z_0)$ and the derivatives of φ with respect to x, y, or z. Thus we obtain

$$\varphi_2 - \varphi_0 = \quad a\left(\frac{\partial\varphi}{\partial x}\right)_0 + \frac{a^2}{2!}\left(\frac{\partial^2\varphi}{\partial x^2}\right)_0 + \frac{a^3}{3!}\left(\frac{\partial^3\varphi}{\partial x^3}\right)_0 + \frac{a^4}{4!}\left(\frac{\partial^4\varphi}{\partial x^4}\right)_0 + \cdots \tag{1.4}$$

$$\varphi_1 - \varphi_0 = -a\left(\frac{\partial\varphi}{\partial x}\right)_0 + \frac{a^2}{2!}\left(\frac{\partial^2\varphi}{\partial x^2}\right)_0 - \frac{a^3}{3!}\left(\frac{\partial^3\varphi}{\partial x^3}\right)_0 + \frac{a^4}{4!}\left(\frac{\partial^4\varphi}{\partial x^4}\right)_0 - \cdots \tag{1.5}$$

Adding these series and solving for $(\partial^2\varphi/\partial x^2)_0$ yields

$$\left(\frac{\partial^2\varphi}{\partial x^2}\right)_0 = \frac{1}{a^2}\{(\varphi_2 - \varphi_0) + (\varphi_1 - \varphi_0)\} - \frac{a^2}{12}\left(\frac{\partial^4\varphi}{\partial x^4}\right)_0 - \cdots \tag{1.6}$$

Similar series can be derived for $(\partial^2\varphi/\partial y^2)_0$ and $(\partial^2\varphi/\partial z^2)_0$.

Adding these series and introducing the finite difference operator L defined by

$$(L\varphi)_0 \equiv \frac{1}{a^2}\sum_{j=1}^{6}(\varphi_j - \varphi_0) \tag{1.7}$$

results in

$$(\Delta\varphi)_0 = (L\varphi)_0 - \frac{a^2}{12}\left(\frac{\partial^4\varphi}{\partial x^4} + \frac{\partial^4\varphi}{\partial y^4} + \frac{\partial^4\varphi}{\partial z^4}\right)_0 - \cdots \tag{1.8}$$

From Eq. (1.8) it can be seen that $(L\varphi)_0$ is an approximation to $(\varDelta\varphi)_0$, approaching it all the more closely as a is made smaller. From Eq. (1.7) it will be clear that $(L\varphi)_0$ contains the finite differences:

$$(\varphi_1 - \varphi_0), \qquad (\varphi_2 - \varphi_0), \qquad (\varphi_3 - \varphi_0), \quad \text{etc.}$$

This procedure for deriving a difference operator from a differential operator can also be used in the more general case in which the differential equation involves the first derivatives as well as the second derivatives. Subtracting Eq. (1.5) from Eq. (1.4) and solving for $(\partial\varphi/\partial x)_0$, we get

$$\left(\frac{\partial\varphi}{\partial x}\right)_0 = \frac{1}{2a}(\varphi_2 - \varphi_1) - \frac{a^2}{6}\left(\frac{\partial^3\varphi}{\partial x^3}\right)_0 - \cdots \tag{1.9}$$

Expression (1.9) will be used when dealing with three-dimensional problems with rotational symmetry (see Section 1.3.2,A,3).

2. Resistance Networks for Two-Dimensional Problems

a. *Solution of the finite difference equation.* In two-dimensional problems, φ is independent of z and, therefore, in Fig: 1.1,

$$\varphi_5 = \varphi_6 = \varphi_0$$

Therefore, Eq. (1.7) reduces to

$$(L\varphi)_0 = \frac{1}{a^2} \sum_{j=1}^{4} (\varphi_j - \varphi_0) \tag{2.1}$$

The Laplace differential equation, in this case, is

$$\frac{\partial^2\varphi}{\partial x^2} + \frac{\partial^2\varphi}{\partial y^2} = 0 \tag{2.2}$$

To take a specific example, let us consider the case of Fig. 2.1. The three closed boundaries, s_1, s_2, and s_3 represent sections taken at right angles through three infinitely long prisms. On the periphery of each prism, φ has a known constant value. The problem is to find a function φ which satisfies (2.2) in the area within s_3, but outside s_1 and s_2, and which assumes the prescribed values along s_1, s_2, and s_3.

Over s_1, s_2, and s_3 we place a square grid of m lines parallel to the x axis and n lines parallel to the y axis, spaced at intervals of a. The position of each line is given by the relations

$$y_i = i \cdot a \qquad x_k = k \cdot a \tag{2.3}$$

with

$$1 \leq i \leq m \qquad 1 \leq k \leq n$$

Intersections (i, k) of grid lines are called *grid points*.

In Fig. 2.1 the boundaries s_1, s_2, and s_3 intersect the grid in grid points only. This is not a necessary limitation but, in the case of the resistance network, a considerable simplification.

Grid points located on the outlines s_1, s_2, and s_3 will be referred to as "boundary" grid points, whereas grid points in the area where φ has to be determined are denoted by "internal grid points." Now to each internal grid point a value φ^*_{ik} is allotted in such a way that the equations

$$(L\varphi^*)_{ik} = 0 \tag{2.4}$$

are satisfied; φ^*_{ik} is called a solution of the finite difference equation (2.4). It will be noted that with (2.4), the "function" φ^*_{ik} is only defined in the internal grid points; its values in the boundary grid points being given. The solution $\varphi(x, y)$ of the differential Eq. (2.2) has the values φ_{ik} in the grid points.

The principle underlying the resistance netword analog is, that φ^*_{ik} approaches φ_{ik} more closely as the meshwidth a is made smaller. Thus,

$$\lim_{a \to 0} \varphi^*_{ik} = \varphi_{ik} \tag{2.5}$$

wherein the boundary conditions are the same in both cases. In Section 1.3.2,A,4 we will enlarge somewhat on this supposition.

The finite difference equation corresponding to the *Poisson equation*

$$\varLambda\varphi = - \varrho/\varepsilon_0 \tag{2.6}$$

is expressed by

$$(L\varphi^*)_{ik} = - \varrho_{ik}/\varepsilon_0 \tag{2.7}$$

b. *The resistance network.* The finite difference equation (2.4) leads to simultaneous equations for all internal grid points $P_0(i, k)$. Using the notation of Fig. 1.1, we have

$$\frac{1}{a^2} \sum_{j=1}^{4} (\varphi_j{}^* - \varphi_0{}^*) = 0 \tag{2.8}$$

In order to find the values φ^*, a network of resistors is built up. The junc-

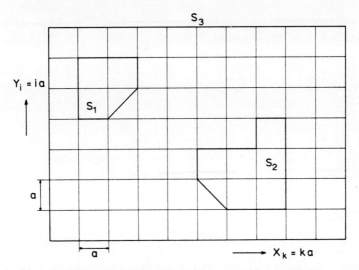

FIG. 2.1. Plane potential problem defined by potentials on the boundaries s_1, s_2, and s_3.

tions of the resistance network will correspond to the grid points in Fig. 2.1. Accordingly, four resistors will meet at each junction $P_0(i, k)$ (Fig. 2.2a).

Boundary junctions, that is, junctions corresponding to boundary grid points are fed with voltages V_b that are proportional to the boundary potentials $\varphi_b^* = \varphi_b$. Thus

$$\varphi_b^* = \beta V_b \tag{2.9}$$

where both V_b and φ_b^* are measured with respect to the lowest boundary voltage or potential value, respectively.

The voltage $V_0 = V(i, k)$ in an arbitrary point of the network is related

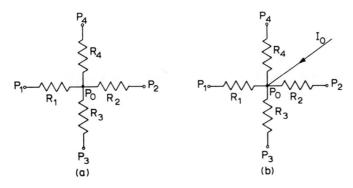

FIG. 2.2. Junction of resistance network. (a) For resolving Laplace's equation; (b) For resolving Poisson's equation.

to the voltages of the surrounding points (see Fig. 2.2a). Thus, applying Kirchhoff's law to the function P_0

$$\sum_{j=1}^{4} \frac{1}{R_j} (V_j - V_0) = 0 \qquad (2.10)$$

Taking the same value for all resistors, thus

$$R_j = R \qquad (j = 1, 2, 3, 4) \qquad (2.11)$$

and putting

$$\varphi^*_{ik} = \beta V_{ik} \qquad (2.12)$$

it follows from (2.10)–(2.12) as well as the boundary conditions (2.9) that

$$(L\varphi^*)_{ik} = \beta(L\,V)_{ik} = 0 \qquad (2.13)$$

The function φ^*_{ik} from (2.12) is, therefore, a solution of the finite difference equation (2.4) and an approximate solution of the Laplace equation for two-dimensional problems [Eq. (2.2)].

An approximate solution of the Poisson equation can be obtained by feeding currents I_0 into the resistor junctions (Fig. 2.2b). Instead of Eq. (2.10), we then obtain, using Eq. (2.11)

$$\sum_{j=1}^{4} (V_j - V_0) = -I_0 R \qquad (2.14)$$

On using the definition of $(LV)_0$ [see Eq. (2.1)], Eq. (2.14) can be written as

$$(L\,V)_0 = -I_0(R/a^2) \qquad (2.15)$$

The solution of Eq. (2.7) can then be obtained, using Eq. (2.12), from

$$(L\,\varphi^*)_{ik} = -\beta I_{ik}(R/a^2) \qquad (2.16)$$

if the injected currents I_{ik} are made equal to

$$I_{ik} = \frac{a^2}{\beta R} \frac{\varrho_{ik}}{\varepsilon_0} \qquad (2.17)$$

In this equation ϱ_{ik} is the average charge density in a square with sides of a centered in the point $0(i, k)$.

In some cases the charge density ϱ in Eq. (2.6) bears a simple relation to the potential φ. Karplus (1955) has shown that the injected currents I_{ik} in (2.17) can be adjusted automatically with the aid of analog computer units, attached to the network junctions. The method is illustrated on the case of the plane diode, where Eq. (2.6) simplifies to

$$d^2\varphi/dx^2 = k\varphi^{-1/2}$$

In general, however, the electron trajectories are not known and therefore, in order to determine ϱ, an iterative method has to be used.

c. *Resistance networks with unequal meshes.* In deriving the finite difference operators (1.7) and (2.1), use is made of points at equal distances from the midpoint P_0. This is not a necessary limitation. In the general

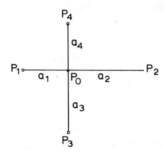

FIG. 2.3. Location of mesh points used in deriving the finite difference equations for unequal meshes.

case (Fig. 2.3), the Taylor series corresponding to Eqs. (1.4) and (1.5) are given by

$$\varphi_1 - \varphi_0 = -a_1\left(\frac{\partial\varphi}{\partial x}\right)_0 + \frac{a_1^2}{2!}\left(\frac{\partial^2\varphi}{\partial x^2}\right)_0 - \frac{a_1^3}{3!}\left(\frac{\partial^3\varphi}{\partial x^3}\right)_0 + \cdots \quad (2.18)$$

$$\varphi_2 - \varphi_0 = a_2\left(\frac{\partial\varphi}{\partial x}\right)_0 + \frac{a_2^2}{2!}\left(\frac{\partial^2\varphi}{\partial x^2}\right)_0 + \frac{a_2^3}{3!}\left(\frac{\partial^3\varphi}{\partial x^3}\right)_0 + \cdots \quad (2.19)$$

Similar series can be given in the y direction. Neglecting terms with a_i^3, the first derivatives can be eliminated from the four equations. Putting

$$\Delta\varphi \equiv \frac{\partial^2\varphi}{\partial x^2} + \frac{\partial^2\varphi}{\partial y^2} \quad (2.20)$$

the resulting expression leads to the approximation,

$$(\varDelta\varphi)_0 \simeq (L_{um}\varphi)_0 = \frac{2}{a_1 + a_2}\left(\frac{\varphi_1}{a_1} + \frac{\varphi_2}{a_2}\right) + \left(\frac{2}{a_3 + a_4}\right)\left(\frac{\varphi_3}{a_3} + \frac{\varphi_4}{a_4}\right)$$

$$- 2\left(\frac{1}{a_1 a_2} + \frac{1}{a_3 a_4}\right)\varphi_0 \qquad (2.21)$$

This equation can be given the alternative form,

$$\frac{(a_1 + a_2)(a_3 + a_4)}{4}(L_{um}\varphi)_0 = \frac{a_3 + a_4}{2a_1}(\varphi_1 - \varphi_0) + \frac{a_3 + a_4}{2a_2}\cdot(\varphi_2 - \varphi_0)$$

$$+ \frac{a_1 + a_2}{2a_3}(\varphi_3 - \varphi_0) + \frac{a_1 + a_2}{2a_4}(\varphi_4 - \varphi_0)$$

$$(2.22)$$

From this equation, a resistance network can be derived (Liebmann, 1954), using resistances (see Fig. 2.2a) related by

$$R_1 = \frac{2a_1}{a_3 + a_4}R_0 \qquad R_2 = \frac{2a_2}{a_3 + a_4}R_0 \qquad R_3 = \frac{2a_3}{a_1 + a_2}R_0$$

$$R_4 = \frac{2a_4}{a_1 + a_2}R_0 \qquad (2.23)$$

A solution of the Poisson equation can, again be obtained by feeding currents I_0 into the resistance junctions (Fig. 2.2b). The junction potential V_0 is then related to the potentials of the surrounding junctions by Eq. (2.14). The application of the finite difference operator as defined in (2.22) to these potentials leads to

$$(L_{um}V)_0 = - \frac{4R_0}{(a_1 + a_2)(a_3 + a_4)}I_0 \qquad (2.24)$$

This equation corresponds to the Poisson equation (2.7) if, using the relation (2.12), the injected currents are given by [see Eq. (2.17)]:

$$I_0 = \frac{(a_1 + a_2)(a_3 + a_4)}{4\beta R_0}\frac{\varrho_0}{\varepsilon_0} \qquad (2.25)$$

A special case of the network defined by Eq. (2.23), arises, if the meshes are equal, but rectangular (Fig. 2.4a):

$$a_1 = a_2 = a' \qquad a_3 = a_4 = a$$

The resistance network then consists of resistors of two different values (Fig. 2.4b):

$$R_H = \frac{a'}{a} R_0 \qquad R_V = \frac{a}{a'} R_0 \qquad (2.26)$$

Liebmann (1954) uses this type of mesh to remove the boundaries of a

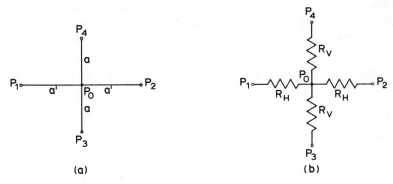

(a)   (b)

FIG. 2.4. Location of mesh points (a) and junction of resistors (b) in case of rectangular meshes.

network a long distance away from the "working part" of the network. This long distance, of course, results from the resistance values and not from the physical sizes of the meshes. The resistance values at the boundary between a square mesh and a rectangular mesh (Fig. 2.5a) are

$$R_1' = R_0 \qquad R_2' = R_H = \frac{a'}{a} R_0 \qquad R_3' = R_4' = \frac{2R_0}{1 + (a'/a)} \qquad (2.27)$$

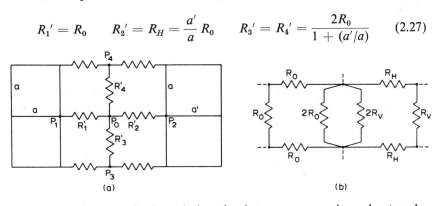

(a)   (b)

FIG. 2.5. Resistors junction P_0 on the boundary between square meshes and rectangular meshes. (a) Situation of junction P_0; (b) Junction resistors R_3' and R_4' in (a) can be obtained from a parallel connection of resistances $2R_0$ and $2R_V$.

This will be clear considering R_3' and R_4' can be obtained by connecting resistors valued $2R_0$ and $2R_V$ in parallel, where R_V is given by Eq. (2.26).

The data for the network *termination at an open boundary* passing through a point P_0 [that is, $(\partial\varphi/\partial n)_0 = 0$] are obtained from Eq. (2.27) by putting $a' = 0$. Thus

$$R_3' = R_4' = 2R_0 \qquad (2.28)$$

d. *Geometrical interpretation.* A useful interpretation of the derived design formulas (2.23) is obtained by considering a shallow electrolytic tank with constant depth h and conductivity σ of the electrolyte. Considering five points located as in Fig. 2.3, it can be easily seen that the resistances R_1 and R_2, given by Eq. (2.23), are proportional to the resistances of elements in the tank of which top views are indicated in Fig. 2.6a. After

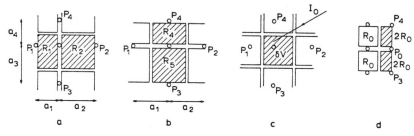

Fig. 2.6. Difference surface elements for: (a) horizontal resistances; (b) vertical resistances; (c) injected current; (d) open boundary.

Tschopp (1961 a,b) these elements will be denoted by *difference surface elements*. The resistances are given by

$$R_1' = \frac{1}{\sigma h}\frac{a_1}{\frac{1}{2}(a_3 + a_4)} \qquad R_2' = \frac{1}{\sigma h}\frac{a_2}{\frac{1}{2}(a_3 + a_4)} \qquad (2.29)$$

Similarly, the resistances R_3 and R_4, given by Eq. (2.23) are proportional to those of the tank elements indicated in Fig. 2.6b. In both cases, a complete analogy is obtained by taking

$$R_0 = 1/\sigma h \qquad (2.30)$$

In cases in which current is injected into the mesh points, it follows from (2.25) and (2.30) that

$$I_0 = \frac{\sigma}{4\beta\varepsilon_0}h(a_1 + a_2)(a_3 + a_4)\varrho_0 \qquad (2.31)$$

Therefore, the injected current is proportional to the *total charge* in the volume element $\delta V = h(a_1 + a_2)(a_3 + a_4)$ (Fig. 2.6c). With the aid of this interpretation, equations such as (2.25) and (2.26) can be easily derived. Fig. 2.6d, for example, illustrates the case of an open boundary, as given by Eq. (2.28).

For networks with *equal square meshes*, Eq. (2.29) simplifies to Eq. (2.30).

3. *Resistance Networks for Axially Symmetrical Problems*

a. *The finite difference equations.* By using cylindrical coordinates r, θ, z (Fig. 3.1), the potential distribution is fully determined by that in a meridional plane, $\varphi(r, z)$.

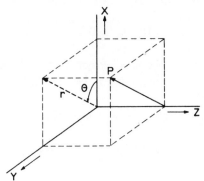

FIG. 3.1. Relation between cylindrical coordinates r, θ, z and orthogonal coordinates x, y, z.

Laplace's equation (1.2), in this case, is expressed by

$$\frac{\partial^2\varphi}{\partial r^2} + \frac{1}{r}\frac{\partial\varphi}{\partial r} + \frac{\partial^2\varphi}{\partial z^2} = 0 \tag{3.1}$$

A finite difference equation, corresponding to this differential equation can be derived, using the method of Subsection 1.3.2,A,1. Substituting z and r for x and y, respectively, in Eqs. (1.6) and (1.9), the partial derivatives for off-axis points r_0, z_0 are, in this case,

$$\left(\frac{\partial^2\varphi}{\partial z^2}\right)_0 = \frac{1}{a^2}\{(\varphi_2 - \varphi_0) + (\varphi_1 - \varphi_0)\} - \frac{a^2}{12}\left(\frac{\partial^4\varphi}{\partial z^4}\right)_0 - \cdots \tag{3.2}$$

$$\left(\frac{\partial^2\varphi}{\partial r^2}\right)_0 = \frac{1}{a^2}\{(\varphi_4 - \varphi_0) + (\varphi_3 - \varphi_0)\} - \frac{a^2}{12}\left(\frac{\partial^4\varphi}{\partial r^4}\right)_0 - \cdots \tag{3.3}$$

$$\frac{1}{r_0}\left(\frac{\partial\varphi}{\partial r}\right)_0 = \frac{1}{a^2}\left(\frac{a}{2r_0}(\varphi_4 - \varphi_3)\right) - \frac{a^2}{6r_0}\left(\frac{\partial^3\varphi}{\partial r^3}\right)_0 - \cdots \tag{3.4}$$

A finite difference operator M can now be defined by

$$(M\varphi)_0 \equiv \frac{1}{a^2}\left\{\varphi_1 + \varphi_2 + \varphi_3\left(1 - \frac{a}{2r_0}\right) + \varphi_4\left(1 + \frac{a}{2r_0}\right) - 4\varphi_0\right\} \quad (3.5)$$

Adding Eqs. (3.2)–(3.4) and subtracting Eq. (3.5) yields

$$(\Delta\varphi)_0 = (M\varphi)_0 - \frac{a^2}{12}\left(\frac{\partial^4\varphi}{\partial z^4} + \frac{\partial^4\varphi}{\partial r^4} + \frac{2}{r_0}\frac{\partial^3\varphi}{\partial r^3}\right)_0 \qquad (3.6)$$

An approximate solution of Eq. (3.1) with given boundary conditions can be obtained, in the same way as has been explained in Subsection 1.3.2,A,1, by solving the finite difference equation

$$(M\varphi^*)_{ik} = 0 \qquad (i > 0) \qquad (3.7)$$

However, for points in the axis of symmetry, the operator M cannot be used because the terms containing r_0 become $0/0$, φ_3 being equal to φ_4 in this case. Therefore, for $r_0 = 0$, the operator L, defined in (1.7), is used. In the axially symmetrical case we have (see Fig. 1.1)

$$\varphi_3 = \varphi_4 = \varphi_5 = \varphi_6$$

Equation (1.7) thus reduces to

$$(L_a\varphi)_0 = \frac{1}{a^2}(\varphi_1 + \varphi_2 + 4\varphi_3 - 6\varphi_0) \qquad (3.8)$$

The meaning of the index a in L_a is that this operator refers to points situated in the axis of symmetry. For these points, instead of Eq. (3.7) we have to solve the finite difference equation

$$(L_a\varphi^*)_{0k} = 0 \qquad (3.9)$$

the index $i = 0$ referring to the axis of symmetry.

b. *A resistance network.* A network of resistors of the type of which an element is shown in Fig. 2.2a can be used to solve Eqs. (3.7) and (3.9), if suitable resistance values R_1, R_2, R_3, and R_4 are chosen. In view of the symmetry of the problem, a network for the upper half of a meridional plane suffices. The row of resistors $i = 0$ then corresponds with the axis of symmetry.

The equations to be solved are, written in full,

$$(\varphi^*_{i,k-1} - \varphi^*_{i,k}) + (\varphi^*_{i,k+1} - \varphi^*_{i,k}) + \left(1 - \frac{1}{2i}\right)(\varphi^*_{i-1,k} - \varphi^*_{i,k})$$
$$+ \left(1 + \frac{1}{2i}\right)(\varphi^*_{i+1,k} - \varphi^*_{i,k}) = 0 \tag{3.10}$$

$$(\varphi^*_{0,k-1} - \varphi^*_{0,k}) + (\varphi^*_{0,k+1} - \varphi^*_{0,k}) + 4(\varphi^*_{1,k} - \varphi^*_{0,k}) = 0 \tag{3.11}$$

From an inspection of these equations it will be clear that the resistance values of the corresponding network will depend on i only, the terms not depending on k explicitly. A network junction can, therefore, be represented as in Fig. 3.2a or 3.2b. From Kirchhoff's law we have in the case of Fig. 3.2a (off-axis points),

$$\frac{1}{R_H[i]} \{V[i, k-1] - V[i, k]\} + \frac{1}{R_H[i]} \{V[i, k+1] - V[i, k]\}$$
$$+ \frac{1}{R_V[i-1]} \{V[i-1, k] - V[i, k]\} + \frac{1}{R_V[i]} \{V[i+1, k] - V[i, k]\} = 0 \tag{3.12}$$

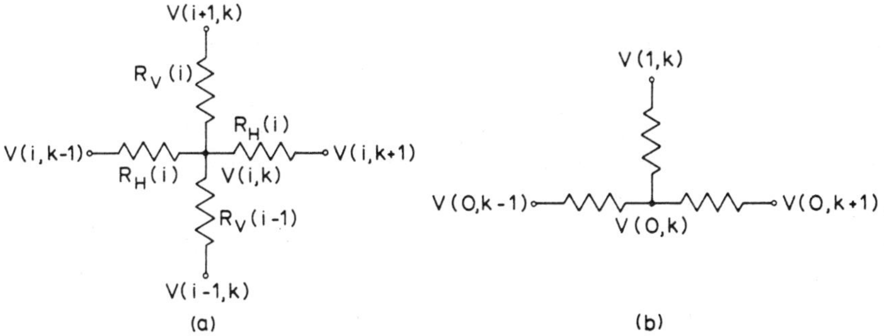

FIG. 3.2. Junctions of resistors at (a) an arbitrary point (i, k) and (b) an axis point $(0, k)$.

Applying boundary conditions as in the case of the network for two-dimensional problems [see Eq. (2.9)], and, again, putting

$$\varphi^*_{ik} = \beta V_{ik} \tag{3.13}$$

it follows from (3.10), (3.12), and (3.13) that

$$(M\varphi^*)_{ik} = \beta(MV)_{ik} = 0 \tag{3.14}$$

provided the resistance values are related by

$$\frac{1}{R_H[i]} : \frac{1}{R_V[i-1]} : \frac{1}{R_V[i]} = 1 : \left(1 - \frac{1}{2i}\right) : \left(1 + \frac{1}{2i}\right) \quad (3.15)$$

Furthermore, to build a coherent network, an "upward" resistance for a junction $[i, k]$ is a "downward" resistance for a point $[i + 1, k]$; namely, we have to satisfy the requirement (Fig. 3.2a)

$$R_V[i] = R_V[(i+1) - 1] \quad (3.16)$$

From Eqs. (3.15) and (3.16) the following relations can be deduced for $i > 0$. Thus

$$R_V[i] = \frac{2i - 1}{2i + 1} R_V[i-1] = \frac{2(i-1) + 1}{2i + 1} R_V[i-1] \quad (3.17)$$

$$R_V[i] = \frac{2i}{2i + 1} R_H[i] = \frac{2(i+1)}{2i + 1} R_H[i+1] \quad (3.18)$$

From these equations we obtain

$$R_H[i+1] = \frac{i}{i+1} R_H[i] \quad (3.19)$$

By starting with a value

$$R_V(0) = R_0 \quad (3.20)$$

Equations (3.17–3.20) result in the following resistance values:

$$R_H[i] = \frac{R_0}{2i} \quad (i > 0) \quad (3.21)$$

$$R_V[i] = \frac{R_0}{2i + 1} \quad (3.22)$$

In the case of on-axis points (Fig. 3.2b), Kirchhoff's law yields

$$\frac{1}{R_H[0]} \{V[0, k-1] - V[0, k]\} + \frac{1}{R_H[0]} \{V[0, k+1] - V[0, k]\}$$
$$+ \frac{1}{R_V[0]} \{V[1, k] - V[0, k]\} = 0 \quad (3.23)$$

After using (3.13), it follows from (3.11) and (3.23) that

$$(L_a\varphi^*)_{ok} = \beta(L_aV)_{ok} = 0 \tag{3.24}$$

if the resistance values in the axis are

$$R_H(0) = 4R_V(0) = 4R_0 \tag{3.25}$$

In Fig. 3.3 a small part of the network near the axis is shown.

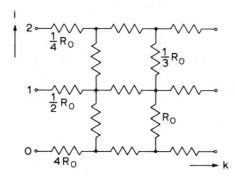

FIG. 3.3. Part of a resistance network for resolving axially symmetrical potential problems.

An analog for solving the Poisson equation can be obtained by feeding currents into the resistor junctions as in the two-dimensional case (Fig. 2.2b). Placing I_{ik} at the right-hand side of Eq. (3.12), using the definitions for the operators M and L_a from (3.5) and (3.8), and using the resistance values from (3.21) and (3.22) or from (3.25), we obtain the following relations:

$$(M\,V)_{ik} = -\frac{R_0}{2i\,a^2}\,I_{ik} \qquad (i > 0) \tag{3.26}$$

$$(L_a\,V)_{ok} = -\frac{4R_0}{a^2}\,I_{ok} \qquad (i = 0) \tag{3.27}$$

From Eq. (2.7) and the relations between φ^* and V from (3.14) or (3.24), it appears that the injected currents have to be made equal to

$$I_{ik} = \frac{2i\,a^2}{\beta R_0}\,\frac{\varrho_{ik}}{\varepsilon_0} \tag{3.28}$$

and

$$I_{0k} = \frac{a^2}{4\beta R_0} \frac{\varrho_{0k}}{\varepsilon_0} \tag{3.29}$$

c. *Geometrical interpretation.* This can be given in a similar way as in the two-dimensional case. However, the electrolytic tank corresponding with the network, has no longer a uniform depth.

As shown in Fig. 3.4, the depth h is a linearly increasing step-function of the distance from the axis. Moreover, the steps are differently located for the simulation of the vertical and the horizontal resistance values. For small values of the mesh width a, the tank approaches a wedge tank.

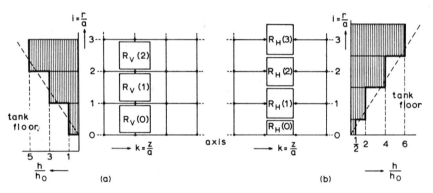

FIG. 3.4. Geometrical interpretation of resistance values in a network for axially symmetrical potential problems. (a) Vertical resistances: $R_v(i) = a/\sigma(2i+1)h_0a$. (b) Horizontal resistances: $R_H(i) = a/\sigma \cdot 2ih_0a \ (i > 0)$; $R_H(0) = a/\sigma \cdot \frac{1}{2}h_0 \cdot \frac{1}{2}a$.

d. *Other resistance networks.* Instead of the finite difference equation $(M\varphi)_0$ as defined in Eq. (3.5), other equations may be derived from the series (3.2)–(3.4), particularly in the vicinity of the axis. To this purpose, we use the series expansion of $\varphi(r, z_0)$ in terms of the potential on the axis $\Phi(z_0) \equiv \varphi(0, z_0)$:

$$\varphi(r, z_0) = \sum_{n=0}^{\infty} \frac{(-1)^n}{(n!)^2} \left(\frac{r}{2}\right)^{2n} \Phi^{(2n)}(z_0) \tag{3.30}$$

This equation can be derived from a Taylor expansion of the function $\varphi(r, z_0)$ near an axis point $(0, z_0)$. The substituting of this series in the Laplace equation (3.1) leads to the values of the coefficients used in Eq. (3.30).

With the aid of this equation, the partial derivatives in Eqs. (3.2)–(3.4) can be expressed in the derivatives of the axis potential $\Phi(z)$. Partial differentiation of (3.30) yields

$$\left(\frac{1}{r}\frac{\partial\varphi}{\partial r}\right)_0 = -\frac{1}{2}\Phi''(z_0) + \frac{1}{16}r_0^2\,\Phi^{IV}(z_0) - \cdots \qquad (3.31)$$

$$\left(\frac{\partial^2\varphi}{\partial r^2}\right)_0 = -\frac{1}{2}\Phi''(z_0) + \frac{3}{16}r_0^2\Phi^{IV}(z_0) - \cdots \qquad (3.32)$$

$$\left(\frac{\partial^2\varphi}{\partial z^2}\right) = \Phi''(z_0) - \frac{1}{4}r_0^2\,\Phi^{IV}(z_0) + \cdots \qquad (3.33)$$

$$\left(\frac{1}{r}\frac{\partial^3\varphi}{\partial r^3}\right)_0 = \frac{3}{8}\Phi^{IV}(z_0) - \cdots \qquad (3.34)$$

$$\left(\frac{\partial^4\varphi}{\partial r^4}\right)_0 = \frac{3}{8}\Phi^{IV}(z_0) - \cdots \qquad (3.35)$$

$$\left(\frac{\partial^4\varphi}{\partial z^4}\right)_0 = \Phi^{IV}(z_0) - \cdots \qquad (3.36)$$

[From these equations it follows, for $r = 0$:

$$\frac{1}{r}\frac{\partial\varphi}{\partial r} = \frac{\partial^2\varphi}{\partial r^2} = -\frac{1}{2}\Phi''(z_0) \qquad (3.1')$$

The substituting of (3.1') in (3.1) results in the relation

$$2\left(\frac{\partial^2\varphi}{\partial r^2}\right)_0 + \left(\frac{\partial^2\varphi}{\partial z^2}\right)_0 = 0 \qquad (3.2')$$

From (3.2'), (3.2), and (3.3) the finite difference equation $(L_a\varphi)$ as given in (3.8) can be found (Francken, 1953).]

Substituting Eqs. (3.31)–(3.36) in Eqs. (3.2)–(3.4), rearranging the terms and putting

$$r_0 = ia \qquad (3.37)$$

we get

$$\frac{1}{a^2}(\varphi_1 + \varphi_2 - 2\varphi_0) = \Phi''(z_0) + a^2\left(-\frac{1}{4}i^2 + \frac{1}{12}\right)\Phi^{IV}(z_0) - \cdots \qquad (3.38)$$

$$\frac{1}{a^2}(\varphi_3 + \varphi_4 - 2\varphi_0) = -\frac{1}{2}\Phi''(z_0) + a^2\left(\frac{3}{16}i^2 + \frac{3}{96}\right)\Phi^{IV}(z_0) - \cdots \qquad (3.39)$$

$$\frac{1}{a^2}\frac{1}{2i}(\varphi_4 - \varphi_3) = -\frac{1}{2}\Phi''(z_0) + a^2\left(\frac{1}{16}i^2 + \frac{1}{16}\right)\Phi^{IV}(z_0) - \cdots \qquad (3.40)$$

From these equations the terms containing $\Phi''(z_0)$ can be eliminated in several ways. A simple addition results in the finite difference equation $(M\varphi)_0$ and a residual with the leading term in a^2:

$$(M \varphi)_0 = \frac{17}{96} a^2 \Phi^{IV}(z_0) - \text{(higher order terms)} \qquad (3.41)$$

A more general result is obtained by the multiplication of Eqs. (3.38)–(3.40) by

$$\frac{1}{(1-A)}, \quad \frac{1-2A}{1-A}, \quad \text{and} \quad \frac{1+2A}{1-A} \qquad (3.42)$$

respectively, and a subsequent addition. This yields (Weber, 1963)

$$
\frac{1}{a^2(1-A)} \left\{ \varphi_1 + \varphi_2 + (1-2A)(\varphi_3+\varphi_4) + \frac{1+2A}{2i}(\varphi_4-\varphi_3) \right.
$$
$$
\left. -4\varphi_0(1-A) \right\} = \frac{17-A(24i^2-6)}{96(1-A)} a^2 \Phi^{IV}(z_0) - \frac{\text{higher order terms}}{1-A} \qquad (3.43)
$$

The left-hand side of this equation, for $A = 0$, obviously yields the finite difference operator $(M\varphi)_0$.

Weber (1963) pointed out that several distinct finite difference operators can be found from (3.43) by substituting different values of the constant A, provided $(1-A)$ does not approach zero. If A is, for example, in the region

$$0.9 < A < 1.1 \qquad (3.44)$$

the higher order terms in (3.43) cannot be neglected. From (3.43), Weber (1963) constructed a resistance network with $A = \infty$ for $i = 1$ and $A = 0$ for $i > 1$. In this case, (3.43)—for $i = 1$—reduces to

$$(W \varphi)_0 \equiv \frac{1}{a^2}(3\varphi_3 + \varphi_4 - 4\varphi_0) = \frac{3}{16} a^2 \Phi^{IV}(z_0) - \cdots \qquad (3.45)$$

A part of the resistance network constructed from Eq. (3.45), is shown in Fig. 3.5. A feature of this network is a considerable reduction in the ratio between the largest and smallest resistance values prescribed by the finite difference equation. It can be seen from Fig. 3.5, in which the "conventional" resistance values (for $A = 0$ throughout) are given between parentheses, that this reduction amounts to a factor of 9.

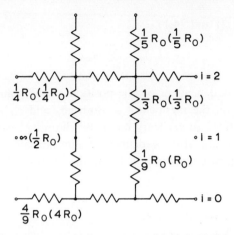

FIG. 3.5. Resistance network for axially symmetrical potential problems after Weber (1963). The resistance values of the network of Fig. 3.3 are given between parenthesis.

4. *Influence of the Meshwidth*

It was stated in Subsection 1.3.2,A, 2a that the solution φ^*_{ik} of a finite difference equation is only an approximation to the solution $\varphi(x, y)$ or $\varphi(r, z)$ of the corresponding partial differential equation. An estimation of the error involved can be made by applying the finite difference operator to the exact solution and calculating the residuals in the points P_0 of the mesh. This will be illustrated for the case of two-dimensional fields. In this case the operator is $(L\varphi)_0$, defined in Eq. (2.1) and it follows from the derivation of Eq. (1.7) that the residual is given by a series:

$$(\Delta\varphi)_0 - (L\varphi)_0 = -(L\varphi)_0 = \sum_{n=1}^{\infty} A_n a^{2n} \tag{4.1}$$

From Eq. (1.8) it can be seen that the leading term of the series is given by

$$(\Delta\varphi)_0 - (L\varphi)_0 \simeq A_1 a^2 = -\frac{a^2}{12}\left(\frac{\partial^4\varphi}{\partial x^4} + \frac{\partial^4\varphi}{\partial y^4}\right)_0 \tag{4.2}$$

In practical cases, however, $\varphi(x, y)$ is unknown and a solution φ^*_{ik} is obtained instead. Defining ε_{ik} by

$$\varepsilon_{ik} = \varphi^*_{ik} - \varphi_{ik} \tag{4.3}$$

it follows from (2.4) that, for $\varphi_{ik} \equiv \varphi_0$,

$$(L\varepsilon)_0 = -(L\varphi)_0$$

and, on using (1.2) and (4.1),

$$(L\varepsilon)_0 = \sum_{n=1}^{\infty} A_n a^{2n} \tag{4.4}$$

The quantity ε_{ik} is dependent on the position of the point P_{ik} where it is determined, and on the meshwidth a. Both φ^*_{ik} and ε_{ik} are only defined in the mesh points of an arbitrary mesh. By choosing a suitable interpolation function, φ^*_{ik} can also be defined in intermediate points and for a continuously variable meshwidth a. Using the relation (4.3), we then have for the "extended" function ε_{ik}:

$$\varepsilon_{ik} = \varepsilon(x, y, a) \tag{4.5}$$

It can be derived from Eqs. (4.4) and (4.5) (Richardson, 1924–1925) that ε_{ik} can also be developed in a power series in a:

$$\varepsilon_{ik} = \sum_{n=1}^{\infty} A_n{}^* a^{2n} \tag{4.6}$$

The constant term of this series is zero in order to fulfill Eq. (2.5). By making measurements at different values of a, ε_{ik} can be estimated from (4.6). This equation can also be used for correcting the potential values measured with the network (Liebmann, 1950a; Culver, 1952).

To this purpose, the potential distribution is measured in both a full-scale and a half-scale model.

Retaining the first term only in the series (4.6), we have from this equation and Eq. (4.3):

$$\varphi(i, k) \simeq \varphi_1{}^*(i, k) - A_1{}^* a^2 \tag{4.7}$$

$$\varphi(i', k') \simeq \varphi_2{}^*(i', k') - A_1{}^* (2a)^2 \tag{4.8}$$

In Eq. (4.8) i' and k' are the row and column numbers in the half-scale model. In corresponding points:

$$i' = \tfrac{1}{2}i \quad \text{and} \quad k' = \tfrac{1}{2}k \tag{4.9}$$

For mesh points where Eq. (4.9) holds we have

$$\varphi(i, k) = \varphi(i', k') \tag{4.10}$$

By using this relation, it follows from (4.7)–(4.10) that

$$\varphi(i, k) \simeq \varphi_1{}^*(i, k) + \tfrac{1}{3} \{\varphi_1{}^*(i, k) - \varphi_2{}^*(i', k')\} \tag{4.11}$$

The right-hand side of this equation is a better approximation to $\varphi(i, k)$ than $\varphi_1^*(i, k)$ is. Even better results can be obtained by making measurements on three scales and using one more term of the series (4.6). From the three resulting equations, corresponding to Eqs. (4.7) and (4.8), the coefficients A_1^* and A_2^* can be eliminated (Culver, 1952), yielding a more accurate approximation. To illustrate the effect of the correction (4.11), its application to the field of a cylindrical condensor with a ratio of the radii 10 : 1 and of the potentials of 1000 : 1, is shown in Table IV (Francken, 1953).

TABLE IV

CYLINDRICAL CONDENSOR $R_2/R_1 = 10$; $V_2/V_1 = 1000$
(Calculated Values)

r/r_1	$\varphi_2^*(i')$ 1 : 2 scale model	$\varphi_1^*(i)$ 1 : 1 scale model	$-\Delta/3$	$\varphi_{corr.}$ corr. value	φ_{exact} ($= \log r/r_1$)	Difference
2	294.14	299.09	1.65	300.74	301.03	−0.29
3	470.62	475.32	1.57	476.89	477.12	−0.23
4	596.68	600.58	1.30	601.88	602.06	−0.18
5	694.73	697.81	1.03	698.84	698.97	−0.13
6	774.94	777.27	0.78	778.05	778.15	−0.10
7	842.82	844.48	0.55	845.03	845.10	−0.07
8	901.65	902.70	0.35	903.05	903.09	−0.04
9	953.56	954.06	0.17	954.23	954.24	−0.01

This type of field is particularly unfavorable for numerical computation with the aid of the operator $(M\varphi)_{ik}$ from (3.5), since it is very nonuniform near the inner electrode. Nevertheless, the corrected potential differs less than $3 \cdot 10^{-4}$ of φ_{max} from the exact value in the critical region.

The potentials in the first rows are calculated using the "exact" values of the vertical resistances as given in Eq. (3.22). The resulting expressions are

$$\varphi_1^*(i) = \left\{ \sum_{j=2}^{i} R_0/(2j+1) \right\} : \left\{ \sum_{j=2}^{19} R_0/(2j+1) \right\} \qquad (i = 2, 4, \ldots, 18)$$

$$\varphi_2^*(i') = \left\{ \sum_{j=1}^{i'} R_0/(2j+1) \right\} : \left\{ \sum_{j=1}^{9} R_0/(2j+1) \right\} \qquad (i' = 1, 2, \ldots, 9)$$

In cases in which this correction method cannot be applied to the whole model because of the geometry, the error due to the meshwidth in *important regions* can be reduced by measuring the potential along a suitable outline of mesh points in a full-scale model, setting up an enlarged model of the region inside that outline, and applying the measured potential values to the new boundary mesh points. The correction (4.11) can now be applied to the selected region.

B. Practical Realization of Resistance Networks.

5. *Experimental Setup and Design*

The first property to decide upon is the size of the network. Obviously, this is largely dependent on the requirements as to accuracy, available space, maximum cost, and so forth. Typical examples of networks for solving problems with rotational symmetry have 20 meshes in the *r* direction and 60 meshes in the *z* direction (Liebmann, 1950a) or 25 meshes in the *r* direction and 50 meshes in the *z* direction (Francken, 1953, 1959–1960). The total number of resistors in the latter case amounts to 2575 and this figure increases quadratically with the linear dimensions of the network.

The resistors are mounted on the back of a sheet of insulating material (laminated paperbased sheet) with a high surface resistivity (for example, 10^{11} ohms per square). The junctions are connected to silver-plated contact pins, protruding through the front panel. In order to determine the potential distribution in a certain case, the electrodes are simulated by linking the junctions corresponding to the electrode outlines by copper wires.

In general, the boundary potentials V_b should be applied such that the relation $\varphi_b = \beta V_b$ (Eq. (2.9) is satisfied. However, a more convenient and accurate way is, if possible, to split up the potential problem into unipotential problems, that is, problems in which the electrode potentials are either zero or equal to an (arbitrary) potential V. Any combination of boundary potentials can then be obtained by applying the superposition principle, as will be illustrated on the following example.

Figure 5.1 shows the analog of the triode part of an electron gun (Francken, 1959–1960). A first plot is made with the electrode G connected to one terminal of a potentiometer (B); all the other electrodes are connected to the other terminal (A), while A and B are connected both to the poles of a voltage source and to the terminals of a potentiometer. The potential of the sliding arm of this potentiometer is made equal to that of an arbitrary point P. The balance condition can be read from a null indicating instrument, such as a dc millivoltmeter and the (relative) potential of P can be read from the position of the sliding arm. By repeating the measurement for other points P, a potential plot $\varphi_1(r, z)$ is obtained, where the respective potentials are given as a proportion of the input voltage $(V_B - V_A)$,

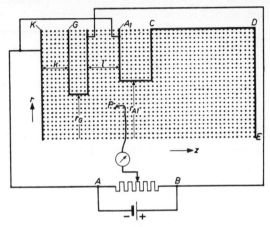

FIG. 5.1. Bridge circuit for measuring the potentials of junctions in the resistance network. K,G and A_1 are electrode models simulating the triode part of an electron gun; AB is potentiometer (1000 Ω.) (Courtesy: *Philips. Tech. Rev*).

which can be taken as unity. If we take $V_A = 0$, we have $V_B = 1$ unit. A second plot $\varphi_2(r, z)$ is obtained by repeating the procedure with the electrode A_1 connected to the terminal $B(V_B = 1$ unit) and all other electrodes to the terminal A $(V_A = 0)$. Since, in the given example, CD and DE are connected to A_1, the number of independent solutions is two and an arbitrary distribution, with boundary voltages

$$V_K = 0, \quad V_G = V_1, \quad \text{and} \quad V_{A1} = V_2$$

is constructed from the measured plots as follows:

$$\varphi(r, z) = V_1\varphi_1(r, z) + V_2\varphi_2(r, z) \tag{5.1}$$

In general, with n different boundary potentials, the measurements have to be repeated $(n - 1)$ times in order to find, by linear combination of the results, a solution for any given set of electrode potentials. The choice of *resistors* and *resistance* values is, again, governed by the accuracy required and the cost involved. In his original network, Liebmann (1950a) used commercial high-stability 1-W resistors of $\pm 1\%$ manufacturing tolerance. Francken (1953, 1959–1960) used 0.5-W wire-wound resistors of 0.2% tolerance, made of manganin wire. It is also possible to use metal film resistors which can be supplied with tolerances of 0.2% or even 0.1%. Apart from the nominal tolerance, attention should be paid to the thermal stability of the resistors. The choice of the *resistance values*, together with the input voltage, needs a careful consideration. In the case of the resistance network for axially symmetrical potential fields, the choice of the arbitrary value R_0 [see Eqs. (3.21) and (3.22)] is limited on one side by the manufacturability of the highest resistance [that is, $R_H(0) = 4R_0$, Eq. (3.25)], on the other side by the requirement that, for large feeding currents, the potential drop along the copper (equipotential) electrodes should be negligible. Moreover, the power dissipation in the smallest resistors [that is, $R_H(n) = R_0/2n$, see Eq. (3.21), for n rows] determines their size, with a given input voltage. The latter should not be too low in order to avoid untolerable loss of sensitivity. With $n = 25$, suitable values are (Francken, 1953, 1959–1960):

$$R_0 = 3600\ \Omega \qquad R_{max} = 14.400\ \Omega \qquad R_{min} = 72\ \Omega$$

By using a voltage source of 2 V, the power dissipation in the most unfavourable case amounts to about 0.05 W. The design of a network on the lines suggested by Weber (Subsection 1.3.2,A, 3d) offers a greater degree of freedom, the ratio between R_{max} and R_{min} being much smaller. Networks for the solution of magnetic field problems can be dimensioned on similar lines as treated above.

6. Experimental accuracy

Apart from the systematic error inherent to the method (that is, the finite size of the meshes), which has been discussed in Subsection 1.3.2,A, 4, the following experimental error sources contribute to the total error.

a. *Influence of the resistor tolerances.* It has been pointed out by Liebmann (1950a) that the *average* precision of the potential measurements should be much greater than would appear from the tolerances of the resistors used, owing to the *error reducing statistical property* of the network. In fact, this observation has opened the road to the acceptance of the resistance network as an accurate tool for determining potential fields. The approximate calculation made by Liebmann shows the average accuracy to be expected with a resistor network consisting of 60 meshes in the z direction and 20 meshes in the r direction, composed of resistors with $\pm 1\%$ tolerance, to be of the order of 1 part in 1000 to 1 part in 10,000, depending on the geometry of the model. A similar experimental result was obtained by Francken (1953), the average error being 0.02–0.002%, of the overall potential, namely, 0.1–0.01 of the (maximum) tolerance of manufacture, depending on the geometry of the model. In general, both the average error and the fluctuations are largest in regions with the greatest field gradient.

b. *Temperature dependency of resistors.* If the resistors are made of manganin wire or film, the temperature coefficient in the region of normal room temperature is

$$\alpha = \pm 0.00001$$

Then, temperature differences of the order of 50°C between individual resistors could be tolerated without the precision of the network being affected seriously. In the example cited in the previous paragraph, using a voltage source of 2 V, overheating with 0.5 W resistors is hardly expected.

In practice a check on the constancy of resistivity can be easily made by repeating a few measurements at the end of each experiment.

c. *Errors in the model of equipotential electrodes.* Any appreciable drop along the electrodes should be avoided. A convenient method to achieve this is to surround the network with two or more heavy "equipotential" strips or bars (Francken, 1953, 1959–1960), made of copper and to make multiple connections between the wires constituting the electrodes and the equipotential bars. With such an arrangement it is possible to confine local potential differences to 1 or 2 parts in 10^5 of the overall potential. Good contact between the electrodes and the contact pins must be ensured.

d. *Tolerances of the potentiometer.* A potentiometer consisting of four pairs of ganged decades, three pairs of ganged decades, and a wirewound potentiometer or a so-called "helipot" can serve the purpose. A tolerance of 1 part in 10^5 should be aimed at. A typical value of the potentiometer input resistance is 1000 ohms (Francken, 1959–1960).

C. Resistance Networks for the Determination of Magnetic Fields

7. *The Scalar Magnetic Potential*

It has been pointed out in Section 1.3.1,C that the scalar magnetic potential $\psi(x, y, z)$ obeys Laplace's equation in current-free regions with constant permeability μ [see Eq. (6.6) in Section 1.3.1]. Such fields, therefore, can be simulated with the electrolytic tank analog, the conductivity of the electrolyte being the analogy of the permeability μ. (See Table II, p. 122).

In cases where μ is very large, pole-pieces can be represented by (ideal) conductors. Regions with smaller values of μ, but with $\mu_r > 1$, can be simulated with the aid of electrolytes with greater conductivity than that of the bulk electrolyte (Section 1.3.1,C 6c).

In Subsection 1.3.2,A,2d it has been pointed out, that the resistance values of a resistance network are inversely proportional to the conductivity of an electrolytic tank [see. Eq. (2.30)]. Therefore, in simulating magnetic fields with different values of μ, with the aid of a network, the resistance values should be taken inversely proportional to μ.

In the following, the discussions will be restricted to two-dimensional fields. A similar reasoning can be used for fields with rotational symmetry, however.

a. *Discontinuous boundary between regions with constant permeability.* Suppose we have two regions of a two-dimensional field with different values of the permeability, μ_I and μ_{II}. From the previous discussion it will be clear, that these regions can be represented by parts of a resistance network with different values of the resistances R_I and R_{II} such that

$$\mu_I R_I = \mu_{II} R_{II} = R_0 \qquad (7.1)$$

The problem is to determine how these two regions should be connected. To take a specific example, let us consider the case of Fig. 7.1a, where the boundary between the two regions is in the y direction, intersecting the grid points P_3, P_0, and P_4.

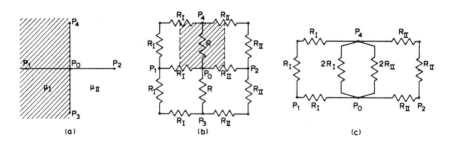

FIG. 7.1. Determination of scalar magnetic potential with the resistance network (plane fields). Discontinuous boundary between regions with different permeability μ_I and μ_{II}. (a) Location of mesh points; (b) Resistance network simulation with difference surface element (hatched); (c) Value of the boundary resistance R.

The corresponding part of the network is shown in Fig. 7.1b. On either side of the boundary the network consists of resistances R_I and R_{II}, respectively, related by Eq. (7.1).

A convenient way to determine the connecting resistances R is to use the geometrical interpretation, given in Subsection 1.3.2,A,2d. The *difference surface element* corresponding with R is indicated in Fig. 7.1b. With a tank of height h and conductivities [see Eq. (2.30)]

$$\sigma_I = 1/hR_I \quad \text{and} \quad \sigma_2 = 1/hR_{II} \tag{7.2}$$

the resistance of the difference surface element can be computed to be

$$R = \frac{2}{h(\sigma_I + \sigma_{II})} \tag{7.3}$$

so that R is related to R_I and R_{II} by

$$\boxed{\frac{1}{R} = \frac{1}{2R_I} + \frac{1}{2R_{II}}} \tag{7.4}$$

the resistance R is, therefore, the equivalence of the resistances $2R_I$ and $2R_{II}$, connected in parallel, as shown in Fig. 7.1c.

b. *Regions with continuously varying permeability.* The determination of magnetic fields in regions with field-dependent permeability has been treated by Liebmann (1953) and, more extensively by Tschopp and Frei (1959; Tschopp, 1961 a, b). In general, the dependence of the magnetic induction **B** on the fieldstrength **H** is *nonlinear* and, with some materials, *anisotropic*. Moreover, in applications where time-dependent fields are involved, the *hysteresis* effect has to be considered as well.

Tschopp (1961 a, b; Tschopp and Frei, 1959) has shown that the effect of nonlinearity and anisotropy can be accounted for by using either a resistance network with variable resistors, or a network consisting of constant resistance values, but with variable current injection into the junctions.

In both cases the process is iterative; the settings of either the resistor or the injection currents have to be adjusted in each iteration cycle. These settings can be computed from the magnetization curve $B = f(H)$ for nonlinear, *isotropic* materials. In the case of anisotropy, the angle between the field vector and a characteristic direction in the material (for example, the direction of rolling) has to be taken into account and a set of magnetization curves $B = f(H, a)$ has to be known.

The hysteresis effect can be simulated for linear, isotropic materials by feeding a network of constant resistances with alternating voltage and feeding alternating currents into the junction through parallel-connected RC-units.

An even more sophisticated approach is the use of inductances with iron cores instead of resistances. If the core material is the same as that of the system to be simulated, the effects of nonlinearity and hysteresis can be taken into account automatically by properly dimensioning the inductances and cores. Anisotropic materials present a greater difficulty, but an approximate solution can be obtained by using different "vertical" and "horizontal" network elements (Tschopp, 1961a).

8. The magnetic vector potential in two-dimensional fields

The magnetic vector potential in this case obeys the Poisson equation [Section, 1.3.1, Eq. (7.8)]:

$$\frac{\partial^2 A}{\partial x^2} + \frac{\partial^2 A}{\partial y^2} = -\mu j_w(x, y) \tag{8.1}$$

where μ is the permeability in the space surrounding the wires with current density $\mathbf{j}_w$, that excite the magnetic field. The term on the right-hand side of Eq. (8.1) differs from zero only in regions containing current-carrying conductors. Outside these regions, the field is determined by the Laplace equation

$$\frac{\partial^2 A}{\partial x^2} + \frac{\partial^2 A}{\partial y^2} = 0 \tag{8.2}$$

and the boundary conditions. These are the same as in the case of the electrolytic tank analog (see Sec. 1.3.1,C,7b). Known **B** lines can be used as equipotentials for A, whereas, at the outlines of polepieces the normal derivative $\partial A/\partial n$ vanishes.

From the similarity of Eqs. (2.20) and (8.2) it will be clear that an approximated solution of $A(x, y)$ can be obtained with the aid of a resistance network with equal resistances, solving the equation of finite differences. Thus

$$(L\,A^*)_{i,k} = 0 \tag{8.3}$$

Pole-pieces with "infinite" permeability are represented by area's with infinite resistance (refer to the tank analogy Section 1.3.1, Table III).

At the outlines of such pole-pieces, the boundary condition is

$$\partial A/\partial n = 0 \tag{8.4}$$

However, this condition cannot be established by simply disconnecting the resistances perpendicular to the outline and pointing inward into the pole pieces but, as shown in Fig. 8.1b, the resistances along the outlines have to be doubled in value at the same time (Tschopp, 1961a). This will be clear from an inspection of the difference surface elements, as hatched in Fig. 8.1

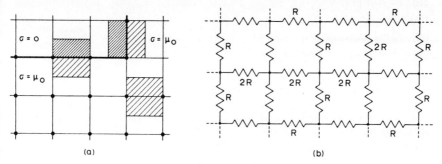

FIG. 8.1. Determination of magnetic vector potential with the resistance network (plane fields). Discontinuous boundary between regions with $\mu = 0$ and $\mu = \mu_0$. (a) Difference surface element on and outside the boundary. (b) Part of the resistance network showing boundary simulation.

(see Subsection 1.3.2,A,2d). The resistivity of the heavily hatched parts inside the pole-pieces is infinite, that of the lightly hatched parts is proportional to $\mu(= \mu_0$ in this case). Obviously, the resistance of a "boundary" difference surface element is twice that of an "inner" surface element.

The practical realization of a resistance network for solving $A(x, y)$ in-

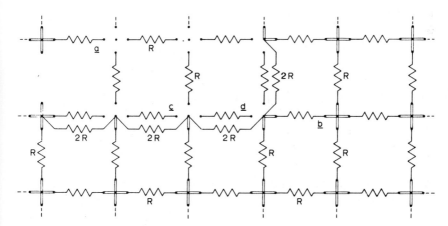

FIG. 8.2. Practical realization of resistance network for determining the magnetic vector potential in plane fields, showing boundary simulation of Fig. 8.1.

volves a construction where resistances can easily be disconnected and substituted. Tschopp (1961b) uses a construction where in each junction the resistances are connected with the aid of a screw-bush. Another possibility is indicated in Fig. 8.2. At each junction the resistances are connected to four bushes which normally are connected to the central bush with the aid of a five-pole plug [fig. 8.2; at (a) disconnected resistors; at (b) connected resistors]. Special plugs can be used to partly connect resistors and connecting other resistors (Fig. 8.2; at (c) two-pole plug with two resistors connected; and at (d) three-pole plug with two-resistors connected).

9. Magnetic Flux in Axially Symmetrical Fields

It has been pointed out in Section 1.3.1,C,7c that in this case the magnetic flux,

$$M(r_0 , z) = 2\pi r_0 A(r_0 , z)$$

has to be determined rather than the vector potential. The partial differential equation has been given in Eq. (7.15) of Section 1.3.1,C,7c. Thus

$$\frac{\partial^2 M}{\partial z^2} - \frac{1}{r} \frac{\partial M}{\partial r} + \frac{\partial^2 M}{\partial r^2} = -2\pi\mu r j_w(r, z) \tag{9.1}$$

From Taylor expansions in four points surrounding a point P_0, expressions for the partial derivatives can be found, yielding Eqs. (3.2)–(3.4) with M substituted for φ. Rearranging the terms of the first equation and of the difference of the two other equations, we have

$$M_2 + M_1 - 2M_0 = a^2 \left(\frac{\partial M^2}{\partial z^2}\right)_0 + \frac{1}{12} a^4 \left(\frac{\partial^4 M}{\partial z^4}\right)_0 + \cdots \tag{9.2}$$

$$(M_4 - M_0)\left(1 - \frac{a}{2r}\right) + (M_3 - M_0)\left(1 + \frac{a}{2r}\right)$$

$$= a^2 \left(\frac{\partial^2 M}{\partial r^2} - \frac{1}{r} \frac{\partial M}{\partial r}\right)_0 - a^4 \left(\frac{1}{6r} \frac{\partial^3 M}{\partial r^3} - \frac{1}{12} \frac{\partial^4 M}{\partial r^4}\right)_0 + \cdots \tag{9.3}$$

Adding these equations leads to a finite difference equation—which is an approximation to Eq. (9.1) with $j_w = 0$—and a residual, given by

$$O(a^2) = a^2 \left(\frac{1}{12} \frac{\partial^4 M}{\partial z^4} - \frac{1}{6r} \frac{\partial^3 M}{\partial r^3} + \frac{1}{12} \frac{\partial^4 M}{\partial r^4}\right) + \cdots \tag{9.4}$$

A better approximation, especially in the vicinity of the axis, can be obtained as follows, using the method described in Subsection 1.3.2,A,3d.

The series expansion of $M(r, z_0)$ in terms of the magnetic induction on the axis, $B(0, z) \equiv B_0$, is given by

$$M(r, z_0) = 2\pi \, rA(r, z_0) = \pi r^2 \sum_{n=0}^{\infty} \frac{(-1)^n}{n!(n+1)!} \left(\frac{r}{2}\right)^{2n} B_0^{(2n)} \qquad (9.5)$$

The first terms of this series and those of the partial derivatives of M are

$$M(r, z_0) = \pi \left(r^2 B_0 - \frac{r^4}{8} B'' + \frac{r^6}{192} B^{IV} - \cdots \right) \qquad (9.5')$$

$$\frac{\partial M}{\partial z} = \pi \left(r^2 B_0' - \frac{1}{8} r^4 B_0''' + \cdots \right)$$

$$\frac{\partial^2 M}{\partial z^2} = \pi \left(r^2 B_0'' - \frac{1}{8} r^4 B_0^{IV} + \cdots \right)$$

$$\frac{\partial^3 M}{\partial z^3} = \pi r^2 B_0''' - \cdots$$

$$\frac{\partial^4 M}{\partial z^4} = \pi r^2 B_0^{IV} - \cdots$$

$$\frac{\partial M}{\partial r} = \pi \left(2r \, B_0 - \frac{1}{2} r^3 B_0'' + \frac{1}{32} r^5 B_0^{IV} - \cdots \right)$$

$$\frac{\partial^2 M}{\partial r^2} = \pi \left(2B_0 - \frac{3}{2} r^2 B_0'' + \frac{5}{32} r^4 B_0^{IV} - \cdots \right)$$

$$\frac{\partial^3 M}{\partial r^3} = \pi \left(-3r \, B_0'' + \frac{5}{8} r^3 B_0^{IV} - \cdots \right)$$

$$\frac{\partial^4 M}{\partial r^4} = \pi \left(-3 \, B_0'' + \frac{15}{8} r^2 B_0^{IV} - \cdots \right)$$

$$\frac{\partial^5 M}{\partial r^5} = \pi \cdot \frac{15}{4} r \, B_0^{IV} - \cdots$$

$$\frac{\partial^6 M}{\partial r^6} = \pi \cdot \frac{15}{4} B_0^{IV} - \cdots \qquad (9.6)$$

Substituting these expressions in Eqs. (9.2) and (9.3) reveals, for small values of r, the contributions of the terms in $\partial^3 M/\partial r^3$ and $\partial^4 M/\partial r^4$ containing B_0'' to be of the same order of magnitude as those in $\partial M/\partial r$, $\partial^2 M/\partial r^2$, and $\partial^2 M/\partial z^2$. For this reason, the residual (9.4) is not the smallest possible one. Multiplying Eq. (9.2) with $(1 + A)$, adding the product to

(9.3), dividing by a^2 and using (9.1) with $j_w = 0$ results in

$$\frac{1}{a^2}\Big\{(1 + A)(M_1 + M_2 - 2M_0) + (M_3 - M_0)\Big(1 + \frac{a}{2r}\Big)$$

$$+ (M_4 - M_0)\Big(1 - \frac{a}{2r}\Big)\Big\}$$

$$= a^2\Big(\frac{A}{a^2}\frac{\partial^2 M}{\partial z^2} + \frac{1 + A}{12}\frac{\partial^4 M}{\partial z^4} - \frac{1}{6r}\frac{\partial^3 M}{\partial r^3} + \frac{1}{12}\frac{\partial^4 M}{\partial r^4}$$

$$- \frac{a^2}{120r}\frac{\partial^5 M}{\partial r^5} + \frac{a^2}{360}\cdot\frac{\partial^6 M}{\partial r^6}\Big) \quad (9.7)$$

The residual on the right-hand side of Eq. (9.7) can be expressed in the derivatives of B_0, using (9.6) and retaining terms with B_0'' and B_0^{IV}:

$$O(a^2) = \pi a^2\Big\{B_0''\Big(A\frac{r^2}{a^2} + \frac{1}{4}\Big) + B_0^{IV}\Big(-\frac{1}{8}A\frac{r^4}{a^2} + \frac{1}{12}Ar^2 + \frac{13}{96}r^2 - \frac{a^2}{48}\Big)\Big\}$$
$$(9.8)$$

The term with B_0'' can now be made to vanish at each value of $r = ia$, by putting

$$A = -1/4i^2 \quad (9.9)$$

The residual, in this case, is given by

$$O(a^4) = \pi a^4\Big(\frac{1}{6}i^2 - \frac{1}{24}\Big)B_0^{IV} + \cdots \quad (9.10)$$

In general, this residual will be smaller than that given in (9.4) and, therefore, resulting from (9.8) for $A = 0$:

$$O(a^2) = \pi a^2\Big\{\frac{1}{4}B_0'' + a^2\Big(\frac{13}{96}i^2 - \frac{1}{48}\Big)B_0^{IV}\Big\} + \cdots \quad (9.11)$$

From (9.7) and (9.9) it follows that a finite difference operator N_i can be defined by

$$a^2(N_iM)_{ik} \equiv \frac{4i^2 - 1}{4i^2}(M_{i,k-1} + M_{i,k+1} - 2M_{i,k}) + \frac{2i+1}{2i}(M_{i-1,k} - M_{i,k})$$

$$+ \frac{2i-1}{2i}(M_{i+1,k} - M_{i,k}) \quad (9.12)$$

This equation can be given the alternative form

$$a^2(N_iM)_{i,k} \equiv \frac{4i^2-1}{2i}\left(\frac{M_{i,k-1}-M_{i,k}}{2i}+\frac{M_{i,k+1}-M_{i,k}}{2i}\right.$$
$$\left.+\frac{M_{i-1,k}-M_{i,k}}{2i-1}+\frac{M_{i+1,k}-M_{i,k}}{2i+1}\right) \tag{9.13}$$

From this equation and Eq. (3.12) it follows that, for solving the equation

$$(N_iM)_{ik}=0 \tag{9.14}$$

a resistance network can be used with resistances[4]

$$R_H(i)=2iR_0 \qquad R_V(i)=(2i+1)R_0 \tag{9.15}$$

These equations are also valid for $i=0$, since the axis is an equipotential for $M(0, z)=0$ (see Fig. 9.1).

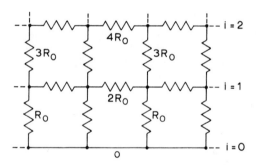

FIG. 9.1. Part of a resistance network for determining the magnetic flux distribution in rotationally symmetrical magnetic fields.

Boundary conditions are established such that a known **B** line is an equipotential [Section 1.3.1, Eq. (7.16)], whereas at a high-permeability polepiece, the normal component of grad M vanishes [Section 1.3.1, Eq. (7.17)]:

$$(\text{grad } M)_n = 0$$

As in the two-dimensional problem for $A(x, y)$, treated previously, this condition is established by disconnecting the resistances pointing inward from the boundary and substituting the resistors along the boundary by others, having twice the resistance. (see Fig. 8.2).

[4] Liebmann (1950b) arrives at the same network with the aid of a mathematically incorrect derivation.

REFERENCES

Bowman-Manifold, M., and Nicoll, F. A. (1938). *Nature* **142**, 39.
Culver, R. (1952). *Brit. J. Appl. Phys.* **3**, 376.
de Beer, A.J.F., Groendijk, H., and Verster, J. L. (1961–1962). *Philips. Tech. Rev.* **23**, 352.
de Packh, D. C. (1947). *Rev. sci. Instr.* **18**, 798.
Einstein, P. A. (1951). *Brit. J. Appl. Phys.* **2**, 49.
Francken, J. C. (1953). Electron optics of the image iconoscope. Thesis, Delft.
Francken, J. C. (1959–1960). *Philips Tech. Rev.* **21**, 10.
Gershgorin, S. A. (1929). *Zh. Prikl. Fiz.* **4**, No. 3–4, 3.
Hogan, T. K. (1943). *J. Inst. Engrs. Australia* **15**, 89.
Hollway, D. L. (1955). *Australian J. Phys.* **8**, 74.
Karplus, W. J. (1955). *Brit. J. Appl. Phys.* **6**, 356.
Karplus, W. J. (1958). "Analog Simulation." McGraw-Hill, New York.
Kennedy, P. A., and Kent, G. (1956). *Rev. Sci. Instr.* **27**, 916.
Kennely, A. E., and Whiting, S. E. (1906). *Elec. World* **48**, 1239.
Kirchhoff, G. R. (1845). *Ann. Physik.* [2] **64**, 497.
Liebmann, G. (1949). *Nature* **146**, 149.
Liebmann, G. (1950a). *Brit. J. Appl. Phys.* **1**, 92.
Liebmann, G. (1950b). *Phil. Mag.* [7] **41**, 1143.
Liebmann, G. (1950c). *Advan. Electron.* **2**, 101.
Liebmann, G. (1953). *Brit. J. Appl. Phys.* **4**, 193.
Liebmann, G. (1954). *Brit. J. Appl. Phys.* **5**, 362.
Loukochkov, V. S. (1956). *Vide* **65**, 328.
Malavard, L. (1956a). *Onde Elec.* **36**, 762.
Malavard, L. (1956b). *Onde Elec.* **36**, 829.
Malavard, L. (1956c). *Onde Elec.* **36**, 1046.
Miroux, J. (1958). *Onde Elec.* **38**, 450.
Musson-Genon, R. (1947). *Ann. Telecommum.* **2**, 298.
Peierls, R. E. (1946). *Nature* **158**, 831.
Peierls, R. E., and Skyrme, T. H. R. (1949). *Phil. Mag.* [7] **40**, 269.
Peters, C. J. (1965), *Rev. Sci. Instr.* **36**, 174.
Richardson, L. F. (1924–1925). *Math. Gaz.* **12**, 415.
Sander, K. F., and Yates, J. G. (1953). *Proc. Inst. Elec. Engrs.* (*London*) *Pt. II* **100**, 167.
Tschopp, P. A. (1961a). *Bull. Schweiz. Elektrotech. Ver.* **6**, 185.
Tschopp, P. A. (1961b). Analogieverfahren zur Bestimmung von magnetischen Feldern in nicht-linearen nicht-isotropen Medien. Thesis, Zurich.
Tschopp, P. A., and Frei, H. A. (1959). *Arch. Elektrotech.* **44**, 441.
van Duzer, T., and Brewer, G. R. (1959). *J. Appl. Phys.* **30**, 291.
van Duzer, T., Buckey, C. R., and Brewer, G. R. (1963). *Rev. Sci. Instr.* **34**, 558.
Verster, J. L. (1960–1961). *Philips Tech. Rev.* **22**, 245.
Verster, J. L. (1963). On the use of gauzes in electron optics. Thesis, Delft.
Weber, C. (1962–1963). *Philips Tech. Rev.* **24**, 130.
Weber, C. (1963). *Proc. Inst. Elec. Engrs.* (*London*) **51**, 1963.

CHAPTER 1.4

MEASUREMENT OF MAGNETIC FIELDS

C. Germain
CERN
GENEVA, SWITZERLAND

1.4.1. Introduction

An increasing demand for exacting measurements of magnetic fields in a wide variety of conditions has resulted from the continuous progress of science and technology. As a consequence, new methods of measurement have been developed and the older ones have been considerably improved in the last two or three decades. This advance in magnetic measurement technique has been stimulated, particularly, by the rapid development of high-energy physics and the associated construction of large accelerators of charged particles whose guiding and focusing are realized with magnetic fields. In constructing such accelerators and their beam transport equipment, the least difficult problem is not that of measuring and adjusting to very stringent conditions a magnetic field that often varies both in space and time.

The general problem of magnetic-field measurements has produced important technical literature, which has been comprehensively reviewed by Symonds (1955) and more recently by other authors (Barber, 1960; Grivet, 1962; Germain, 1963; Hermann, 1964). In order to stay within the frame of the present chapter we shall only examine the methods of magnetic-field measurement pertaining to the problem of charged-particle focusing; and furthermore, for the benefit of a more detailed discussion of the practical applications of these methods, we have selected some typical examples in the list of recent references. The reader will easily find more references about a given problem of magnetic-field measurement in the literature reviews mentioned in the foregoing.

The principles of the usual methods of measuring magnetic fields in magnets or magnetic lenses are presented in Section 1.4.2. We have devoted Section 1.4.3 to the measurement of the magnetic field itself. In Section 1.4.4 we discuss the problem of measuring the gradient of the magnetic field in inhomogeneous field magnets and in quadrupole magnetic lenses.

We have subdivided both Sections 1.4.3 and 1.4.4 to consider separately the cases when the magnetic field is constant with time as in permanent magnets and in dc electromagnets, or variable with time as in pulsed and ac electromagnets. In a further subdivision we distinguish between point measurements and averaging measurements: The former give the value of the magnetic field, or its projection on a direction, at a well-defined point and the latter provide the integral, along a direction, of the magnetic-field component that is perpendicular to that direction. A similar distinction is made for magnetic-field gradient measurements. Such spatially integrated measurements are often required for computing the trajectories and focusing of high-energy charged particles.

The MKSA system of units is used and therefore the magnetic induction B is expressed in teslas ($1\ T = 1\ Wb/m^2 = 10^4\ G$). When magnetic field is implied in a method of measurement, we have given the corresponding induction value in vacuum for the sake of simplicity in comparing different methods. The frequency is expressed in hertz ($1\ Hz = 1\ cps$).

1.4.2. Principles of the Most Usual Methods of Measuring Magnetic Fields

A. MAGNETIC RESONANCE

1. *Introduction*

Since 1946 when the first researches were published on nuclear magnetic resonance, this phenomenon has been widely used for the measurement and stabilization of magnetic fields: Examples of such application in the relevant literature have been rapidly multiplying as the technique of nuclear magnetic resonance overflowed the domain of specialized laboratories to become almost an industrial practice; different types of apparatus for measuring magnetic fields are now commercially available. The main feature of this method of measuring magnetic fields is the high degree of accuracy that can be readily obtained (Pople *et al.*, 1959).

From a phenomenological point of view, when a sample of paramagnetic material is placed in a steady magnetic field H_0, it becomes magnetized and its magnetic moment $\mathbf{M}_0$ is directed along $\mathbf{H}_0$ with $\mathbf{M}_0 = \chi \mathbf{H}_0$, where χ is the susceptibility of the material. If a sinusoidal magnetic field $\mathbf{H}_1$, with $H_1 \ll H_0$, is now applied at right angles to $\mathbf{H}_0$ and if its angular frequency is equal to the Larmor frequency $\omega = \gamma H_0$, where γ is the gyromagnetic ratio of the material, the magnetic-resonance phenomenon appears: The resulting magnetic moment $\mathbf{M}$ is no longer parallel to $\mathbf{H}_0$ but makes an angle with $\mathbf{H}_0$. The component $\mathbf{M}_p$ of the magnetic moment that is perpendicular to $\mathbf{H}_0$ can be decomposed into two vectors; one in phase with the field $\mathbf{H}_1$ and the other in quadrature with $\mathbf{H}_1$. The latter component is proportional to the power required by the sample when its magnetic moment is being deflected from alignment with $\mathbf{H}_0$. The necessary power is provided by the ac field H_1.

In the quantum mechanics picture of this phenomenon the charged particles responsible for the magnetic properties of the material are considered. They all have an angular momentum vector $\mathbf{p}$ and a corresponding magnetic moment vector $\boldsymbol{\mu} = \gamma \mathbf{p}$, which become orientated with respect to the magnetic field $\mathbf{H}_0$. This orientation can only be such that the component of $\mathbf{p}$ along $\mathbf{H}_0$ is equal to $mh/2\pi$. Here h is Planck's constant; $m = \pm (I - k)$, where I is the spin of the particle, and k any integer smaller or equal to I, so that m can take on several discrete values, each giving a different orientation for $\mathbf{p}$. To each of these orientations of $\boldsymbol{\mu}$ in the magnetic field there corresponds a different energy level. In the paramagnetic material employed for measuring magnetic fields, only electrons or protons are made to resonate. These particles both have a spin $I = \frac{1}{2}$, and therefore only two possible energy levels between which the difference in energy is $\Delta E = H_0 \gamma h / 2\pi$.

Upon irradiating a sample of paramagnetic material with photons of the right frequency ν_0, such that $h\nu_0 = \Delta E$, an exchange of energy is produced between the electromagnetic wave and the sample material: Photons are absorbed by the protons (or electrons) in the sample when they jump from the lower to the higher energy level. The principle of the measuring method is to determine the Larmor frequency corresponding to the resonance in the magnetic field to be measured. It is an absolute measurement that can be made with a very great accuracy.

A magnetic-field measuring apparatus based on magnetic resonance consists of two parts: one for producing the irradiating electromagnetic wave of frequency ν_0 and the other for detecting the energy exchange between this wave and the paramagnetic material sample. The electromagnetic wave is usually generated in a small rf emitting coil, containing the sample, orientated so that its axis is perpendicular to the magnetic field $\mathbf{H}_0$. The resonance is detected in the absorption method by the variation of the emitting coil impedance: In this case the apparatus has only one coil. The resonance can also be detected by the induction method with the help of another rf coil orientated at right angles to both the magnetic field $\mathbf{H}_0$ and the emitting coil so as to reduce inductive coupling as much as possible: In this case the resonance signals are induced in the detection coil by the photons emitted when particle spins return from the high to the low-energy level.

2. Nuclear Resonance

For the magnetic resonance the sample generally consists of water or mineral oil in a small ampul to provide the necessary free protons of hydrogen atoms. At resonance the induction value B_0 of the magnetic field is related to the electromagnetic-wave frequency. Thus

$$B_0 = 2.3487 \times 10^{-2}\, \nu_0 \quad \text{or} \quad \nu_0 = 42.576\, B_0 \tag{1}$$

where B_0 is in teslas and ν_0 in *megahertzs*. The measuring range of this method is from about one hundredth of a tesla to a few tens of teslas without definite limits. For higher field values the measurements would be possible with the use of centimetric wavelength technique if such magnetic fields could exist in steady enough conditions. Stretching to the low values presents some difficulty, as the magnitude of the signal depends on the measured magnetic field and then becomes very small. This difficulty may be overcome in different ways for measuring magnetic fields down to the microtesla range, as can be found in the literature, but this will not be discussed here as it would take us rather far afield. The proton magnetic resonance can easily provide a method of measuring magnetic fields to one part in 10^4 with simple apparatus. About two orders of magnitude in the accuracy can still be gained with extensive precautions and refined equipment.

For observing the resonance signal, produced by the exchange of energy between the electromagnetic wave and the spins of the protons, the resonance conditions are crossed by varying either the magnetic field intensity or the electromagnetic-wave frequency. In steady fields a small local sweeping ac field is produced by coils in the volume of the sample or alternatively the frequency ν can be modulated around the resonance value ν_0. The longitudinal relaxation time T_1 of the protons in the sample comes into account in the conditions under which pulsed or ac magnetic fields can be measured. The time T_1 can be reduced by dissolving a paramagnetic catalyst, such as ferric nitrate or sulphate, in the water sample and adjusted to any value between 1 sec and 10^{-4} sec (or even 10^{-5} sec at saturation). By increasing the concentration of paramagnetic catalyst the relaxation time T_1 can be made shorter than the duration of the magnetic-field pulse, within the limits given above, thus still allowing a large enough polarization of the protons for producing an intense signal.

Decreasing the longitudinal relaxation time T_1 causes a progressive increase of the width of the signal from a fraction of a microtesla to a few tenths of a millitesla because of the corresponding reduction of the spin–spin relaxation time T_2. The accuracy is then limited by the shape of the signal, the center of which can hardly be determined to better than about one tenth of the linewidth without special line-narrowing devices. However, if the magnetic field is not very homogeneous, a corresponding instrumental relaxation time $\bar{T}_2 = (\gamma \overline{\Delta H})^{-1}$, related to the rms value $\overline{\Delta H}$ of the field inhomogeneity over the volume of the sample, has to be introduced. The optimum conditions are obtained for the resonance signal when this time $\bar{T}_2$ is equal to the natural time T_2 of the probe. The longitudinal relaxation time T_1, which is equal to T_2 in practical cases of field measurements, is then determined.

3. Electron Resonance

If electrons instead of protons are involved in the magnetic-resonance phenomenon, the range of frequency in the same magnetic field is displaced upwards by nearly three orders of magnitude. The gyromagnetic ratio for electrons is about 660 times larger than for protons and at resonance the magnetic induction is

$$B_0 = 3.5657 \times 10^{-2} \nu_0 \quad \text{or} \quad \nu_0 = 28.045 \, B_0 \qquad (2)$$

where B_0 is in teslas and ν_0 in *gigahertzs*. When contemplating the use of electron magnetic resonance the first problem is providing the necessary single electrons because electrons usually go in pairs with opposite spins that have a null resultant moment. The second problem is the limitation of the magnetic-field measurement accuracy due to the width of the resonance signal, which is now more important compared to the resonance field because of the strong coupling between neighboring electron magnetic moments. For a long time the best material in this respect has been diphenyl-picryl-hydrazyl (DPPH) which contains a free radical with an unpaired electron and provides a remarkable resonance line with a total width at half amplitude of only about 85 μT (Lothe and Eia, 1958). Now even better materials are used also, for instance, carbonized dextrose or solutions of potassium in liquid ammonia which have a linewidth of, respectively, 20 μT (Gabillard, 1956) and 3 μT (Hutchinson and Pastor, 1953).

The practical range of field measurements by electron magnetic resonance is from the millitesla region, where the accuracy is limited by the linewidth, up to the tesla region in which centimetric wavelength techniques have to be used. Thus a large overlap exists with proton magnetic resonance methods. The main interest of electron magnetic resonance in high fields is the small relaxation time values ($T_1 \sim T_2 \sim 50$ nsec for DPPH) that make it possible to measure pulsed magnetic fields in the millisecond range (Grivet, 1962).

B. HALL EFFECT

1. *Introduction*

When a magnetic field is applied at right angles to a current-carrying strip of thickness d, a transversal difference of potential V_H appears across the strip at right angles to both the current I and the magnetic field B (Fig. 1).

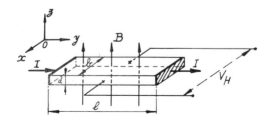

FIG. 1. Hall plate in a magnetic field.

This phenomenon, discovered by Hall in 1879, has produced a simple and increasingly popular means of measuring magnetic fields with a good accuracy. The Hall voltage V_H is expressed by

$$V_H = R_H I B/d \tag{3}$$

All the quantities are in MKSA units, that is, the Hall coefficient R_H is in cubic meters per coulomb.

The Hall effect is a consequence of the Lorentz force acting on electrical charges that move in a magnetic field (Fig. 2). When the charges are constrained to move within the boundaries of a conductor they accumulate on one of the sides of the conductor, building up a repulsive force that balances the Lorentz force $\mathbf{F}_L$ and restores the steady-state current flow conditions that were perturbed upon applying the magnetic field. The mean charge motion is now again parallel to the conductor axis but the equipotentials are no longer perpendicular to the axis: They have been rotated by an angle θ, called the Hall angle (Fig. 3). Hall voltage results from the rotation of the equipotentials by the angle θ or—if one prefers—it is the potential variation created by the accumulation of charges on one of the sides of the conductor.

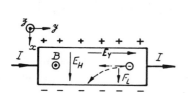

FIG. 2. Hall field E_H in a strip.

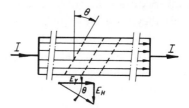

FIG. 3. Hall angle θ in a strip.

2. Mechanism of Solid-State Electrical Conduction

Let us recall briefly the model of electrical conduction in the quantum theory before examining a little more in detail the mechanism of the Hall effect.

In a crystal structure the energy levels of the electrons in all the atoms of the sample are arranged in bands of preferred energies that are separated by gaps containing no permitted levels. The number of levels in each band is discrete and corresponds to the total number of atoms in the sample. A state of motion in the crystal is represented by each level but the distribution is such that there is no net motion of the electrons in the absence of electric field or other force, created by a temperature gradient, for instance. When such a force is applied the electron distribution in the possible levels may be modified and give rise to a current.

The electrons fill all the possible levels of the lower energy bands and the highest band may be either completely filled or not. The energy gap between bands is usually much greater than the mean thermal energy kT or the energy the applied field can provide to the electrons: In these conditions the electrons cannot move from one band to the next higher one, and the crystal is insulating if the highest band is completely filled because the distribution cannot be modified. This is the situation for quartz. If the highest band is not completely filled, a current is produced upon applying an electric field. This band is called the conduction band and the highest completely filled band is the valence band.

Semiconductors are considered "intrinsic" when they are absolutely pure, that is without any imperfection in the crystal structure. At low temperatures they have no electrons in the conduction band. However, the energy gap ε_g with respect to the valence band is small enough so that at room temperature an observable number of electrons are moved into the conduction band, leaving holes in the valence band that contribute also to conduction, for each missing electron is equivalent to a positive charge moving in the opposite direction. The concentrations of electrons and holes are equal; they both are rapidly increasing functions of the temperature, proportional to a Boltzmann factor $\exp(-\varepsilon_g/kT)$.

In practice there are always impurity atoms that we have to take into account in the conduction process even if their proportion appears to be completely negligible by the usual standards of chemical purity. In fact, the intrinsic conduction at moderate or low temperatures has long been masked by impurity or "extrinsic" conduction and only recently has it been possible to study it with the advent of new techniques of refining. In turn semiconducting materials are now produced with an extremely high degree of purity or with controlled minute impurity rates. (The technique of injecting impurities into the purer materials is called "doping".)

In the extrinsic semiconductors the impurity atoms perturb the periodic structure of normal crystal atoms and the corresponding energy levels may now fall in the gap between the valence and conduction bands, requiring a much smaller energy to be excited than for intrinsic conduction. The concentration of charge carriers is then practically independent of an increasing temperature until the onset of intrinsic conduction, which results first in a transition or mixed conduction, and later in dominant intrinsic conduction. There are different ways of introducing impurity atoms and the resulting extrinsic semiconductors may be n-type or p-type, corresponding, respectively, to electron and hole conduction below the transition temperature.

3. *Hall Effect in Metals and Extrinsic Semiconductors*

Let us consider a strip of metal or an n-type semiconductor in which a volume density n of charges is available in the conduction band. All the vector quantities are determined by their components in the coordinate system of Fig. 1: The control current $I(0, I, 0)$ is produced by the charges q ($q = -e$ for electrons) that move with a velocity $v(0, v, 0)$ under the action of the applied electric field $E(0, E_y, 0)$. When the magnetic induction $B(0, 0, B)$ is applied, the Lorentz force F_L acts on the moving charges until it is balanced out by the Hall field E_H. We can thus write

$$F_L + qE_H = 0 \tag{4}$$

$$F_L = qv \times B = (qvB, 0, 0) \tag{5}$$

$$I = qnvbd \tag{6}$$

and by definition of the Hall voltage, we have

$$V_H = R_H IB/d = E_H b \tag{7}$$

so that

$$V_H = -IB/qnd \tag{8}$$

We obtain from these relations the following expressions for the Hall coefficient in metals or n-type extrinsic semiconductors:

$$R_H = -1/qn = 1/en \tag{9}$$

By introducing the electrical conductivity $\sigma = nev/E_y = ne\mu$ where μ is the mobility of the charge carriers we have the simple relation,

$$R_H \sigma = \mu \tag{10}$$

in which μ is expressed in $(\text{teslas})^{-1} = \text{coulomb} \cdot \text{second} \cdot (\text{kilogram})^{-1}$.

From Eq. (8) we see that the Hall voltage polarity is reversed if holes are considered instead of electrons, I and B having stayed unchanged. Equation (9) is valid for metals but the corresponding values for R_H are very small because of the high density of electrons in the conduction band. In semiconducting materials Eq. (9) is only approximative and in the exact derivation of the R_H expression a coefficient $3\pi/8$ is introduced, due to the large dispersion of values of v about the mean velocity.

$$R_H = \frac{3\pi}{8} \cdot \frac{1}{en} \tag{11}$$

The Hall angle θ is given by

$$\tan \theta = \frac{E_H}{E_y} = \frac{3\pi}{8} \mu B \tag{12}$$

When connecting the Hall voltage to a measuring circuit, the transversal resistance R_2 of the Hall plate has to be considered as well as the longitudinal resistance R_1 of the plate, which is related to the control voltage V_c by $R_1 I = V_c$. The efficiency η of the Hall plate is, in the case of maximum output power,

$$\eta = \frac{\text{output power}}{\text{input power}} = \frac{V_H{}^2}{4R_2} \cdot \frac{R_1}{V_c{}^2} = \frac{R_1}{4R_2} \cdot \frac{b^2}{l^2} \cdot \left[\frac{E_H}{E_y} \right]^2 \tag{13}$$

thus

$$\eta = \frac{R_1}{4R_2} \cdot \frac{b^2}{l^2} \tan^2 \theta \sim (\mu B)^2 \tag{14}$$

On the other hand, since the control current I is generally limited by the power dissipation of the Hall plate, we should compare materials under conditions of equal power dissipation p per unit volume. Thus

$$V_H = R_H(IB/d) = R_H b B \sqrt{\sigma} \sqrt{p} = \mu b B \sqrt{\varrho} \sqrt{p} \tag{15}$$

in which ϱ is the resistivity in ohm $\cdot$ meters.

It is clear from these relations that the carrier concentration should be small for obtaining a high value of Hall coefficient. Semiconductors are largely superior to metals in this respect. The mobility should also be high and for this reason n-type are preferred to p-type semiconductors. For measuring magnetic fields it is convenient to have a Hall voltage as independent of temperature as is possible: intrinsic semiconductors are therefore much less favorable than extrinsic semiconductors. Besides, the mechanism of the Hall effect in intrinsic semiconductors is slightly different, because in this case both types of charge carriers exist with the same concentration, and they both are deflected in the same direction. Thus a differential effect only appears because of the difference in the mobility values. This type of semiconductor turns out to be interesting mostly for its large magnetoresistance effect, which is another means of measuring magnetic field, which will not, however, be discussed here because it is rather less popular presently. Continuous progress in that field may however change this situation (Weiss and Wilhelm, 1963).

Details of the theory of the Hall effect can be found either in books (Welker and Weiss, 1956; Jan, 1957; Putley, 1960) or in several articles (Ross *et al.*, 1957; Kuhrt, 1960, 1961; other references in the bibliography given by Kuhrt).

4. *Choice of the Material for Hall Generators*

We have seen that, in order for a material to be suitable to make a good Hall generator, the concentration of charges should be small and their mobility high. Metals are ruled out because of their high value of electron concentration in the conduction band, but within the last decade semiconducting materials have been specially developed for the purpose of measuring magnetic fields. Typical characteristics are given for some materials in Table I.

TABLE I

SOME CHARACTERISTICS OF SEMICONDUCTING MATERIALS

Material	Resistivity, ϱ (Ωm)	Hall coeffic., R_H (m³/C)	Temperature dependence, $\dfrac{1}{R_H} \cdot \dfrac{dR_H}{dt}\left(\dfrac{\%}{°C}\right)$	Mobility, μ (T⁻¹)	Factor of merit, $\mu \sqrt{\varrho}$
Cu	1.7×10^{-8}	5.2×10^{-11}		2.7×10^{-3}	3.5×10^{-7}
n-type Ge	0.1	3.6×10^{-2}	0.2	0.36	0.11
InSb	6×10^{-5}	2.4×10^{-4}	1.5	4	0.031
InAs	5×10^{-5}	1×10^{-4}	0.07	2	0.014
InPAs	2.5×10^{-4}	2×10^{-4}	0.07	0.8	0.013

Hall generators have been made from germanium crystals but new materials with a higher mobility have been developed, especially at Siemens Laboratories, that present a smaller dependence upon temperature. In that respect, InAs is much better than InSb. The latter has a particularly high mobility but is in the mixed conduction state at room temperature. Commercially available Hall generators (Siemens types) may have Hall coefficients as high as 3×10^{-3} m³/C but values of the order 1–2×10^{-4} m³/C are obtained if an operating temperature range 20–100 °C is required (Breunersreuther, 1965). We shall see in Section 1.4.3,A, 1b that a good temperature stabilization is often necessary in the Hall probe when a high degree of accuracy in the magnetic-field measurements is required; thus such a large operating temperature range is not necessary. InAs, as well as $InAs_yP_{1-y}$, with $0.6 < y < 0.85$ (Weiss, 1956), is used for making Hall plates; in the latter material the transition temperature is slightly increased by the presence of phosphorus compared to pure InAs.

Typical Hall generators are made with thin rectangular semiconducting plates—a fraction of a millimeter thick, in dimensions of a few millimeters to one or two centimeters—embedded in epoxy resin to provide enough mechanical strength to the plate and electrode connections. The length is from two to four times the width in order not to be too far away from the infinite length case that was considered in the theory. For a given power dissipation per unit area on the plate, corresponding to satisfactory operating temperature, the current density, and therefore the sensitivity of the Hall generator, varies as $d^{-0.5}$; when a high sensitivity is required the plates are made as thin as possible, but it is hardly possible to go below 0.05 mm without having troubles with instabilities in the residual resistance (that is, the ratio of zero field Hall voltage over control current).

Other Hall generators have been obtained with thin films of vacuum-evaporated semiconducting materials, such as HgSe, and their sensitivity is comparable to that of good Hall generators of the thin plate type (Elpatevskaya and Regel, 1956). Among other advantages of this type of construction, there is the possibility of miniaturizing the probe by using printed circuit techniques: Evaporated Bi film microprobes with a sensitive area as small as 10^{-4} mm² have been described by Roshon (1962).

5. Characteristics of Hall Generators

The Hall generator is a transducer that changes a magnetic field into voltage for measurement, but the Hall voltage is proportional to the magnetic induction to the first approximation only. Detailed studies of the properties of the Hall generator and ways to optimize them have been presented by Kuhrt (1954), Schwaibold (1956), and Kuhrt and Hartel (1957). Several subsidiary effects have to be taken into account for precise measurements:

(a)   Both the Hall coefficient and the resistivity of the plate vary greatly with temperature, if this is not kept below the transition temperature by appropriate cooling of the Hall generator.

(b)   The longitudinal resistance R_1 of the plate varies with the applied magnetic field as a result of the magnetoresistance effect so that a constant-current and not a constant-voltage supply is required for the control circuit of the Hall generator. For instance, R_1 may be more than doubled when the induction is increased from 0 to 1 tesla.

(c)   The size of the pickup electrodes used to measure the Hall voltage enters into the relationship between V_H and B if they are not point-like.

(d)   At the plate-electrode contacts thermoelectric voltages, as well as contact resistance variations, and some rectifying effects in the case of alternating current, result in small errors in the field measurement. For instance, the copper electrode–InAs plate contact gives a thermoelectric voltage of $0.3 \, mV/°C$.

(e)   If some power is provided by the Hall voltage into the measurement circuit, the magnetoresistance effect on the transversal resistance R_2 of the plate produces a magnetic-field-dependent voltage across the plate.

It is possible both to optimize the characteristics of Hall generators and to obtain a good enough linearity in the relationship between the measured voltage and the applied magnetic field, by selecting the plate material and by choosing appropriate values for the free parameters: dimensions of the plate and of the electrodes, external resistance in the measurement circuit. For example, mutual cancellation of effects c and e is easily obtainable (Kuhrt, 1954).

The Hall voltage produced by commercially available plates (Siemens types) is usually in the range 0.1–1 V/T in normal operating conditions, but it could even be made higher by simply increasing the control current and at the same time providing the cooling power necessary to keep the plate temperature at the right level. Pulsing the control current is another way of enhancing the Hall voltage by one order of magnitude or more without increasing the plate temperature (Shirer, 1960).

C. INDUCTION LAW

1. *Introduction*

The voltage induced in a coil, when the magnetic flux Φ that it links varies, is expressed by the Faraday induction law:

$$e = - \, d\Phi/dt \tag{16}$$

From this law are derived a great variety of methods for measuring magnetic

fields and gradients in all the possible cases examined in Section 1.4.3 and
1.4.4: point or integrating measurements of steady or dynamic fields.

The magnetic flux Φ is obtained by integrating over the entire volume of
the coil the expression of the elementary flux linkage

$$d\Phi = \mathbf{B} \cdot d\mathbf{S} \tag{17}$$

where $\mathbf{B}$ is the magnetic induction at this point and $d\mathbf{S}$, representing an
element of a turn of the coil, is a vector whose direction is perpendicular
to the plane of that element and whose length is equal to its area. With a
solenoidal coil in a homogeneous field parallel to the axis of the coil, the
flux linked is

$$\Phi = NB_0 S \tag{18}$$

where N is the total number of turns and S is the mean area of each turn,
with B_0 the constant value of the induction. When the field is not homogeneous, Φ can be expressed as a function of the coefficients in the Fourier
series expansion of the field around the center of the coil: It will be shown
that the induction value at this point, B_0, can still be obtained with a given
accuracy by using a coil of appropriate dimensions and shape.

From Eq. (17) it is clear that only the component of the induction that
is parallel to the coil axis contributes to the flux linkage. Thus it is possible
to determine the direction of the flux lines.

For point measurements small coils can be used. If, on the contrary,
the integral $\int_c B_n \cdot ds$ is required, where B_n is the component of the induction perpendicular to a plane P containing the line C along which the
integration is to be performed, then a set of identical coils connected in
series and aligned on the curve C, or just a long coil if C is a straight line,
can provide the flux linkage necessary for the measurement.

The voltage induced in the coil is proportional to the time derivative of
the induction. If the field varies in time, the coil can be kept at a fixed position and an integration in time of the output voltage provides a measurement
of the induction B. If the field is steady it is necessary to change the orientation or the position of the coil in order to induce a voltage which again
by integration gives the measurement of B. Integration of the induced
voltage can easily be obtained with suitable electronic circuits, such as a
simple RC circuit or a Miller integrator for longer time constants. In another way, a direct measurement of steady fields is given by the first deviation
of a ballistic galvanometer upon flipping or removing the coil from the
field. Instead of directly measuring the integral of the voltage it is also
possible to balance it against an equivalent quantity calibrated in advance

or more easily measurable, as in the well-known Grassot-type fluxmeter or in the photoelectric fluxmeters.

For steady magnetic fields it is possible to avoid the integration stage by varying the flux linkage periodically. By rotating a coil at a constant speed or vibrating it with constant amplitude and frequency, a sinusoidal voltage is induced whose amplitude is proportional to the induction B or its gradient, respectively. For accurate measurements the influence of fluctuations in the parameters of the coil motion can be eliminated by using, here also, a balance method as will be shown later.

In the following subsections the various methods of measurement and the related apparatus will be examined in more detail.

2. Measuring Coils

The first aim in designing coils for precise magnetic field measurements is to keep the measurement errors that are introduced by their imperfections much below those introduced by the measuring apparatus itself. Secondly, very stable coils are wanted in order to reduce as much as possible the number of recalibrations necessary if the program of measurements is expected to last many months; this was the case at CERN for measuring all the 1000 individual magnet blocks that were afterwards assembled into the 100 magnet units of the 25 GeV proton synchrotron. In order to achieve the accuracy of 0.1% required for the gradient measurements in these magnets, the pairs of coils employed had to stay matched to much better than 1 part in 10^4 (de Raad, 1958; Reich, 1963).

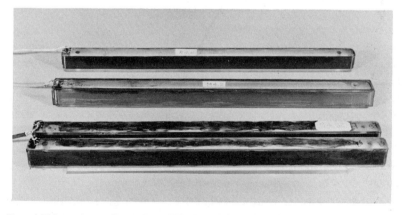

FIG. 4. Measuring coils used at CERN. G for gradient, M for median plane, R for remanent field.

a. *Construction of measuring coils.* For the magnet-block measuring program of the synchrotron at CERN about 50 precision coils were made to the same standards of construction. The coils are about as long as a block to provide an integrating measurement along the azimuthal direction, which is the direction of the protons in the synchrotron. There are three types of coils, shown in Fig. 4: gradient coils, assembled in pairs, glued to a common glass base plate; remanent field coils; and lastly median plane coils, similar to field coils but with an auxiliary winding for determining the plane of symmetry of the magnetic field. Their characteristics are given in Table II.

TABLE II

CHARACTERISTICS OF SOME MEASURING COILS USED AT CERN

Type of coils	Length (mm)	Width (mm)	Wire φ (mm)	Number turns	Total area (m²)
Gradient coils	415	20	0.17	290	2.45
Remanent field coils	415	20	0.10	2350	20.9

Glass was chosen as an insulating, homogeneous, and low-thermal-expansion material which proved suitable for making the coil formers. For ready comparison of experimental results and for ease of replacing damaged coils, all the formers were ground to the same dimensions to within 0.1 mm, except for the width which was ground to 0.01 mm. These tight tolerances ensured that any pair of gradient coils had the same distance between their axes to about 0.02 mm.

Each coil was wound very regularly by hand on the glass former sandwiched between two suitable metal plates. The first layer of the copper wire was firmly bonded to the former by a thin film of Araldite 103. The thermoplastic insulation of the copper wire was cured at 125°C with a gentle pressure applied by two straight bars on the free long sides of the winding. After their calibration all coils were protected by a few layers of shellac and the ends of the wires were soldered to tags glued to the former.

Shorter coils were also made by the same methods of construction for measuring the field in the center of magnet blocks. Instead of using rectangular coils, the simplest construction for point measurements is achieved by winding the coil on a cylindrical former; this is generally the case for rotating coils and, if thermoplastic insulation is used, coils can be made without formers, since the winding constitutes a solid core.

b. *Matching and calibration of the coils.* All coils of a given type were made equal in area turns by comparison with a standard coil in the homogeneous field of a standard magnet, shown in Fig. 5. The same magnet served to match even more accurately the two coils of a pair used for gradient measurements; the exact value of the distance d between the magnetic axes of the two coils was measured in a gradhelm coil, that is, a quadrupolar coil arrangement generating a field with a constant gradient in its central region (de Raad, 1958).

FIG. 5. Coil area calibration in a standard magnet.

For the calibration of the effective area, a large single layer standard coil was used. Its former is a steatite slab 49×13 cm^2, the central part of which is cut away so that a free space is available wherein the coil to be calibrated is placed. The effective area of the standard coil is known from its dimensions to within 0.02%. The common support of the two coils can slide on two brass rails that extend on one end about 1.5 meters beyond the end of the standard magnet as shown in Fig. 5. For calibration the coil support is withdrawn from the central region of the magnet where the field is uniform to within 2 parts in 10^4 at 0.7 T, and the voltage thus induced in the measuring coil is subtracted from a fraction of the voltage induced in the standard coil; the difference is fed into an electronic integrator and the signal from the standard coil is adjusted until the integrated output voltage is zero. The adjustment is made on a potentiometer, shunting three independent turns that are connected in series with part of the main coil, which was tapped at every five turns. The overall precision of this calibration was better than 0.04%.

The gradient in the CERN synchrotron is about 4% per cm and the distance between the axes of the standard type of gradient coil pair is 3 cm. Consequently, the areas of the two coils were matched to a few parts in 10^5, so that the corresponding error in gradient measurement would be much smaller than the desired accuracy. The area matching is achieved by flipping the pair of coils through an angle of 180° in the homogeneous field of the standard magnet. A correction is made on the auxiliary loop, glued on top of the former of every coil and connected in series with the main winding for adjusting the area, until the integrator output voltage is again zero after flipping. The matching of the coils is not affected by the residual inhomogeneity of the field of the standard magnet because the flipping brings each coil exactly to the place occupied by the other one before the operation. The precision obtained in the matching for coils of 2.5 m² effective area was about 1 part in 10^5. The coils were checked regularly and the mismatch seldom exceeded 2 parts in 10^5 over periods of several months.

In order to measure accurately the distance between the magnetic axes of the two coils, the gradient coil pair was placed on a small carriage inside

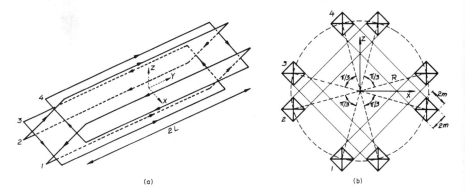

(a)   (b)

FIG. 6. Schematic drawing of gradhelm coil.

the quadrupolar field of a gradhelm coil supplied with 50 Hz current, shown in Figs. 6 and 7. The measuring carriage is displaced until the voltage induced in each coil has vanished successively upon crossing the gradhelm coil axis. The distance between the corresponding carriage positions is read on a micrometer knob and provides the distance of the coil axes to better than 0.01 mm. The parallelism of the two axes is checked by measuring the distance d in this way as a function of the distance z above the median plane of the gradhelm coil. For parallel axes one has $\partial d/\partial z = 0$ and otherwise a small auxiliary loop, glued on one side of the coil at right angles to the main winding and connected in series with it, is adjusted until this condition is fulfilled.

FIG. 7. Gradhelm coil with its measuring carriage.

c. *Optimum shape of the coils.* The field can be expressed in the volume of the coil by a Fourier series expansion at the center of the coil. The total flux linked by the coil is then expressed as a function of the coil dimensions and the Fourier series coefficients of the field expansion. For measuring the field at the center of the coil the contribution in the flux of all the terms of order higher than the fundamental one should be made very small with respect to this one. This result can be achieved by proper shaping of the coil; there even exists a shape for which the unwanted terms completely vanish, the so-called fluxball coil whose winding can be well enough approximated by a series of coaxial cylindrical coils (Brown and Sweer, 1945; Williamson, 1947). Several authors have studied this problem and given recommended coil shapes which depend on the field configuration and what term of the field expansion is to be measured: Garrett (1951), Laslett (1954a,b), Elmore and Garrett (1954), Gautier (1956), and other works mentioned by Symonds (1955).

In two-dimensional fields, where the induction is a function of the coordinates x and z and yet is independent of the azimuth y, measurements integrated in the azimuthal direction that yield a given term of the field expansion, can be achieved with the "harmonic coil" (Grekov et al., 1956; de Raad, 1958). Let us use both rectangular x, z and polar r, φ coordinate systems, with $x = r \cos \varphi$ and $z = r \sin \varphi$, to express the flux linking the

182 C. GERMAIN

harmonic coil which is wound on a cylinder of radius R whose axis is on the oy axis. The winding consists of wires placed as generators of the cylinder, with a density $dN/d\varphi = 0.5 \cos m\varphi$, where $N =$ total number of wires, the wires being connected in such a way that every turn is made with one positive direction wire ($\cos m\varphi > 0$) and one negative direction wire ($\cos m\varphi < 0$).

The general expression of B_z in this two-dimensional field is

$$B_z = \sum_{k=1}^{\infty} [a_k r^{k-1} \cos(k-1)\varphi - b_k r^{k-1} \sin(k-1)\varphi] \tag{19}$$

where all the b_k are zero if the plane $z = 0$ is a plane of symmetry. On the central line ($x = z = 0$) the field and its derivatives are

$$(B_z)_0 = a_1, \quad (\partial B_z/\partial x)_0 = a_2, \cdots, (\partial^n B_z/\partial x^n)_0 = n!a_{n+1} \tag{20}$$

The magnetic flux Φ_m linking the harmonic coil is

$$\Phi_m = \frac{\pi N a_m R^m}{2m} \tag{21}$$

Any Fourier coefficient a_q of the field expansion can be directly measured without any contribution from other coefficients by using the corresponding harmonic coil with $m = q$. The actual construction of the harmonic coil, its degree of approximation with respect to the theoretical shape, and the corresponding error term have been discussed by Grekov et al. (1956) and Daniltsev and Plotnikov (1963).

The problem of the coil shape, for the different types of field measurements in the CERN synchrotron magnet with static or rotating coils, has been discussed by de Raad (1958) as well as the influence of the electrical characteristics of the coil concerning the accuracy of the measurements (for the latter point, see also Grivet, 1962). It is shown, for instance, that in two-dimensional fields the best shape for long field measuring coils is width is equal to height if the winding is thin. More generally, the relation can be written

$$W_i^2 + W_o^2 = 2h^2 \tag{22}$$

where $W_i =$ width of the former, $h =$ height and $W_o =$ outside width of the winding. If the gradient is to be measured in such a field with a pair of coils ($d =$ distance of the two coil axes) the optimum shape is obtained for

$$W_i^2 + W_o^2 + 2d^2 = 2h^2 \tag{23}$$

For point measurements of the field with a rotating coil the best shape is given by

$$h = 1.34 \, r_0 \left(\frac{1 - a^5}{1 - a^3} \right)^{0.5} \quad \text{with} \quad a = r_i/r_o \qquad (24)$$

where h is the height, r_i the inner radius, and r_o the outer radius of the winding.

3. Rotating Coil

In this method for measuring a steady magnetic field, the flux linkage is varied periodically by rotating the coil at a constant angular velocity ω. The voltage V, induced in the coil as a direct application of the generator principle, is proportional to the induction and provides a simple means for measuring magnetic fields, which was very popular until its recent overtaking by the new high quality Hall plates. The rotating coil is, however, still preferred in some cases, especially in its most elaborate two-coil type used in an opposition method for very accurate measurements. The range of measurement is practically unlimited: In very weak fields large coils can usually be employed to obtain an appreciable voltage; and there is no limit for measuring the highest possible steady fields, in which very small coils can be used. The induced voltage is proportional to the induction and requires only one point of calibration. Many examples of magnetic field and gradient measurements with rotating coils can be found in the literature (Symonds, 1955; Germain, 1963).

The coil used in this method is generally wound on a small cylindrical former with a diameter of a few millimeters and a ratio length/outer winding diameter of about 0.7 (see Section 1.4.2,C, 2c). The coil is fixed at one end of a hollow shaft so that the axis of rotation goes through the center of the coil at right angles to its axis. The coil is connected by a pair of wires running through the shaft to slip-rings on which the induced voltage V is picked up by brushes. In the most usual case, a synchronous motor is coupled at the other end of the shaft that is long enough for the motor not to interact with the magnetic field. The rotating shaft is generally insulating, a bakelite tube for instance, and is guided by bearings inside a metal tube that provides the mechanical strength required to avoid, as much as possible, the vibrating and bending of the shaft. The alternating voltage generated in the coil has an amplitude V:

$$V = NS\omega B_n \qquad (25)$$

where NS = total area-turns of the coil, ω = angular velocity in radian/ second, and B_n = component of induction perpendicular to the axis of rotation.

a. *The one-coil system.* If a simple and quick field measurement only is required, the voltage is usually collected by slip-rings and directly measured with an electronic voltmeter, providing an accuracy of about 2% for the measurement. As an example, at the moderate revolution speed of 1500 rpm a 25-Hz voltage is induced with an amplitude of $5\pi = 15.7$ V/T in a rotating coil of 0.1 m² area-turns; if the rms value is given by the voltmeter, the coefficient of measurement for the same coil is $5\pi/\sqrt{2} = 11.1$ V_{rms}/T. Different types of coils have been used at CERN: the usual type was wound with 0.09-mm copper wire, area-turns = 0.33 m², resistance = 300 Ω, length = 10.7 mm, inner diameter = 7 mm, and the outer diameter = 14 mm. The thermoplastic insulation of the wire made it possible to achieve a solid winding without former after curing. The smallest coils were made with 0.025-mm wire on an ivory former, length = 3.8 mm, inner diameter = 2 mm, outer diameter = 5 mm, area-turns = 4 dm², and resistance = 1800 Ω.

The accuracy of the simple one-coil system can be increased by several possible improvements to about 0.1%, by

(1)   Stabilizing the revolution speed to a precision better than the above mentioned accuracy by feeding the synchronous coil-driving motor with a constant frequency power supply; or alternatively correcting the influence of fluctuations of ω by passing the voltage V through an amplifier whose gain is inversely proportional to the frequency around ω_0 (Wills, 1952; Lozingot *et al.*, 1953).

(2)   Using the output electronic voltmeter at a constant deflection position by attenuating the induced voltage with a precision potentiometer.

(3)   Collecting a dc instead of an ac voltage with the brushes by using a commutator instead of slip-rings. This dc voltage can be accurately measured by a potentiometric method but the coil must then be specially constructed as the winding of a dc generator and have regularly distributed connections to the many commutation segments (Cork *et al.*, 1947). In a simpler way, it is possible to rectify half periods of the voltage with a simple two-segment commutator (Peters, 1950; Müller, 1955) or with electronic circuits to amplify and rectify the voltage collected on slip-rings (Jürgens, 1953).

Different methods have been proposed in order to avoid use of slip-rings and brushes:

(1)   Oscillating or vibrating the coil in a rotary fashion instead of simply rotating it at a constant angular velocity (Klemperer and Miller, 1939; Caldecourt and Adler, 1954; Spighel, 1957).

(2)   Using a rotating transformer that consists essentially of two coaxial solenoidal windings; the inner one is the rotating primary connected to the measuring coil and the outer one is the steady secondary providing the output voltage (Langer and Scott, 1950; Wills, 1952; Alon, 1962).

(3)   Generating the measuring voltage in a fixed search coil by the influence of the eddy currents the unknown field induces in a fast rotating copper vane, as in the "turbo-inductor magnetometer" (McCutchen, 1959; Arnaud and Cahen, 1960).

However, trouble-free connections with slip-rings and brushes can be realized: for instance, a noise level of about $10\mu V$ has been obtained at CERN with silver graphite brushes on silver slip-rings at a relative velocity of 2.4 meters/sec for a rotation speed of 1500 rpm (Reich, 1963). Thus it does not seem often worthwhile to avoid slipping contacts by using the different means which have been proposed above, all the more as their efficiency is smaller and their complication greater than the combination of brushes and slip-rings.

   b. *The two-coil system and balance method.*   When a greater accuracy is required, a balance method has to be used, whereby the voltage induced in the measuring coil is opposed to a well-determined reference voltage. For this purpose, in the two-coil system, the voltage induced in the search coil is exactly counterbalanced by the voltage induced in the second coil that is located in a reference magnetic field. The accuracy obtained by this method strongly depends on the degree of synchronization of the two voltages that are opposed to one another. The best synchronization is still obtained when the two coils are rigidly fixed to the same shaft, and in this case the accuracy of the measurement can be pushed to 0.01% or even better. But since the unknown field and the reference field have to be shielded from each other, a rather long shaft is required and the apparatus is cumbersome.

The reference field may be produced in air-cored coils whose current is adjusted for balancing the two voltages or it may be a fixed field, provided by a permanent magnet for instance; whereupon, the two voltages are balanced against each other with the help of a precision attenuation potentiometer. In both cases, the intensity of the unknown field is determined with respect to the reference field by a potentiometer reading and a single point calibration of the system only is required (Lamb and Retherford, 1951;

Hedgran, 1952; van der Walt, 1953; Alon, 1962). The balance method has also been used for vibrating coils (in a rotary fashion), in just the same manner as for rotating two-coil systems (Caldecourt and Adler, 1954), or by counterbalancing the search coil voltage with a voltage generated by a condenser vibrating in phase with the coil (Spighel, 1957).

The balance method has also been used with rotating coil systems much more flexible and less cumbersome than the one described in the foregoing. In a two-coil system a long shaft is required for the fixed-reference field to be at a large enough distance from the field to be measured; but a possible way of reducing the length of the shaft is by using a rotating reference field that induces the reference voltage in static coils as described in Section 1.4.3,A,1,c(i). Alternatively, the reference voltage may be produced by other means, provided the condition of rigorous synchronisation is fulfilled. As in the case mentioned above for vibrating coils, a reference voltage can be generated electrostatically: A rotating condenser with rotor plates shaped to produce a sinusoidal voltage, whose amplitude is determined by a dc voltage applied to the stator, replaces the reference field and coil system in the apparatus described by Burson et al. (1959). As the problem of shielding different magnetic fields vanishes the instrument can be made more compact and more easy to handle. The accuracy achieved with these magnetometers is about 0.1%.

4. Vibrating Coil

The variation of the magnetic flux linking a search coil in a steady magnetic field can be obtained by a periodical motion of the coil. Translatory motions only will be considered under this heading since we have discussed vibrating the coil in a rotary fashion in the rotating coil method. The vibrating coil method is also a direct application of the generator principle: The periodic variation of the magnetic flux linking the coil induces in it an alternative voltage, whose amplitude is proportional to the flux variation and to its frequency, that can be used for measuring at a given point either a steady magnetic field or its gradient according to the shape of the search coil.

When the coil is a long solenoid that is vibrated in the direction of its axis, a direct measurement of the axial field component at one end of the coil is obtained, provided the field at the other end is negligible. This method is useful in measuring the axial field of magnetic electron lenses (Gautier, 1954). If a single-turn coil consisting of a narrow rectangular loop that is fastened to a support strip is vibrated in the longitudinal direction, a measurement of the field component perpendicular to the loop at one end is obtained, provided the field value at the other end is negligible. The resolving power of such a vibrating coil is very good and with a frequency of vibration in the ultrasonic range a sensitivity of a fraction of a millitesla has been achieved (Radus, 1960).

The most frequent use of this method is for measuring the gradient of steady magnetic fields by vibrating a small coil. The direction of the vibrating motion with respect to the coil axis may be chosen freely, making it possible to measure the gradient of a field component in any direction.

The accuracy of a gradient measurement is generally limited to 2 or 3% by the stability in the frequency and in the amplitude of the coil motion, which is usually obtained with the help of an electrodynamic vibrator. The frequency of the vibration is often in the range 0.1–1 kHz. For better accuracy, the balance method with a two-coil system similar to what has been described for rotating coils is also possible and cancels out the influence of frequency and amplitude fluctuations in the coil motion.

5. Fluxmeter and Ballistic Galvanometer

One of the most usual, long-established methods of measuring magnetic fields is based on measuring with a galvanometer the magnetic flux linking a search coil. For this purpose, special types of galvanometers are used: either fluxmeters or ballistic galvanometers. In spite of the recent development of more modern methods of measurement, fluxmeters and ballistic galvanometers remain popular measuring instruments, even for accurate magnetic-field measurements. Significant improvements have, however, been introduced in the construction of these instruments since the early days.

a. *Grassot fluxmeter.* A Grassot-type fluxmeter is a galvanometer in which the suspension torque and the mechanical coil damping are both made as small as possible. It consists essentially of a coil that can freely orientate in the magnetic field of a permanent magnet. Therefore this coil, when connected to an external circuit that undergoes a variation $\Delta\Phi$ in the magnetic flux linking it, will orientate under the electromagnetic torque then generated so that its own linking flux varies by $-\Delta\Phi$. The total flux linking the fluxmeter coil and external circuit will thus stay constant, provided the total resistance of this circuit is small enough. The fluxmeter can be looked at as a mechanical voltage integrator: The deviation of the fluxmeter pointer gives the value of $\Delta\Phi = \int(d\Phi/dt)\,dt$ which is the integrated value of the voltage applied at its connections.

A discussion of the use of the fluxmeter and of the various improvements different authors have brought to this apparatus is found in Symonds' review (1955). By compensating the residual suspension torque and the thermal emf, the drift can be reduced to a negligible level; direct-reading field measurements have been performed with an accuracy of about 0.1%. The sensitivity of usual Grassot fluxmeters is of the order of 100 μWb per division.

b. *Ballistic galvanometer.* A ballistic galvanometer is nothing else other than a conventional galvanometer in which a special attention is given to the value of the oscillation period. An ordinary galvanometer can be used for measuring a magnetic-flux variation, provided its period of free oscillation is at least ten times longer than the duration of the flux variation. In as much as it may take about one second to flip a coil through 180° or remove it from a magnet gap, galvanometers having periods of ten seconds, or more, are called ballistic when used for measuring magnetic fields in this condition. When the duration of the flux variation in the search coil is short enough, the first maximum deviation of the galvanometer coil is proportional to the flux variation and, in a first approximation, independent of the exact duration of the flux change. The condition of low total-circuit resistance which applies to the Grassot fluxmeter is irrelevant for ballistic galvanometers, thereby making the construction of search coils easier.

After the works mentioned in Symond's review (1955), other authors have also studied possible improvements in using ballistic galvanometers. When the total resistance of the galvanometer and search coil circuit is much less than the critical damping resistance, the sensitivity of the instrument is greatly decreased. However, it is still possible to leave the galvanometer either critically or lightly damped, as desired, by mechanically opening a switch in the circuit a short time after the flux change is completed (Pohm and Rubens, 1956). The effect of the duration of the voltage pulse, applied to the ballistic galvanometer, on the maximum coil deflection obtained has been analyzed mathematically by Fragstein (1957) and a curve of the corresponding error of measurement is given. The same problem has been studied experimentally by Woodbridge and Warner (1958), and from the curves they give, it can be concluded that the pulse duration should not exceed 1/16 of the galvanometer period.

Direct-reading field measurements are easily performed by this method with a simple equipment available in any laboratory, and if the galvanometer is used in a correct manner an accuracy of even 0.1% can be achieved, provided the angle of rotation of the search coil and the duration of the rotation are reproducible enough. A sensitivity of the order of one micro-weber per division is obtained with good galvanometers.

c. *The balance method and the servofluxmeter.* If instead of a direct measurement a balance method is used with a fluxmeter or ballistic galvanometer for measuring the magnetic flux, the accuracy is improved and can reach 0.01%. The opposition can be obtained by comparing the magnetic flux values in two separate coils, either both in the same magnet, when plotting a relative field map, or with one in a reference field when absolute values are required. A precisely known adjustable fraction of the coil voltage is readily obtained from a potentiometer connected across the coil terminals. The reference flux can also be produced in the secondary of a mutual inductance whose primary current is adjusted to counterbalance exactly the unknown field and is then proportional to it. Other possibilities of achieving a balance measurement exist and have been examined systematically by Neumann (1954).

The current producing the reference flux in the mutual inductance secondary, instead of being adjusted manually to balance the search coil flux, can be controlled by suitable electronic circuits. The galvanometer mirror is used to detect any small unbalance by deflecting a light beam from its zero position on a photoelectric device controlling the mutual inductance primary current. This apparatus, called a photoelectric fluxmeter or servofluxmeter, was invented in 1937 by Edgar and has been studied further by Dicke (1948), Cioffi (1950), Kapitsa (1955), and Grivet *et al.* (1961).

The principle of operation of the photoelectric fluxmeter is the same as that of the electronic integrator, but its dc amplifier, instead of being purely electronic, incorporates a galvanometer whose input voltage, related to the small flux unbalance in the secondary circuit, is amplified of the order of one million times in the photoelectric device as a result of the light-beam deflection. The value of the primary current, which is proportional to the magnetic flux being measured, can be recorded by a self-balancing potentiometer or read off directly on it with a great accuracy. The time constant of this apparatus with respect to the possible duration of the voltage integration is very large compared to the corresponding values of the period in ballistic galvanometers, whereas the time of response of the current in the mutual inductance primary circuit is short, possibly as short as 0.01 sec.

A CdS split photocell and a galvanometer of the immersed coil type (the galvanometer coil is immersed in a liquid which, by its damping action, greatly reduces the influence of mechanical vibrations) have been used by Grivet *et al.* (1961) in order to obtain a servofluxmeter both highly sensitive and reliable, with a drift of 5 nWb/sec and a sensitivity of 100 nWb. When a high sensitivity is required, precautions should be taken against thermal and contact emf's. Magnetic-flux measurements taking advantage of the high performances of servofluxmeters have often been reported in the literature and relevant references can be found in the reviews mentioned in the introduction (Section 1.4.1).

6. Electronic Integrator

a. *Basic integrating circuits.* When the field is changing in time, the corresponding variation of the induction can be obtained by integrating the voltage induced in a search coil kept at a fixed position in the magnet gap. If the value of the induction at any instant is required, the integration process should give its result continuously and as far as possible instantaneously. A simple resistance R and capacity C circuit has suitable integrating properties if the measurement can be completed in a time that is much shorter than the circuit time constant RC, so that the potential of point A (Fig. 8a) varies only by an amount that must be a negligible fraction of

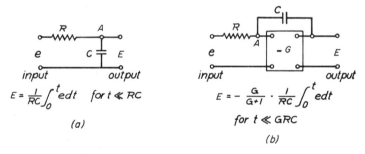

$$E = \frac{1}{RC} \int_0^t e\,dt \quad \text{for } t \ll RC$$

(a)

$$E = -\frac{G}{G+1} \cdot \frac{1}{RC} \int_0^t e\,dt$$

$$\text{for } t \ll GRC$$

(b)

FIG. 8. Integrating circuits. (a) Simple RC circuit; (b) Miller integrator.

the input voltage e. The maximum admissible value of this fraction is determined by the accuracy required in the integration and sets a limit to the possible duration of the integration. A slight modification of the circuit block diagram by using a dc amplifier with a high gain G, as shown in Fig. 8b, makes it possible to maintain the potential of point A within the admissible limit during a time which is G times longer than in the simple RC circuit. In both cases the output voltage E has been assumed to be zero when starting the integration ($E = 0$ at $t = 0$). The latter circuit, described by Chance (1949), is known as the Miller integrator; very large values of the integrating time GRC can be achieved with this circuit, of the order of one hour, for instance.

The advantage of the Miller integrator with respect to the simple RC circuit, for a given value of the error in the integration, is to allow a much longer duration T of integration with the same components R and C (thus eq. $T \ll GRC$ is easily achieved for usual measurements) or alternatively, with the same value of the integrating time ($RC = GR'C'$) to yield a much greater sensitivity by using a correspondingly smaller $R'C'$ value. The high value of G makes the Miller integrator output voltage E almost insensitive to gain fluctuations and the expression of E is the same as with a simple RC circuit.

b. *Effects of electrical transients.* The integrated value of the input voltage e is given by the output voltage E at any instant after a short period of transient phenomena that depend on the characteristics of the input coil circuit. This problem, of particular importance when considering the integration of signal voltages induced by short magnetic-field pulses, has been reviewed by Grivet (1962). For measuring the field and its gradient in the CERN synchrotron magnet with a precision of 0.1% even at injection level, that is, 10 msec after the start of the field cycle, the transient electrical behavior of the different types of input coil circuits was studied in detail (de Raad, 1958). In field measurements it was shown that the error introduced by the transients, which were completely damped in about 0.5 msec, was equivalent to a delay of about 2 μsec in the start of a perfect integration, which is negligible compared to the time $t = 10$ msec of the first measurement. In gradient measurements it was important to avoid the parasitic capacities of the circuit introducing different delays in the signals from the two coils. By suitable connections a difference in delay smaller than one microsecond was achieved. The same problem has been studied by Vizir *et al.* (1963) and solved in a rather more complicated manner to reduce the duration of the transients to 50 μsec.

c. *Drift-correction and calibration methods.* The main problem in the design of an electronic integrator is the construction of a dc amplifier with a low-enough drift. The different types of dc amplifiers have been reviewed by Kandiah and Brown (1952) and the best one for the purpose of reducing the drift incorporates a modulated channel with a chopper amplifier in parallel with the first stage of dc amplification (Goldberg, 1950; Robinson, 1956). This type of dc amplifier was chosen for the integrator used at CERN during the program of magnetic field measurements of the 25 Gev synchrotron magnet (de Raad, 1958; Reich, 1963). A servosystem has been added to reset the integrator output voltage to zero between the measurements (Blewett *et al.*, 1953).

The basic diagram of this integrator is shown in Fig. 9. The ac channel has two identical synchronous choppers CH-1 and CH-2 working at 50 Hz and the ac amplifier consists of two push-pull stages in cascade so that there is in principle no 50 Hz ripple at the point F. A smoothing filter is however used but its time constant $R_m C_m$ can be smaller by a factor of 10 than in Goldberg's diagram, thus improving the frequency response of the ac channel. The fluctuations of this channel referred to its input are about one microvolt, and its effective gain is $A_m = -1000$.

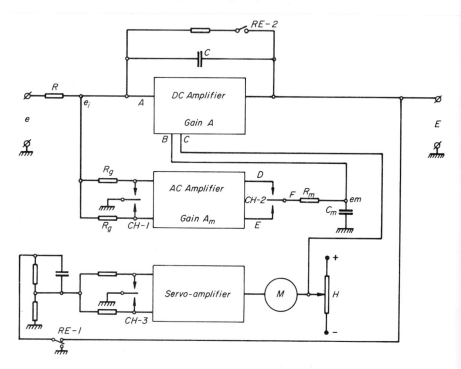

FIG. 9. Chopper stabilized integrator with additional drift correction.

The servoamplifier drives the servomotor M that is coupled to a potentiometer, thus controling a variable voltage that is introduced into the first stage of the dc amplifier: The output voltage E is then maintained to zero to within ± 0.5 mV. The servosystem is stopped with the relay RE-1 just before a measurement is made and the integrator starts drifting slowly but only during the short time required for the measurement. Thereafter the condenser C is short-circuited by the relay RE-2 and then the servosystem is switched on again and RE-2 is opened.

The detailed circuit of the integrator is shown in Fig. 10. The dc amplifier consists of two differential stages and a cathode follower output stage. Its drift voltage, referred to the input, is smaller than one microweber per second and its gain is $A = -6000$. The dominant time constant in the dc amplifier filters used to avoid oscillations is $\tau_c = 0.5$ msec and the resulting error in the integration is negligible, corresponding to a delay of about 0.1 μsec in the integration. Polarized relays that have a small power consumption were used for RE-1, RE-2, and RE-3 in order to avoid inducing a drift voltage in the integrator when they were operated by the timing circuit that was controlled by the master timer of the magnet power supply.

The integrating circuit RC is mounted in a separate box, that is thermally insulated from the integrator, in order to keep the calibration more constant. The resistor R is wirewound and its value is one megohm. The condensor C is selected in a set of six polystyrene capacitors ranging from 1 to 300 nanofarads; their leakage resistance is about 10^7 megohms. The linearity and stability of the integrator have been measured with an electronic switch that was also used for calibrating the time constant τ of the integration. The voltage e across a potential divider is applied at the input of the integrator by a switching circuit during a time T that is derived from an electronic counter fitted with a quartz oscillator, and the output voltage $E = eT/\tau$ is measured with a digital voltmeter. It was found that the integrator output voltage E was proportional to the time integral of the input voltage e within the error of measurement (0.1%) for periods of integration between 0.01 and 5 sec, and the change of calibration constant during a period of six months was about 0.2%.

d. *Integrator output quantity: voltage or frequency.* The integrator can be used either as a null-reading instrument in a balance method, for relative gradient measurement for instance, or as a direct-reading instrument with a digital voltmeter, for remanent field measurement with a flipping coil for instance. In another method [first used on the Birmingham synchrotron (Fuller and Hibbard, 1954) and later on the Saclay synchrotron (Taieb *et al.*, 1955; Chonez *et al.*, 1959), the ITEP synchrotron in Moscow (Vasilev *et al.*, 1962), and the Harwell synchrotron (Gray *et al.*, 1963)] the integrator output voltage is allowed to vary in a well-defined range only before being reset to zero and started again; the instrument thus generates a frequency that is proportional to its input voltage, and the total number of oscillations, which can be read on a scaler, is a measurement of the integral of the input voltage. In fact, two identical integrators are used alternately and at the preset limit of the output voltage, a commutating circuit starts the other

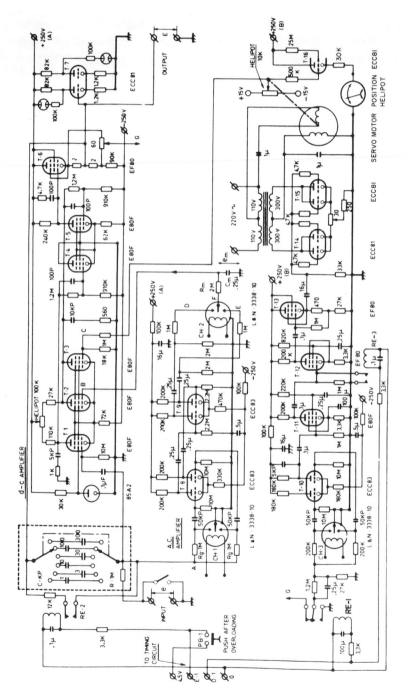

FIG. 10. Detailed circuit of the integrator.

integrator and returns the first one to its initial voltage, waiting for the next commutation. A linearity of 1 part in 10^3 can be achieved in that way, provided the input voltage remains in a rather narrow range: This was the case in the dynamic field measurement of the Birmingham synchrotron. If a larger range of possible input voltage values is to be allowed, a dc amplifier is inserted between the measuring coils and the integrator. This dc amplifier should have a drift-correction circuit, stable and adjustable gain, and be calibrated together with the integrator (Taieb *et al.*, 1955). If the first measurement is to be done after a duration of integration of 10 msec, for instance, the period of oscillation should not exceed 10 μsec if an accuracy of 0.1% is required; but, on the other hand, for such short periods, the time it takes to switch from one integrator to the other must be taken into account, and the circuit made to correct for this delay in a first approximation.

D. PEAKING STRIP

1. *Basic Principle and Operation in Dynamic Fields*

The peaking strip is essentially a zero-field detector based on the properties of some modern ferromagnetic materials, such as mumetal or permalloy, which are saturated in very weak magnetic fields and have a small coercive force H_c. As a consequence, the induction of saturation B_s in the strip may easily be reversed by a small variation of the magnetic field about zero value, and the corresponding large variation of induction, $2 B_s$, is detected by a pick up coil. The ferromagnetic materials specially prepared for peaking strips have almost rectangular hysteresis loops with characteristic values in the ranges 0.5–1 T for B_s, 5–10 A/meter for H_c, and 10^4–10^5 for the maximum relative permeability.

The principle of operation of a peaking strip in a magnetic field H increasing with time t is illustrated in Fig. 11. A piece of fine wire of mumetal, for instance, is placed along a field line and a small search coil is wound closely round the central section of the strip, where the B-H loop is represented in Fig. 11. When the rising magnetic field H goes through the value H_c, the induction B in the strip rapidly changes from $- B_s$ to $+ B_s$ and a voltage V is induced in the search coil. Practically all the induction flux in the coil is concentrated within the strip, so that the expression of the voltage V is

$$V(t) = NA \frac{dB}{dH} \cdot \frac{dH}{dt} \tag{26}$$

where N is the number of turns of the coil, A is the cross-section area of the strip, and dB/dH is read on the hysteresis curve of the strip material. As a rough approximation, $(dB/dH)_{max} \sim 2B_s/H_c$ can be used to estimate the order of magnitude of the voltage pulse. The voltage V, represented in Fig. 11 in full line when H increases with time, can be displayed on a cathode-ray oscilloscope for detecting the time when the rising field H goes

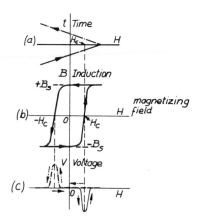

FIG. 11. Principle of the peaking strip.

through zero. The small displacement from exactly zero-field conditions, due to H_c not being completely negligible, can easily be taken into account in calibrating the apparatus.

If the reference value of the rising field H is not zero but H_0, a biasing field $-H_0$ is provided locally around the strip by a small solenoid so that the voltage pulse, generated when the actual field in the strip $H_t \equiv H - H_0$ is zero, determines the time when H goes through the value H_0. Here again the small deviation due to H_c can be absorbed into the calibration. The peaking strip does not perturb the field distribution in the electromagnet where it is placed when no biasing field is required, because the magnetic flux which is concentrated into the fine strip can usually be considered as completely negligible with respect to the useful flux in the gap. If a biasing field is required it is, however, still possible to make the related perturbation negligible by using a self-contained magnetic-flux solenoid. For that purpose Voelker and Leavitt (1955) have placed around the bias solenoid of radius r another one of equal length and radius $r \sqrt{2}$. The outer solenoid has half the number of turns of the inner one and carries the same current but produces a field of opposite direction so that the stray fields of the two coils nearly cancel. The reluctance and the magnetomotive force for the flux path in the inner coil are the same as for the return of this flux between the two coils. Thus in this approximate picture there is no magnetic potential difference available externally for perturbing the field distribution in the gap. In these conditions, at the peaking strip position where the biasing field H_0 is applied, there exists in the absence of the measuring apparatus an induction $B_0 = \mu_0 H_0$, where $\mu_0 = 4\pi \cdot 10^{-7}$ is the permeability of vacuum.

The characteristic features of a typical peaking strip have been given by Kelly (1951). After annealing in hydrogen to optimize its magnetic properties, a piece of permalloy wire, 0.05 mm in diameter and 5 cm long is sealed in a quartz-tube of 1.25 mm o.d. A search coil of 2000 turns of 0.025-mm enamelled copper wire, directly wound on the quartz tube, produces in a magnetic field varying at a rate of 0.27 T/sec, a voltage pulse of 0.1 V with a width of 20 μT at half amplitude. Such peaking strips have been used in these conditions at Brookhaven to produce marker pulses at predetermined magnetic-field strengths in the Cosmotron magnet for injection studies (Giordano et al., 1953). The output pulse is fed into suitable electronic circuits that give a timing pulse with an accuracy of 0.5 μT at a biasing field level of 5 mT, that is, a 0.01% precision in the marker pulse, which also proved to have a long-term stability of the same order over one year. A similar use is reported by Radkevich et al. (1962). Kelly has investigated the influence of several factors on the performance of peaking strips:

(1) The length to diameter ratio should be at least 1000 to prevent the demagnetizing field of the strip ends reducing the effective permeability too much.

(2) The search coil should be wound at the center of the strip and its length should not exceed 10% of the strip length to take advantage of the maximum effective permeability at this position.

(3) A magnetic field gradient in the strip deteriorates the shape of the voltage pulse: broadening in time and reducing its amplitude.

(4) If the field rate of change with time is much higher than 1 T/sec, eddy currents in the strip result into broadening of the pulse. For instance, a pulse width of 19 μT at 1 T/sec becomes 210 μT at 100 T/sec. More recently, Huber and Rogers (1964) have carried out studies at field rates of change from 8×10^2 to 3×10^6 T/sec with oscillating magnetic fields created by the discharge of a condenser in a loop.

2. Operation in Steady Fields

If the sweeping magnetic field is reversed after the first voltage pulse, as in the dotted line in Fig. 11, another pulse is generated with the opposite polarity. If there is no steady field component H_t, the whole pattern is symmetrical with respect to the zero field position, and with a sinusoidal sweeping field H_s the output signal retains its symmetry when displayed on an oscilloscope with the horizontal sweep in phase with H_s. For measuring a steady field H the biasing field H_0 is adjusted until the symmetry of the signal is restored, which implies that the actual field in the strip H_t is again zero. A more sensitive cancelling of H_t is achieved by using a suitable horizontal sweep that makes the two pulses coincide when H_t is zero; this correct sweep is easily obtained by placing the strip inside a multilayer magnetic shielding.

The value $H = H_0$ is obtained from the calibration curve of H_0 as a function of the biasing current, which is in principle a straight line requiring only one calibrated point. The sensitivity achieved in measuring steady magnetic fields in such a way is of the order of one microtesla and the accuracy is 0.1% or better. The range depends on the type of biasing coils used, that is, up to a few centiteslas for air-cooled coils and to one decitesla for water-cooled coils.

For measuring steady magnetic fields, the peaking strip can be considered from another viewpoint as a saturable or biased transformer. As we have seen in the foregoing, the zero-field point is a center of symmetry for the B-H loop of the strip and for the $V = f(H)$ pattern when there is no steady field component H_t in the strip. If the sweeping field H_s is sinusoidal, with a frequency f_0, we always have $V(t + 0.5/f_0) = -V(t)$ and therefore the Fourier-series expansion of $V(t)$ can only contain odd harmonics of f_0. When the external steady field H is not exactly balanced by the biasing field H_0, so that $H_t \neq 0$, the voltage $V(t)$ also contains even harmonics of f_0. The cancelling of the second harmonic of f_0 in the search coil voltage V is used for adjusting the biasing field H_0 or, alternatively, the amplitude of the second harmonic can directly provide a measurement of the steady field H.

Now the main problem is that of eliminating in the voltage V the fundamental frequency f_0 which is generated by the inductive coupling of the search coil with the sweeping coil. This problem can be solved by different means:

(1) A filtering circuit connected to the output of the search coil (Voelker and Leavitt, 1955).

(2) Two identical peaking strips connected in series in such a way that in the two search coils the effects of H are in the same direction whereas the voltages induced by H_s are in opposition (Förster, 1955).

(3) Completely decoupling the sweeping coil and the search coil. This is obtained by any arrangement such that the sweeping field H_s is everywhere at right angles to the steady field H. Twice per cycle of H_s the magnetic core is saturated transversely and thereby the longitudinal permeability, parallel to H, is modulated with a frequency $2f_0$. For this purpose, Palmer (1955) feeds a 5-kHz current along a mumetal peaking strip, thus creating a circular magnetic field H_θ that acts as a sweeping field H_s and can saturate the strip except very near its axis; the steady field H, parallel to the strip, is at right angles to H_θ. Magnetic cores with a shape other than a strip may be used: for instance, Montague (1955) has wound on a small piece of ferrite tube a toroidal coil for the sweeping field whereas the search coil has the usual shape of a small solenoid placed on the tube, which is orientated along the field H.

The sensitivity obtained with such second-harmonic transformers is very high, and can reach one nanotesla or even better. The accuracy, as for peaking strips, usually of the order of 0.1%, is determined by the characteristics and calibration of the biasing coil and its current supply. For a more detailed study, many references to magnetic field and gradient measurements in high-energy particle accelerators at relatively low field intensities, most often in dynamic conditions, can be found in the bibliography (Germain, 1963). As a typical example, the studies on the field and gradient errors in the ac magnet of the 1 GeV electron synchrotron at Tokyo have been done both in static and dynamic conditions at low field levels with peaking strips (Sasaki, 1962).

1.4.3. Measurement of Magnetic Fields

A. STEADY FIELDS

1. *Point Measurements*

a. *Magnetic resonance.* Starting with the easy measurement of a steady uniform magnetic field, we can use a simple apparatus consisting of a transistor marginal oscillator, as described by Donnally and Sanders (1960). This circuit is a Q-multiplier with increased feedback to provide sustained oscillations. The magnetic resonance modulates the oscillations, which are detected by a diode and fed to an audio amplifier. For each measurement the frequency is adjusted by a variable capacitor and the feedback is reduced by a control resistance until oscillations are barely sustained. The circuit is nonmicrophonic and noncritical; the short-term frequency stability is a few parts in 10^6. The magnetic field is modulated by small coils providing an adjustable ac field over the volume of the resonating sample. The frequency can be measured with a digitized frequency meter if quick direct-reading information is needed.

The problem of easy handling of the measuring probe in the magnet gap sometimes requires that the oscillator be connected to the probe by a few meters of coaxial cable. In this condition the oscillator must be designed in such a way as not to be perturbed by the cable, which is moved about. Buss and Bogart (1960) have adapted the desirable features of a Colpitts oscillator to a Franklin oscillator, resulting in a simple wide-range marginal oscillator with excellent frequency stability and a high signal-to-noise ratio. The apparatus described can be used in the range 0.05–2 T.

The effect of inhomogeneity in the magnetic field is to broaden the resonance line and decrease its amplitude. As we have seen, the situation can be improved by adding paramagnetic ions in the water sample and reducing the size of the sample so that the instrumental relaxation time $\overline{T_2}$ corresponding to the field inhomogeneity within the sample is equal to the spin–spin relaxation time T_2 of the water with added ions. In that way it was possible to Denisov (1958) to measure accurately fields with a relative gradient $(1/B) \cdot \partial B/\partial x$ up to 5 (meter)$^{-1}$ in the range 3 mT–2 T. The sample had a volume of 0.2–10 mm^3 according to the range of field and was as thin as 0.1 mm in radial extent in the case of high-gradient azimuth-symmetrical fields.

It is, however, simple to measure fields with even larger relative gradients by cancelling the gradient in the volume of the sample with a small quadrupole winding that is mounted on the measuring probe as was proposed by Denisov (1960). The water sample is placed on the axis of the quadrupole so that the field to be measured is not altered by the compensating gradient. An application of this method has been presented by Vasilevskaya et al. (1963) for measuring magnetic fields with a gradient up to 10–35 T/m with an accuracy of 0.01%. The gradient value can be determined from the current value in the quadrupole winding. As an example of the gradient compensation in this apparatus, a resonance linewidth of 0.2 mT was obtained with a gradient of 10 T/m in a field of 1 T.

When a steady magnetic field has to be measured with a high accuracy at a given instant, one may be bothered by the residual fluctuations of the field due to some imperfection of the magnetizing current stabilization. A solution to this problem is to control the oscillator frequency by the magnetic field so that the requirements on the magnetic field stabilization are greatly reduced. The frequency measurement can be obtained directly in a short time with a digitized frequency meter. Two methods have been proposed for controlling the oscillator frequency by using the nuclear magnetic resonance signal.

In the first method, described by Pound and Freeman (1960), a super-regenerative oscillator produces a coherent signal at a frequency close to the magnetic resonance frequency of the sample which is placed in its coil. A cw (continuous wave) oscillator is phase-locked to the super-regenerative oscillator and acts as a very sharp filter to select the right signal amongst the many others produced in the latter oscillator. The resulting signal follows variations in the resonance frequency that are caused by variations of the magnetic field. This method is based on the property of a super-regenerative circuit, of natural frequency ω_0, to get coupled to a small external signal of frequency ω close to ω_0, and to some extent the device can be regarded as being locked to the Larmor frequency in the same way that an oscillator is locked by a cw signal. The super-regenerative oscillator is on during a period τ (a few microseconds for instance) once every period of duration $T(\Omega$ is the corresponding angular frequency) and a quenching device is made to damp the ringing, at frequency ω_0, of the tuned circuit at the end of the on period for the remainder of the period T.

Under the influence of the magnetic resonance the super-regenerative circuit frequencies will consist of

$$\omega_{SR} = \omega_L' \pm (\tfrac{1}{2} + n)\Omega \tag{27}$$

where $\omega_L' = \omega_L + \Delta\omega$. The error $\Delta\omega$ is small with respect to ω_L and is virtually constant for fixed operating conditions. The frequency which is nearest the Larmor frequency can be extracted by coupling to a Colpitts oscillator. It is possible to calculate the error made in measuring the Larmor frequencies or to eliminate it by taking the mean of the frequencies at the two tuning points which give maximum response, but this is not necessary for plotting a field map, for instance. The range over which the frequency will automatically follow the Larmor frequency is about $\pi/\gamma\tau$, which was 2 mT in the example given.

Another method of frequency control is using a variable capacitor in the oscillator tank circuit and getting from the magnetic resonance line an error signal that is fed to a control loop adjusting the capacitor value. Feldman (1960) has described a marginal oscillator controlled in such a way by using a voltage-sensitive capacitor upon which the correcting dc voltage is applied. The field modulation is small compared to the resonance linewidth so that the output of the phase-sensitive detector (reference lock-in) is proportional to the derivative of the resonance signal shape and provides the correcting dc voltage which acts on the variable capacitor so as to bring the oscillator frequency to the center of the resonance. This apparatus was able to stay within 3 Hz of the center of a sodium resonance line at 10 MHz and could follow variations of the magnetic field of 1 part in 10^4.

A device based on the same principle has been designed at CERN by Brown (1966) for measuring to better than 1 part in 10^5 the absolute value of the magnetic field in a muon storage ring at an induction of about 1.7 T. The oscillator, whose frequency can be varied through the range 72–74 MHz, drives a fixed tuned probe, tuned to the center of this frequency range.

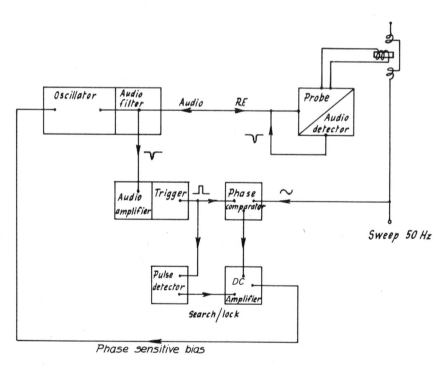

FIG. 12. Nuclear resonance servomagnetometer.

Local detection at the probe is used and the audio signal is returned through the rf feed cable (see Fig. 12). The audio signal is then filtered and amplified, and used to fire a trigger circuit which discriminates against the noise. The standard pulses thus generated are compared in phase with the sweep field, giving a phase sensitive dc signal which is amplified and fed back to control the bias of the oscillator voltage-sensitive capacitor. In the absence of the resonance signal, a search routine is automatically started across all the locking range. The frequency is automatically displayed at a given instant on a 6-digit scaler that counts the number of oscillations during a fixed time of about 100 msec. When measuring an induction of 1.7 T stabilized to 1 part in 10^4 the jitter on the measurement is about: $\pm 6\ \mu T$. A difference on the level of the previous bias corresponding to $\pm$ 10 mT results in a steady error on the next measurement of: $\mp 3\ \mu T$. The locking range is ± 25 mT and the maximum dB/dt when locked is about 5 mT/sec.

b. *Hall effect.* Hall generators are commonly used for measuring steady magnetic fields in high-energy accelerators or beam transport magnets, most frequently in the range 0.1–2 T. With some precautions and appropriate calibration the accuracy is readily 0.1% and can even reach 0.01%. Point measurements in inhomogeneous fields are always possible by using probes of a small enough size, and sizes down to $10 \times 10\ \mu m^2$ have been achieved (Roshon, 1962). The Hall effect gives a very flexible method of measurement whose range covers the practical values of field in magnets; it is adaptable to any requirement from direct-reading measurements with a 1% accuracy to high-precision measurements related with laboratory techniques.

In contrast to nuclear magnetic resonance it is not necessary to cancel the gradient over the volume of the measuring probe or to reduce the size of the probe to very small dimensions when a Hall generator is used for measuring magnetic fields. Zingery (1961) has studied the long-term stability of the inhomogeneous field of a permanent magnet with a precision of 3 parts in 10^5 by balancing the Hall voltage against a low-voltage reference source. A short-time component stability only was required when the Hall generator was placed in the uniform field of an auxiliary magnet that was varied to obtain again null without changing the reference voltage. The auxiliary field was then measured by a nuclear magnetic resonance method. The requirements were a high-component stability during times of one minute and accurate positioning of the Hall probe, whose temperature was controlled to better than $\pm 0.01°C$.

The magnetic field of the large synchro-cyclotron at CERN has been measured by Braunersreuther (1956) with InAs Hall generator prototypes provided by Siemens. Whereas the maximum permissible control current was 400 mA, the normal current value was fixed at 100 mA only, thus making it possible to achieve a reproducibility better than 0.02% in the measurements. The Hall voltage was 0.16 V/T at 100 mA, with a temperature coefficient $-0.07\%/°C$. Between 1.4 and 2.4 T the measured voltage was linear with the magnetic field to better than 0.05%, but a good linearity is not indispensable since the calibration curve against nuclear magnetic resonance can be plotted with as many points as one needs, the important condition being a good enough reproducibility in the Hall voltage. The temperature coefficient of the longitudinal resistance R_1 of the Hall plate, in the control current circuit, was 0.2%/°C in the range 10–50°C, and the variation of R_1 as a function of the magnetic field for this plate is given in Table III.

TABLE III

VARIATION OF LONGITUDINAL RESISTANCE AS A FUNCTION OF MAGNETIC FIELD IN
A PARTICULAR HALL PLATE USED AT CERN

Applied magnetic field B, (teslas):	0	0.5	1	1.5	2
Resistance R_1 normalized at $B = 0$:	100	110	132	160	200

The accuracy required in measuring the magnetic field of this cyclotron was 0.1%, so the Hall voltage and control current were measured with a potentiometer by a balance method. The magnetic-field configuration was studied with two Hall generators connected in opposition so that the resulting voltage was null when both plates were in the same field. One of the probes was kept at a fixed reference position and the other one was used to plot on a recording potentiometer the field configuration along one coordinate of the cyclotron magnet gap. The control currents, provided by two independent current sources, are not necessarily equal but must be very stable.

With the Hall plates, type SBV 544, recently developed by Siemens, the temperature coefficient of the Hall voltage is in the region of $10^{-4}/°C$ at room temperature, thus relaxing the temperature stabilization requirements (Braunersreuther, 1965).

A detailed study of the use of Hall generators for measuring magnetic fields has been carried out by Dahlstrom *et al.* (1962) in order to exploit the new advances in cyclotron development using the sector-focusing method. The following requirements are met:

(1) Routine accuracy of 1 part in 10^4;

(2)   A small sensitive volume for accurate knowledge of measurement position;

(3)   Applicability in high-gradient magnetic fields;

(4)   Convenient when used in automatic recording equipment.

The performances of four types of Hall generators manufactured by Siemens are discussed: types FA 21 in InAs (size 1×2 mm²), FA 23 in InAs (3.5×7 mm²), FC 32 in InAsP (2×4 mm²), and FC 33 in InAsP (3.5×7 mm²), the plate thickness being 0.1 mm. The Hall voltage is about 0.1 V/T with a temperature coefficient β which varies relatively as much as 50% as a function of the field but is always smaller than 0.1%/°C. The transversal resistance R_2 of the plate, in the Hall voltage circuit, increases with the field and its temperature coefficient α is a decreasing function of the field (for example, R_2 varies from 3 to 6.5 Ω and α from 0.3 to 0.04%/°C when the induction is raised from 0 to 2.4 T in FC 33 plate). The long-term stability is good enough to allow accurate measurements since the variation measured over a year was about 1 part in 10^4, but the temperature coefficient β of the Hall voltage is still too large for a direct utilization of the Hall generator.

A reduction by an order of magnitude of the temperature dependence of the measuring voltage V_0 has been obtained with a compensating network in the Hall voltage circuit (Fig. 13). The compensating network consists

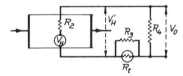

FIG. 13. Temperature compensating network for Hall plates.

of a thermistor R_t, with a negative temperature coefficient $-4\%/°C$, in parallel with a shunt resistance R_3 so that a larger fraction of the Hall voltage V_H appears across the output resistance R_4 when the temperature is increased, thereby compensating, in the measuring voltage V_0, the influence of the negative temperature coefficient β of the Hall voltage. The values of R_3 and R_4 are determined by the following conditions, providing optimum compensation at a given field level:

$$\beta_0 \equiv \beta - \frac{R_2}{r}\alpha - \frac{R^2}{rR_t}\alpha_t = 0 \rightarrow \frac{R^2}{rR_t} = \frac{\beta'}{\alpha_t} \qquad (28)$$

$$\frac{d\beta_0}{dt} = 0 \rightarrow R_t \sim R_3 \qquad (29)$$

where

$$\beta_0 = \frac{1}{V_0}\frac{dV_0}{dt} \qquad \beta = \frac{1}{V_H}\frac{dV_H}{dt} \qquad \beta' = \frac{1}{V_H'}\frac{dV_H'}{dt} \qquad \alpha = \frac{1}{R_2}\frac{dR_2}{dt}$$

$$\alpha_t = \frac{1}{R_t}\frac{dR_t}{dt} \qquad R = \frac{R_t R_3}{R_t + R_3} \qquad r = R_2 + R_4 + R$$

The thermistor is placed in the measuring probe in thermal contact with the Hall generator case. The range of magnetic field to be covered determines the value of R_4 since β is a function of the field. With an FC 33 Hall generator a β_0 value smaller than $4 \times 10^{-5}/°C$ was obtained in the range 0.6–2.4 T since for this Hall plate the curve of β is flat enough in that range. The temperature difference between the Hall generator and the thermistor must not vary by more than 0.1°C, as a result of change of ambient temperature or applied magnetic field, so that a stabilization time of about 0.7 sec is needed to reach thermal equilibrium for proper temperature compensation in the field measurement.

Several other examples of magnetic field measurements in accelerators are presented in the Proceedings of the International Conference on Sector-Focused Cyclotrons, Los Angeles, April 1962, which are published in Nuclear Instruments and Methods, Volume 18–19. We shall not examine them all in detail but only bring out the essential features. The objective was to resolve, to a few parts in 10^5, geometrically complex fields in the range 1–2 T and to measure automatically a large number of points with an absolute accuracy of about 0.01%. Siemens types FC 33 or FC 34 ($7 \times 14mm^2$) Hall generators, fed by current sources stabilized to a few parts in 10^5, have been used. The Hall voltage after amplification by a fixed gain amplifier was either directly measured with a digital voltmeter or converted to frequency by an integrating voltage-to-frequency converter and read on a scaler. The data pertaining to each point, that is, the Hall voltage, probe location etc., can easily be punched on one card. The Hall voltage is converted later to field value in a computer by a program utilizing the probe calibration. Temperature stabilization of the Hall plate or temperature compensation in the Hall voltage circuit, or both, are essential conditions to meet the high degree of accuracy requested. Most of the equipment required by this elaborate method of measurement is commercially available. Let us now discuss some of the problems involved in it.

(1)   The control current.   The control current for accurate measurements should be kept below one half of the permissible maximum to avoid errors due to heating of the Hall generator; for instance, de Forest (1962) has

used at 99 mA a Siemens plate, type FC 34, yielding a Hall voltage of 0.22 V/T. The control current must be kept constant to at least a few parts in 10^5 even when the longitudinal resistance R_1 of the Hall plate varies by a factor of 2 or 3 under the influence of the magnetic field. This result is achieved by some commercially available constant current supplies. It is also possible to adjust the level at which the control current is kept constant so as to compensate for the temperature variation of the Hall voltage: Alon (1962) has used the variation of resistance with temperature of a fine copper wire wound over the total length of the Hall plate to control the current of the generator, so that the Hall voltage measurement was independent of temperature in the range of utilization (30–35°C).

If the Hall voltage is measured by opposition with the voltage across a resistor fed by a current proportional to the control current in the Hall generator, the balance conditions are then independent of the control current fluctuations: the Hall generator can therefore be incorporated in a bridge circuit so that a high-grade current stabilization is no longer necessary for accurate field measurements. Jousselin (1962) has described such an apparatus, commercially available, in which a temperature-compensating circuit is also used so that the overall temperature effect in the calibration is only $2.10^{-5}/°C$ and the accuracy of measurement obtained is 1 part in 10^4.

(2) The temperature effect. The characteristics of a Hall generator vary with the temperature and though some materials like InAs and InAsP have rather low-temperature coefficients, it is necessary to compensate for that effect if a high accuracy of measurement is required. We have described above the temperature compensating network that was used by Dahlstrom *et al.* (1962) to achieve with an FC 33 Hall plate an overall temperature coefficient smaller than $4 \times 10^{-5}/°C$ in the measuring voltage. With this method, only a rudimentary temperature stabilization of the Hall generator is required if an accuracy of measurement of 0.01% is aimed at. Hudson *et al.* (1962) have kept the probe temperature constant to 0.5°C at about 35°C by a control system consisting of a detecting thermistor, a transistor-controlled power source and a heating element capable of supplying 5 W to the probe holder. The Hall generator and thermistor responses are matched.

If a high-grade temperature stabilization of the Hall generator is achieved the temperature compensating circuit is not necessary. De Forest (1962) has placed in a small oven the Hall plate, a temperature-sensitive platinum resistor and an inert resistor (both of which formed the legs of a bridge), and two heating pads. The Hall generator and resistors are between copper plates to avoid thermal gradients and this assembly is placed between two heating pads. The whole thing is surrounded by thermal insulation and placed in a brass box at ground potential for electric shielding. The two probe resistors were parts of an ac bridge feeding a temperature controller, which could hold the temperature stable to $\pm 0.01°C$ in the usual operation when the input power variation with changes in magnetic field was the only perturbing influence. The resulting error in field measurement due to temperature was smaller than 1 part in 10^5.

(3) The Hall voltage measurement. For accurate measurements it is convenient to bring the Hall voltage into the 1–10 V range by amplification with a constant gain amplifier, that is, a negative feedback amplifier whose gain is determined by the ratio of two stable resistors. Resistor values of 1 MΩ and 50 kΩ were used by de Forest (1962) to amplify 20 times the Hall voltage with a commercially available amplifier, which was stable to $\pm 2\mu V$/day, referred to the input. The Hall voltage was then measured with a digital voltmeter linked to an IBM card punch that encoded on one card for each data point the Hall voltage, probe location, and other pertinent information. Alternatively, the Hall voltage can be converted to frequency, after amplification, by an integrating voltage–frequency converter (Tickle, 1962; Lind, 1962), read on a gated scaler and recorded on punched cards or paper tape, for processing on a computer by a program utilizing the probe calibration over the whole operating range. The probe is calibrated against nuclear magnetic resonance signals at various field values chosen over the operating range so that a cubic interpolation formula, in sets of four successive points, would give a maximum error of 10 μT (Dahlstrom et al., 1962). A slow drift in calibration, smaller than 0.1% in 56 days has been reported by de Forest (1962), whereas Dahlstrom et al. (1962) claim shifts less than 2 parts in 10^4 over a year's time for one of their probes.

c. *Induction law.*

(i) *Rotating Coil.* The rotating coil method provides a point measurement of steady magnetic fields from practically zero to the highest values and the output voltage is proportional to the induction, making the calibration procedure very simple. As was shown in Section (1.4.2,C,3) this method is suitable both for direct-reading measurements to about 1% and for very accurate measurements with a balance method. In order to give another example of application of this method, an accuracy of 0.1% has been achieved at CERN with a two-coil system of compact design that proved to be of a flexible and convenient use for point steady field measurements (Reich, 1963). In this apparatus, shown schematically in Fig. 14a, the reference voltage is induced in static coils by a local rotating reference field that is generated by a set of coils attached to the rotating shaft. This reference voltage generator can be placed comparatively close to the field to be measured and the total length of the shaft is only 60 cm. The rotor and stator coils are split in two halves and the stator connected in series aiding with respect to the voltage induced by the rotor but in series opposition with respect to an external field varying in time, in order to cancel out to the first order any inductive coupling with such a field.

The balancing of amplitude and phase of the reference and measurement voltages was carried out with the help of a selective amplifier and an oscilloscope. The phase matching of the two signals was obtained by rotating the stator coils to the correct angle and for the fine setting, by adjusting a variable condenser connected across these coils; the balancing of the amplitude of the two signals was made either with the aid of a voltage divider or by varying the magnetizing current of the reference field that is fed to the rotor by slip-rings. The search coil is wound with a 0.09-mm copper wire whose thermoplastic insulation is cured so that the winding makes a solid core without a former; the coil outer diameter is 14 mm, the total area is 0.33 m²; at a revolution speed of 1500 rpm, the coil constant is 2.71×10^{-2} T/V_{rms} . The interest of the rotating coil as a field measuring instrument has certainly not vanished completely in spite of the rapid progress in Hall plate technique and it is even commercially available (Lush, 1964), though most often it has been constructed in the laboratories that use it.

(ii) *Fluxmeter, Ballistic Galvanometer, and Electronic Integrator.* It is shown in Section 1.4.2,C,2c that by shaping a coil in the right way the magnetic flux it links in an inhomogeneous field can be made proportional to the induction at its center with a given accuracy, the constant of proportionality, or effective area, being the same as in a homogeneous field. The application of the fluxmeter or ballistic galvanometer method is then straightforward for point measurement of magnetic fields. For direct-reading measurements with an accuracy of a few per cent, the Grassot fluxmeter or the ballistic galvanometer is still popular, because these relatively inexpensive and simple instruments can be found in almost any laboratory. The accuracy in direct-reading measurements may be pushed to the 0.1% region with some precautions (Inozemtzev and Latyshev, 1949), but it is simpler and also more accurate to use a balance method.

For studying the field shape in accelerator magnet models, Finlay *et al.* (1950) have opposed the search coil voltage to an adjustable fraction of the voltage of the reference coil which was kept at a fixed position in the gap. The reference coil potentiometer setting was adjusted until a null deviation of the ballistic galvanometer was obtained upon reversing the magnetizing current in the magnet, making it possible to achieve a relative precision better than 0.05% in the field shape. With the same potentiometric method and by flipping the two coils through 180°, Fechter and Rubin (1955) have precisely measured the ratio of two separate magnetic fields,

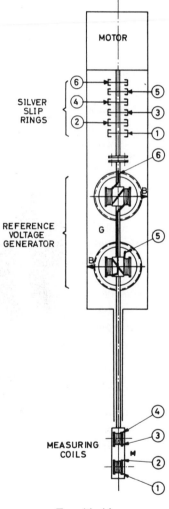

FIG. 14. (a)

FIG. 14. Rotating coil for field and gradient measurements by a balance method. (a) Schematic drawing; (b) Rotating coil on its motorized carriage.

one in a mass spectrometer and the other in an analyzing magnet. The readings were reproducible to within 0.01% when the two coils were rotated by a common shaft and to about 0.02% when they were driven by separate Selsyn motors for more flexibility in using the apparatus. An accuracy of 0.01% is claimed by Tenzer (1955) in measuring the stability of permanent magnets with a similar method but using the voltage induced in the secondary of a mutual inductor as a reference voltage.

FIG. 14. (b)

The application of the photoelectric fluxmeter to these measurements is straightforward and the sensitivity obtained is very high (Grivet *et al.*, 1961) With an electronic integrator the sensitivity, in a measurement requiring one second to be completed, may be limited by the drift to about one microweber, which is similar to what is obtained with a good ballistic galvanometer, but it has the advantage of making it possible to use a coil with a much higher resistance (de Raad, 1958). The application of the electronic integrator is also straightforward; it is discussed in more detail for integrating steady field measurements, but could be done in the same way for point measurements with suitable coils.

2. Integrating Measurements

a. *Hall effect.* Integrating measurement of the magnetic field along a particle trajectory in a magnet is possible indirectly by using point measurements, with a Hall plate, for instance, in conjunction with a computer (Braunersreuther, 1965). The method of computation is similar to that used for calibrating the Hall voltage over the entire operating range with a set of discrete calibration points given by nuclear magnetic-resonance signals. A special program in the computer gives to a polynomial expression of the field, of high enough degree as a function of the coordinate along the trajectory, the best possible fit with the point field measurements at selected coordinate values. The integral, along the trajectory, of the field expression is also computed by this program. When a powerful electronic computer is available, there is thus no need to make a distinction between point measurement methods and integrating measurement methods since the former ones can easily yield the results of the latter ones.

b. *Fluxmeter, ballistic galvanometer, and electronic integrator.* An integrating measurement can be done in exactly the same way as a point measurement with these instruments, the only difference being the shape of the search coil. For studying the field nonlinearity in linac magnetic quadrupole lenses, Daniltsev and Plotnikov (1963) have used long harmonic coils whose linking flux was measured with a ballistic galvanometer upon switching off the magnetizing current. Systematic measurements of magnet elements for the construction of large particle accelerators have been performed with servofluxmeters, with which an automatic recording of the data is easily obtained (Green *et al.*, 1953; Gray *et al.*, 1963).

At CERN, since it was convenient to use the same instrument for both steady and dynamic field measurements on the synchrotron magnet, the electronic integrator was preferred, though it required larger coils (de Raad, 1958; Reich, 1963). Remanent field coils, described in Section 1.4.2. C, 2a, with an effective area of about 20 meters^2 and a resistance of about 4 kΩ, were displaced or flipped through 180° and the integrator output voltage, measured with a digital voltmeter, provided a measurement of the remanent field to 0.1%. Birss and Fry (1960) have described a general purpose electronic integrator whose output voltage is read directly on a meter; there are four operating ranges and provision is made for self-calibration of both the sensitivity and the linearity. With this instrument the accuracy is about $\pm 0.5\%$ if the measurement is completed within three seconds and the full-scale deviation is obtained for a flux change of 0.05 Wb on the most sensitive range.

B. FIELDS VARYING WITH TIME

1. *Point Measurements*

a. *Electron magnetic resonance.* Using magnetic resonance for pulsed or ac magnetic-field measurements often requires relaxation times shorter than those possible with nuclear magnetic resonance, even with high concentrations of paramagnetic ions in the water sample. Furthermore, with such concentrations the resonance linewidth becomes prohibitive for accurate measurements. Fortunately, the relaxation time of electron magnetic resonance in DPPH is much shorter, and several authors have used it for measuring inductions with an accuracy of 0.1% in the decitesla range with time derivatives approaching one millitesla per microsecond (Spokas and Danos, 1962; Stahlke, 1962; Muray and Scholl, 1964). Resonance frequencies are in the microwave region, the range 0.3–0.44 T, for instance, is in the X band (8.2–12.4 GHz). Klystron oscillators are required and conventional microwave equipment and technique are used.

For measuring the time at which the induction crosses a given value, it is possible to use a fixed frequency cw system provided the dB/dt is fast enough to produce a well-defined resonance signal. If the resonance is to be observed near the peak of an alternating field where dB/dt is small, the frequency of the klystron can be modulated at a few megahertzs rate. The field-measuring apparatus basically consists of a klystron oscillator supplying energy to a sample of DPPH placed at the end of a waveguide. A microwave bridge technique is used so that the unbalance of the bridge when crossing the resonance conditions results into a large relative signal on the detector. The klystron is connected to a hybrid tee which divides the rf power between the probe and an adjustable termination set so as to reflect a balance signal (180° out of phase with the reflection from the probe and of suitable amplitude to minimize the energy level in the fourth arm of the tee where the detector is placed). At resonance the signal reflected from the probe is changed, the bridge is unbalanced and the resulting rf pulse sent to the crystal detector is demodulated, amplified, and displayed on an oscilloscope. A suitable signal triggers the oscilloscope and a time-reference pulse can be displayed together with the resonance signal if a second channel is available.

No metal element can be placed in the varying magnetic field without perturbing it and to make the measurement possible at any frequency within the range of the klystron and waveguide system no cavity should be used. The air-filled metal guide is connected by a transition section to a dielectric guide at the end of which the DPPH sample is placed. Muray and Scholl (1964) have used a travelling-wave helix in the measuring arm, at a frequency of about 4.8 GHz, when measuring the amplitude jitter of pulsed magnetic-field waveforms. The corresponding time resolution of the resonance signal was a fraction of a microsecond and the accuracy of the field measurement better than 0.1%.

b. *Hall effect.* Some Hall probes have a good frequency response up to the megahertz region and can be as useful as peaking strips for detecting rapidly changing magnetic fields. In contrast to peaking strips they do not require a compensating field and are therefore suitable for measuring high fields but, on the other hand, their sensitivity is smaller and their use in low fields does not currently appear interesting. It is, however, possible to pulse the control current in order to increase the Hall voltage; Shirer (1960) has fed, with no apparent harm, 1 or 2 A at a duty cycle of less than 1% into Siemens FC 33 probes, which are rated 100 mA, measuring a field of 40 mT changing at a rate of 3 T/sec with a reproducibility of 0.03%. A mercury relay was used to switch the output of a stabilized current source from a bypass resistor R_b to the current leads of the Hall plate, and upon doing so the current regulation recovered within 1 or 2 msec if R_b had the right value, that is, if the current did not change upon switching. During the remaining useful time (20 to 50 msec), the Hall voltage was measured by applying it after amplification to an amplitude comparator.

The problem of measuring with Hall probes pulsed magnetic fields of millisecond or microsecond duration and of a few tens of teslas intensity, has been discussed by Grivet (1962). In pulsed magnetic fields the connection lines for the control current and the Hall voltage circuits must be carefully arranged to keep the induced parasitic signals at a low enough level, but if the field is to be measured at its maximum value the induced parasitic voltages are then at their minimum level. The relative importance of induced signals is moreover easy to determine by a measurement with a null control current. A 20-T pulsed magnet for high-energy physics emulsion experiments has been measured in such a way at CERN (Braunersreuther *et al.*, 1960, 1962).

c. *Electronic integrator.* Since the methods of dynamic field measurements with electronic integrators are identical in both cases, except for the shape of the coils, some examples of application are given here and others in the section dealing with integrating measurements. Penfold and Garwin (1960) have discussed the problem of accurately measuring the unbiased 60 Hz magnetic field of a betatron, for a precise energy determination, with an integrator that was good for frequencies at least up to 200 kHz. When the duration of the pulsed field becomes very short, or its frequency high, the problem becomes similar to that reviewed by Grivet (1962) and encountered in plasma physics for instance. Probes consisting of small coils giving a flat response, to within 1%, up to 20 MHz have been used to study the magnetic field in plasma (Segre and Allen, 1960), and, to mention only one other example, Berglund *et al.* (1963) have described an integrator + coil system with which an overall sensitivity of 10 V/T was achieved with an *RC* time constant of 8 msec and a bandwidth extending to 2 MHz.

2. *Integrating Measurements*

Electronic integrator. The magnet of large synchrotrons consists of many magnet elements, called blocks or units, which are aligned with respect to the particle ideal trajectory, or "equilibrium orbit." For the beam optics the integral of the field in the direction of the particles has to be measured in each individual magnet element, so that all the magnet elements can be distributed azimuthally in the optimum way to minimize the closed orbit perturbations. At CERN the coils used for individual block measurements were 415 mm long and sets of the same coils in series were employed for measuring, one by one, the 100 magnet units, each consisting of 10 magnet blocks fixed on a common girder. The techniques described below for integrating field measurements can be directly applied to point measurements with suitable coils.

The field measurement may be "absolute," when the intensity of the magnetic field as a function of the other parameters is required, or relative, when plotting the map of the field in a magnet gap or measuring successively the field of a series of magnet blocks as a function of that in a reference block. Relative measurements can be made easily an order of magnitude more accurate than "absolute" measurements since the integrator is used as a null indicator and the coils can be matched with high precision; furthermore, in this case small variations of the common excitation current do not perturb the comparison. Relative field measurements are made just as relative gradient measurements by exactly opposing fractions of the coil voltages that are defined with the help of precision resistance boxes used as potentiometers. The best azimuthal position for a magnet element is determined from relative measurements with respect to a reference element.

The simplest solution for recording the instantaneous value of the rapidly varying integrator output voltage in "absolute" dynamic field measurements is to display it on a cathode-ray oscilloscope (CRO) and to photograph the screen. Green *et al.* (1953) have used this method with sets of calibrating lines obtained by applying known voltages to the CRO but the overall accuracy was only about 2%. A somewhat improved accuracy has been obtained by Lambertson (1954) in biasing the voltage applied to the CRO so that a higher amplification was possible on the small upper part of the signal appearing on the screen; there were seven successive steps of attenuation and biasing during the field rise and calibrating signals were also used.

A different approach to the problem is recording only the necessary information by counting the number of oscillations that are generated in the integrator described by Fuller and Hibbard (1954), at discrete values of time, instead of making a continuous record. The linearity of the relation, obtained by this method, between the frequency of oscillation and the input voltage was good enough since a 0.1% accuracy was required only over a range of input voltage from 5 to 6.5 V, for the coil area chosen and the actual time rate of change of the field in their measurements; the corresponding frequency of the oscillation was about 100 kHz. The induction rate of rise was from remanent value to 1.2 T in 1 sec, and the field was measured to 0.1% at 10 msec time intervals, with the number of counts displayed on a CRO and recorded on a 35-mm film.

Taieb *et al.* (1955) have used a similar method and measured the number of oscillations at ten selected values of the time with a memory circuit consisting of ten counting channels that were locked successively. The frequency of oscillation was stable to 0.1% during 5 min for a constant input voltage. The frequency/voltage ratio was 1.8 kHz/V and remained constant to 0.1% when the input voltage remained in the range 3–6 V.

Gray *et al.* (1963) have made comparative measurements of the 336 magnet sectors, to 1 part in 10^4 with respect to a reference sector, for the synchrotron Nimrod, with a method similar to that used at Saclay by Taieb *et al.*, but the gating of the scalers was obtained from either the magnetizing current or the field in the reference sector. An auxiliary search coil placed in the reference sector gap was connected to a simple Miller integrator that provided the control voltage for the gates to operate at the chosen field levels. The coil in the test sector and that in the reference sector were made with slightly different areas so that the difference signal fed to the input of the integrator always had the same polarity and a large enough amplitude. A similar method has been used for the magnetic-field sensor in the control system of the ITEP synchrotron in Moscow (Vasilev *et al.*, 1962).

The magnetic measurements on the ITEP synchrotron magnet have been done partly with a conventional electronic integrator and partly with a ballistic galvanometer that was switched from the measuring coils to an equivalent resistance circuit at a specified value of the field (Borisov *et al.*, 1962). The switching device was a thyratron relay circuit, controlled by a ferrite pickup unit acting as a field sensor, that provided a good reproducibility in the switching-field level (Alekseev *et al.*, 1962 b).

The induction can be accurately measured, as a function of the magnetizing current intensity, by a balance method as in Fig. 15 where the integrator

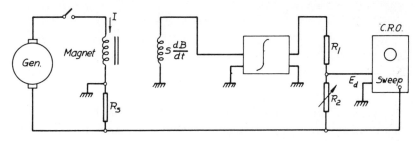

FIG. 15. Circuit for the measurement of $B = f(I)$.

output voltage is opposed to the voltage drop produced by the current across a measuring resistor R_s (de Raad, 1958). The resulting voltage E_d is applied to the vertical amplifier of a CRO and the resistance R_2 is adjusted until E_d is zero. Thus

$$E_d = \frac{1}{R_1 + R_2} \left[R_2 \frac{BS}{RC} - IR_sR_1 \right] \qquad (30)$$

The horizontal deflection of the CRO beam is made proportional to I and, for an exact determination of R_2 at the selected current value I_0, the beam is suppressed when $I = I_0$ by applying to the z axis of the CRO the voltage pulse of a current marker. The balance condition $E_d = 0$ is obtained when the end of the spot trace on the screen just falls on the horizontal reference line corresponding to the rest position of the spot immediately before starting the measurement, as in Fig. 16. A storage oscilloscope, on which the trace remains visible until it is deliberately erased, was used to obtain an accurate zero setting. A block diagram of the current marker is given in Fig. 17; this instrument produces a voltage pulse when the current I

passes through a value that can be chosen at will. The overall accuracy achieved in determining the ratio B/I was better than 0.2%.

Relative field measurements with a precision of the order of 1 part in 10^4 were also made with the same apparatus, by using the integrator as a null

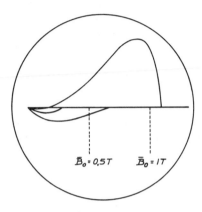

$\bar{B}_0 = 0.5T$ $\bar{B}_0 = 1T$

FIG. 16. Traces on the CRO screen for relative measurements at different field levels.

detector for balancing the voltages of the field coils with the help of potentiometers at any given current value I_0. A similar method has been described, using a gated integrator and a recorder instead of blanking the spot on the CRO; the theory and performance of the apparatus are given in detail (Palmer, 1959).

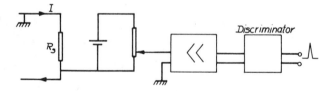

FIG. 17. Block diagram of the current marker.

1.4.4. Measurement of the Gradient of Magnetic Fields

A. STEADY FIELDS

1. *Point Measurements*

a. *Nuclear magnetic resonance.* Very accurate field measurements are possible by using the nuclear magnetic resonance method; thus variation of the magnetic field in space can be obtained as a difference of the values given by two probes, and the gradient can be computed from the separation of the two probes. However, since the distance between the two probes cannot be very small, a mean value only is obtained for the gradient in that way. A point measurement is nevertheless possible with the help of a quadrupole winding providing a compensating gradient: as was shown in Section 1.4.3, A, 1a, the current in this winding is adjusted so as to cancel the resulting gradient in the probe sample and thus to obtain the best resonance signal shape. The gradient is then determined from the value of the compensating current with an accuracy of 1% in good conditions (Denisov, 1960; Vasilevskaya *et al.*, 1963).

If a compensating field coil is added to one of a pair of resonance probes fed from the same oscillator, the other one being used at a fixed point as a

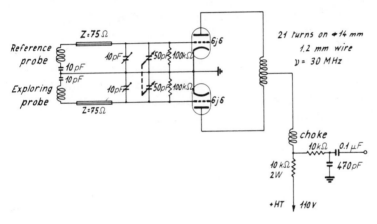

FIG. 18. Oscillator for differential field measurements by the nuclear resonance method.

reference for instance, the resonance signals from the two probes can be superposed by adjusting the current in the compensating coil, and the value of the difference in the field intensity at the two probes is readily obtained from the current value. Field contours in a calibration homogeneous field magnet were plotted at CERN in 1957 at about 0.7 T, using the simple push–pull oscillator presented in Fig. 18 (Bleeker, 1957).

A differential magnetometer in which two probes are fed in parallel in a marginal oscillator has been described by Silver (1964). Two identical pairs of coils in Helmholtz configuration, connected in series, are used for the field modulation at 50 Hz on the two probes and in addition on the exploring probe there is another pair of Helmholtz coils providing the compensat-

ing field. This apparatus was used for checking the field homogeneity in the
6-cm gap of a 1-meter radius isotope-separator magnet, with a sensitivity
of 10 μT at an induction value of 0.5 T. The compensating field calibration
as a function of current is valid even in different magnet gaps if a self-contained magnetic flux solenoid is used for providing that field, as in the case
of peaking strips (see Section 1.4.2,D).

A direct measurement of the difference of the field intensities at two
different points in the gap of a magnet, which is fed by a standard power
supply, is possible in spite of erratic field fluctuations by using a differential
magnetometer based on the properties of the super-regenerative oscillator
described by Pound and Freeman (1960) and presented in Section 1.4.3,A,1a.
The difference between the frequencies of the two super-regenerative oscillators is

$$\omega_1 - \omega_2 = \gamma(H_1 - H_2) + \Delta\omega' \tag{31}$$

The term $\Delta\omega'$, which is due to the inevitable difference in the tuning of the
two oscillators, is very nearly constant and can be accounted for in setting
the zero of the apparatus.

The field contours of a magnet for use in a high-resolution nuclear magnetic resonance spectrometer have been plotted with a sensitivity of 0.1 μT
at an induction level of 1.4 T without requiring a superlatively high current
stabilization by using two super-regenerative oscillators in that way (Freeman, 1961). The problem of shielding the two oscillators from one another
has been avoided elegantly: A master multivibrator controls two out-of-step
pulse generators, which in turn fire the two super-regenerative oscillators
alternately. The beat frequency between the two oscillators is picked up on
a communication receiver and the audio-frequency output, after filtering
to remove components at the quench frequency, is fed to an electronic
counter for a direct-reading measurement. The apparatus was mounted on a
motorized carriage and a marker pen attached to the exploring probe was
driven every time the probe crossed a point where the magnetic field difference with respect to the reference probe had a predetermined value defining a field contour. Automatic plotting of the field contours at given intervals was thus achieved with a high degree of accuracy.

Firsov et al. (1962) have also used this method for measuring differences
in magnetic field intensity of 1 part in 10^3 to 1 part in 10^6, in the presence of
fluctuations with time of the order of 0.1%. The minimum distance between
the probes was 10 mm. The beat frequency was determined from the Lis-
sajous figures produced on a CRO which was connected to the output of
the mixer into which the two frequencies were fed, the horizontal sweep
being provided by an audio generator.

b. *Hall effect.* We have seen in Section 1.4.3,A,1b that the configuration of a magnetic field can be determined precisely with a pair of Hall generators connected in opposition, one of them staying at the reference position and the other one exploring the magnetic field. If now the two Hall plates are mounted on the same probe at a fixed distance the resulting voltage is proportional to the gradient if the terms of higher order in the expansion of the magnetic field are negligible. It is practically always possible to arrange such conditions because the Hall plates and their separation can be made very small (Roshon, 1962). The magnetic center of each Hall generator can be determined accurately, to a few hundreths of a millimeter, by driving the Hall plate through a symmetrical magnetic field bump whose axis of symmetry is accurately known. The magnetic field bump may be produced by placing an accurately machined thin iron strip on the pole face of a magnet (Dahlstrom *et al.*, 1962). By this means the distance between the two Hall plates is also determined with the same precision. The electrical connections of the two Hall generators for measuring magnetic gradients, represented in Fig. 19, have been used at CERN by Braunersreuther.

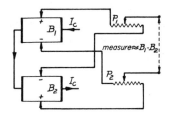

FIG. 19. Circuit for gradient measurements with a pair of Hall plates.

c. *Induction law.*

(i) *Rotating coil.* The rotating coil is generally used for measuring the field component which is perpendicular to the shaft tube, though other possibilities exist with special devices (Müller, 1955; Arnaud and Cahen, 1960). If this field component has a gradient in the direction of the shaft, a straightforward measurement of the gradient can be obtained from the difference in the voltages induced in two identical parallel coils attached at a known distance on the rotating shaft. The two coils can be matched in area and angle by the methods discussed in Section 1.4.2,C,2b. With the rotating coil system described by Reich (1963) and presented in Section 1.4.3,A,1c (i) a pair of parallel coils at a distance of 2 cm was used, instead of the single search coil, for gradient measurements which could thus be performed with an accuracy of a few parts in 10^3.

(ii) *Vibrating coil.* A real point measurement of the gradient in steady fields is obtained with a vibrating coil even in strongly inhomogeneous magnetic fields since the amplitude of the coil motion can be made very small, much smaller than the distance between two field-sensing probes. Magnetic lenses for electron microscopes or other types of charged-particle focusing devices, in which strongly inhomogeneous fields are present in rather small volumes, have often been studied with vibrating coils when the magnetic field and its first derivatives were to be measured with a good spatial resolution (Frazer *et al.*, 1955; Giertz, 1956; Spighel, 1957).

(iii) *Fluxmeter, ballistic galvanometer, and electronic integrator.* If a gradient coil, that is, a pair of parallel identical coils connected in opposition, is used instead of a field coil the measurement can be performed exactly at a point as field measurement. The gradient coil is removed from the magnet gap to provide the necessary flux variation to the measuring instrument. Another possibility is to give a small and well-known translatory displacement to a point field coil. The accuracy of a gradient measurement depends on the relative value of the gradient: It is more difficult to measure a given $\Delta B = B_1 - B_2$ when B is high since errors in matching and orientation of the coils then give larger errors in this differential field measurement. If the relative value of the gradient is required the usual potentiometric method shown in Fig. 20a is recommended. The examples given for integrating measurements of steady gradients are clearly valid also for point measurements.

2. Integrating Measurements

Fluxmeter, ballistic galvanometer, and electronic integrator. Integrating gradient measurements can be performed with a pair of long, parallel identical coils of suitable geometry connected in opposition or with a long harmonic coil as shown in Section 1.4.2,C,2c. The flux variation is obtained by removing the coil from the magnet gap and the differential field measurement is made just as a simple field measurement, with one of these instruments chosen according to the particular conditions of the measurement: sensitivity, accuracy, duration of the coil displacement, characteristics of the coil, and so forth.

For measuring the gradient of the remanent field in the CERN synchrotron magnet the search coil, with an area of 20 m², described in Section 1.4.2.C,2a, was displaced by successive steps of 2 cm and the change in flux linkage, about 4 mWb for each step, was measured with the integrator described in section 1.4.2,C,6c. The output voltage of the integrator is directly read on a digital voltmeter and since the motion of the coil is completed in about one second the drift of the integrator gives an error smaller than 0.1%. A similar method, but with a servofluxmeter instead of an electronic integrator, was used by Gray *et al.* (1963) for measuring the remanent field gradient in the synchrotron Nimrod.

At Brookhaven the magnetic lenses designed for beam transport have been measured with harmonic coils and their magnetic field analyzed into their harmonic content to provide the nonlinear coefficients for beam optics computation (Danby and Jackson, 1963). With a ballistic galvanometer and a harmonic coil, Daniltsev and Plotnikov (1963) have measured the field nonlinearity in magnetic lenses by switching off the excitation current to provide the necessary flux variation in a short enough time.

When a pair of coils is used the potentiometric connections shown in Fig. 20a can provide the relative value of the gradient with a good accuracy.

B. FIELDS VARYING WITH TIME

1. *Point Measurements*

a. *Hall effect.* As a direct application of Sections 1.4.3,B,1b and 1.4.4,A,1b in which we discussed, respectively, the use of the Hall effect for measuring time-varying magnetic fields and steady magnetic gradients, we can see that measuring gradients in time-varying magnetic fields is possible by this method provided precautions are taken to keep the parasitic induced voltages at a negligible level.

b. *Electronic integrator.* With a gradient coil, consisting of a pair of parallel identical coils connected in opposition, instead of a field coil, the measurement can be performed exactly as a point-field measurement with an electronic integrator. The gradient coil can be maintained at a fixed position in the gap since the field is varying in time. The problem of point measurements of the gradient in the ITEP synchrotron has been discussed in detail by Borisov *et al.* (1962), using a potentiometric method with an integrator or with a ballistic galvanometer switched by a special circuit. All the other methods that are described in Section 1.4.4,B,2, dealing with integrating measurements of time-varying gradients, can be applied here with small coils: In such measurements an accuracy of 0.1% is often achieved, especially when the relative value of the gradient is required, which makes it possible to use a potentiometric method.

2. *Integrating Measurements*

Electronic integrator. Integrating gradient measurements can be performed, with an integrator and a pair of long, parallel identical coils connected in opposition, in just the same way as a field is measured with a simple field coil. The gradient is determined from the differential flux measurement and the gradient coil calibration, that is, the area and distance between axes, provided the coil has a proper geometry as discussed in Section 1.4.2,C,2c.

For determining the focusing properties of the magnet elements for the CERN synchrotron the relative value of the gradient—or more precisely the field index $n = - (R_0/B_0) \cdot (\partial B/\partial r)$, where R_0 is the radius of curvature of the equilibrium orbit—was required rather than the absolute value of the gradient. The value of n on the equilibrium orbit was obtained with the pair of coils placed as shown in Fig. 20b and the integrator used as a null reading instrument in the circuit of Fig. 20a. The integrator output voltage is

$$E = \frac{S}{RC} B_0 \left[k \left(1 - \frac{d}{2R_0} n_0 \right) - \left(1 + \frac{d}{2R_0} n_0 \right) \right] \qquad (32)$$

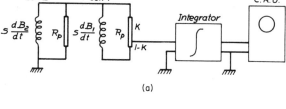

(a)

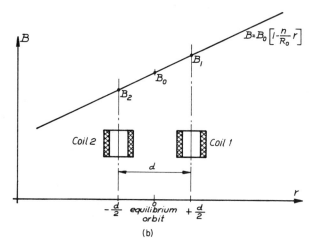

(b)

FIG. 20. Measurement of relative gradient g. (a) Connections with the two coils across potentiometers; (b) Location of the two coils for absolute measurements of n.

When the fraction k of the potentiometer setting is adjusted for exact balance condition at a given excitation current value, in the way described in Section 1.4.3,B,2 for the measurement of $B = f(I)$ at CERN, the corresponding value $n_0(I)$ is obtained

$$n_0(I) = \frac{k(I) - 1}{k(I) + 1} \cdot \frac{2}{d} R_0 \qquad (33)$$

With the coils and the integrator described, respectively, in Sections 1.4.2,C,2a and 1.4.2,C,6c the accuracy of the measurement of n_0 was 0.1% (de Raad, 1958; Reich, 1963). The coils had to be positioned very accurately since an error of 0.1 mm with respect to the equilibrium orbit corresponds to an error of 0.04% in the value of n_0.

To determine the range over which the focusing properties of the magnet elements were acceptable the relative value of the gradient $g = (1/B) \cdot (\partial B/\partial r)$ as a function of r at different currents was also required. To obtain this measurement the voltage in the gradient coil is compared with the voltage in a reference coil as shown in Fig. 21. The reference coil is maintained at

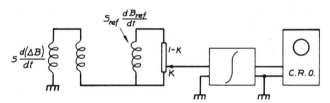

FIG. 21. Circuit for the measurement of $g = f(r)$.

a fixed position, whereas the gradient coil is translated step by step along a radius. At each position the potentiometer setting $k(r, I)$ corresponding to balance conditions is recorded, and the curve $k(r)$ can be translated into $g(r)$ or $n(r)$ by multiplying its values with a constant factor derived from the value of $k(r_0)$ at the equilibrium orbit and that of n_0 known from earlier measurements. The shape of these curves was obtained with an accuracy of 4 parts in 10^4 at injection field level ($B_0 = 15$ mT) progressively improving to 1 part in 10^4 with increasing field values.

The gradient in weak-focusing synchrotrons, such as Saturne at Saclay and Nimrod at Harwell, need not be measured with such a high degree of accuracy; but, on the other hand, its value is much smaller. The measurement programs on these synchrotrons, which have been reported, respectively, by Taieb et al. (1955) and by Gray et al. (1963), were performed with pairs of long coils and the oscillating integrators described in Section

1.4.2,C,6d. By using a harmonic coil, instead of a pair of coils, for measuring the gradient, the error due to higher order terms vanishes in principle. Thus a better accuracy should be obtained in nonlinear fields (Grekov et al., 1956). Complementary information on gradient measurements can be found in other reports dealing with the measurement of magnets for accelerators or beam transport systems (Palmer, 1959; Desy, 1961; Borisov et al., 1962; Danby and Jackson, 1963).

REFERENCES

Alekseev, A. G., Gorelkin, A. S., Mozalevskii, I. A., Mozin, I. V., Tarasov, B. I. and Trokhachev, G. V. (1962a). Instr. Exptl Tech. (USSR) (English Transl.) 4, 797–802.

Alekseev, A. G., Veselov, M. D., Mozalevskii, I. A., Rozhdestvenskii, B. V., and Trokhachev, G. V. (1962b). Instr. Exptl. Tech. (USSR) (English Transl.) 4, 790–797.

Alon, G. I. (1962). Onde Elec. 42, 330–353.

Arnaud, M., and Cahen, O. (1960). Rev. Tech. Compag. Franc. Thomson-Houston 32, 43–50.

Barber, E. (1960). Astronautics Information Literature Search No. 195. California Institute of Technology, Pasadena, California.

Berglund, S., Westerlund, S., and Svennerstedt, S. (1963). J. Sci. Instr. 40, 250–252.

Birss, R. R., and Fry, J. P. (1960). J. Sci. Instr. 37, 31–32.

Bleeker, J. J. (1957). Private communication.

Blewett, J. P., Rogers, E. J., and Swartz, C. E. (1953). Rev. Sci. Instr. 24, 782–788.

Borisov, V. S., Goldin, L. L., Goryachev, Y. M., Grekov, N. N., Ryabov, A. P., Skachkov, S. V., and Talyzin, A. N. (1962). Instr. Exptl. Tech. (USSR) (English Transl.) 4, 824–830.

Braunersreuther, E. (1956). Private communication.

Braunersreuther, E. (1965). Private communication.

Braunersreuther, E., Kuhrt, F., and Lippmann, H. J. (1960). Z. Naturforsch. 15A, 795–799.

Braunersreuther, E., Combe, J. C., Hoffmann, L., and Morpurgo, M. (1962). CERN Rept. 62-7.

Brown, R. A. (1966). CERN Rept. 66–15.

Brown, W., and Sweer, J. H. (1945). Rev. Sci. Instr. 16, 276–279.

Burson, S. B., Martin, D. W., and Schmid, L. C. (1959). Rev. Sci. Instr. 30, 513–521.

Buss, L., and Bogart, L. (1960). Rev. Sci. Instr. 31, 204–205.

Caldecourt, V. J., and Adler, S. E. (1954). Rev. Sci. Instr. 25, 953–955.

Chance, B. (1949). "Waveforms," M.I.T. Radiation Lab. Ser., Vol. 19, p. 664. McGraw-Hill, New York.

Chonez, A., Gabet, A., Labois, E., Lerond, P., Rastoix, G., and Taillet, J. (1959). Onde Elec. 39, 473–479.

Cioffi, P. P. (1950). Rev. Sci. Instr. 21, 624–628.

Cork, J. M., Shreffler, R. G., and Shull, F. B. (1947). Rev. Sci. Instr. 18, 315–316.

Dahlstrom, T. S., Howe, H. A., Mallet, W. M., and Smith, W. E. (1962). Rept. USNRDL–TR–559. U.S. Naval Radiological Defense Laboratory, San Francisco, California.

Danby, G. T., and Jackson, J. W. (1963). Rept. BNL-7700. Brookhaven National Laboratory, Upton, New York.

Daniltsev, E. N., and Plotnikov, V. K. (1963). *Instr. Exptl. Tech. (USSR) (English Transl.)* **3**, 387–392.

de Forest, R. (1962). *Nucl. Instr. Methods* **18-19**, 584–587.

Denisov, I. N. (1958). *Instr. Exptl. Tech. (USSR) (English Transl.)* **5**, 658–661.

Denisov, I. N. (1960). *Instr. Exptl. Tech. (USSR) (English Transl.)* **1**, 89–90.

de Raad, B. (1958). Dynamic and static measurements of strongly inhomogeneous magnetic fields. Ph. D. Thesis, Delft Institute of Technology, Holland.

Desy (1961). Rept. DESY-A 2.84. Deutsches Electronen-Synchrotron, Hamburg, Germany.

Dicke, R. H. (1948). *Rev. Sci. Instr.* **19**, 533–534.

Donnally, B., and Sanders, T. M., Jr. (1960). *Rev. Sci. Instr.* **31**, 977–978.

Edgar, R. F. (1937). *Trans. AIEE* **56**, 805.

Elmore, W. C., and Garrett, M. W. (1954). *Rev. Sci. Instr.* **25**, 480–485.

Elpatevskaya, O. D., and Regel, A. R. (1956). *Soviet Phys.—Tech. Phys. (English Transl.)* **1**, 2350–2356.

Fechter, H. R., and Rubin, S. (1955). *Rev. Sci. Instr.* **26**, 1108–1111.

Feldman, D. W. (1960). *Rev. Sci. Instr.* **31**, 72–73.

Finlay, E. A., Fowler, J. F., and Smee, J. F. (1950). *J. Sci. Instr.* **27**, 264–270.

Firsov, E. P., Pivovarov, S. P., and Latyshev, G. D. (1962). *Bull. Acad. Sci. USSR, Phys. Ser. (English Transl.)* **26**, 1094–1096.

Förster, F. (1955). *Z. Metallk.* **46**, 358–370.

Fragstein, C. V. (1957). *Z. Angew. Phys.* **9**, 268–272.

Frazer, J. F., Hofmann, J. A., Livingston, M. S., and Vash, A. M. (1955). *Rev. Sci. Instr.* **26**, 475–476.

Freeman, R. (1961). *J. Sci. Instr.* **38**, 318–321.

Fuller, W., and Hibbard, L. U. (1954). *J. Sci. Instr.* **31**, 36–42.

Gabillard, R. (1956). *Arch. Sci. (Geneva)* **9**, 316–318.

Garrett, M. W. (1951). *J. Appl. Phys.* **22**, 1091–1107.

Gautier, P. (1954). *J. Phys. Radium* **15**, 684–691.

Gautier, P. (1956). *Compt. Rend.* **242**, 1707–1710.

Germain, C. (1963). *Nucl. Instr. Methods* **21**, 17–46.

Giertz, W. (1956). *Allgem. Elek.-Ges. (Berlin)* **46**, 133–136.

Giordano, S., Green, G. K., and Rogers, E. J. (1953). *Rev. Sci. Instr.* **24**, 848–850.

Goldberg, E. A. (1950). *RCA Rev.* **11**, 296–300.

Goldin, L. L., Skachkov, S. V., and Shorin, K. N. (1962). "Magnetic Measurements in Charged-particle Accelerators." Gosatomizdat, Moscow (in Russian).

Gray, D. A., Harold, M. R., Jones, P. F., Morgan, R.H.C., Partington, J. E., and Pyle, I.C. (1963). Rept. NIRL/R/4. Rutherford High Energy Laboratory, England.

Green, G. K., Kassner, R. R., More, W. H., and Smith, L. W. (1953). *Rev. Sci. Instr.* **24**, 743–754.

Grekov, N. N., Ryabov, A. P., and Goldin, L. L. (1956). *Pribory i Tekhn. Eksperim.* **2**, 29–37.

Grivet, P. (1962). In "High Magnetic Fields; Proceedings of the International Conference on High Magnetic Fields, M.I.T., Cambridge, Massachusetts, 1961" (H. Kolm, B. Lax, F. Bitter, and R. Mills, eds.), pp. 54–84. M.I.T. Press, Cambridge, Massachusetts and Wiley, New York.

Grivet, P., Sauzade, M., and Stefant, R. (1961). *Rev. Gen. Elec.* **70**, 317–328.

Hedgran, A. (1952). *Arkiv Fysik* **5**, 1–27.

Hermann, P. K. (1964). *Arch. Tech. Messen, Lfg.* **345**, 237–240.

Huber, H. J., and Rogers, K. C. (1964). *Rev. Sci. Instr.* **35**, 801–802.

Hudson, E. D., Lord, R. S., Marshall, M. B., Smith, W. R., and Richardson, E. G., Jr (1962). *Nucl. Instr. Methods* **18–19**, 159–169.

Hutchinson, C. A., and Pastor, R. (1953). *Rev. Mod. Phys.* **25**, 285–290.

Inozemtzev, K. V., and Latyshev, G. D. (1949). *Izv. Akad. Nauk. SSSR, Ser. Fiz.* **13**, 453–455.

Jan, J. P. (1957). *Solid State Phys.* **5**, 1–96.

Jousselin, J. (1962). *Atompraxis* **8**, 140–144.

Jürgens, B. F. (1953). *Philips Tech. Rev.* **15**, 49–62.

Kandiah, K., and Brown, D. E. (1952). *Proc. Inst. Elec. Engrs. Pt. II* **99**, 314–326 and 344–348.

Kapitsa, S. P. (1955). *Zh. Tekhn. Fiz.* **25**, 1307–1315.

Kelly, J. M. (1951). *Rev. Sci. Instr.* **22**, 256–258.

Klemperer, O., and Miller, H. (1939). *J. Sci. Instr.* **16**, 121–122.

Kuhrt, F. (1954). *Siemens-Z.* **28**, 370–376.

Kuhrt, F. (1960). *Elektron. Rundschau* **14**, 10–13.

Kuhrt, F. (1961). In "Halbleiter Probleme" (W. Schottky, ed.), Vol. 6, pp. 186–205. Vieweg and Sohn, Braunschweig.

Kuhrt, F., and Hartel, W. (1957). *Arch. Elektrotech.* **43**, 1–15.

Lamb, W. E., Jr., and Retherford, R. C. (1951). *Phys. Rev.* **81**, 222–232.

Lambertson, G. R. (1954). Rept. UCRL 2818. Radiation Laboratory, University of California, Berkeley, California.

Langer, L. M., and Scott, F. R., (1950). *Rev. Sci. Instr.* **21**, 522–523.

Laslett, L. J. (1954a). Rept. LJL–1. Brookhaven National Laboratory, Upton, New York.

Laslett, L. J. (1954b). Rept. LJL–2. Brookhaven National Laboratory, Upton, New York.

Lind, D. A. Rickey, M. E., and Bardin, B. M. (1962). *Nucl. Instr. Methods* **18–19**, 129–134.

Lozingot, J., Pinet, D., and Taieb, J. (1953). Note CEA No. 7 Service des Accelerateurs, Saclay, France.

Lothe, J. J., and Eia, G. (1958). *Acta Chem. Scand.* **12**, 1535–1537.

Lush, M. J. (1964). *Instr. Control Systems* **37**, 111–113.

McCutchen, C. W. (1959). *J. Sci. Instr.* **36**, 471–474.

Montague, B. W. (1955). *Mullard Tech. Commun.* **2**, 64–73.

Müller, M. (1955). *St. Elek.-Ges. (Stuttgart)* **3**, No. 2, 96–98.

Muray, J. J., and Scholl, R. A. (1964). Rept. SLAC–26. Stanford Linear Accelerator Center, Stanford University, California.

Neumann, H. (1954). *Arch. Tech. Messen, Lfg.* **222**, 161–164.

Palmer, J. P. (1959). Rept. BNL–4657. Brookhaven National Laboratory, Upton New York.

Palmer, T. M. (1955). In "Precision electrical measurements," Paper 9. H. M. Stationery Office, London.

Penfold, A. S., and Garwin, E. L. (1960). *Rev. Sci. Instr.* **31**, 155–163.

Peters, W.A.E. (1950). *Elektrotech. Z., A* **71**, 193–194.

Pohm, A. V., and Rubens, S. N. (1956). *Rev. Sci. Instr.* **27**, 306–308.

Pople, J. A., Schneider, W. G., and Bernstein, H. J. (1959). "High Resolution Nuclear Magnetic Resonance." McGraw-Hill, New York.

Pound, R. V., and Freeman, R. (1960). *Rev. Sci. Instr.* **31**, 96–102.

Putley, E. H. (1960). "The Hall Effect and Related Phenomena." Butterworth, London and Washington, D.C.

Radkevich, I. A., Sokolovskii, V. V., Talyzin, A. N., Goldin, L. L., Bysheva, G. K. and Goryachev, Y. M. (1962). *Instr. Exptl. Tech.* (*USSR*) (*English Transl.*) **4**, 848–854.

Radus, R. J. (1960). *J. Appl. Phys.* **31**, Suppl., 186S–187S.

Reich, K. H. (1963). The CERN proton synchrotron magnet. Internal report: MPS/Int. DL 63–13. CERN, Geneva, Switzerland.

Robinson, D. A. (1956). *Electronics* **29**, No. 9, 182–185.

Roshon, D. D., Jr. (1962). *Rev. Sci. Instr.* **33**, 201–206.

Ross, I. M., Saker, E. W., and Thompson, N.A.C. (1957). *J. Sci. Instr.* **34**, 479–484.

Sasaki, H. (1962). *Nucl. Instr. Methods* **14**, 252–262.

Schwaibold, E. (1956). *Arch. Tech. Messen, Lfg.* **246**, 153–156.

Segre, S.E., and Allen, J. E. (1960). *J. Sci. Instr.* **37**, 369–371.

Shirer, D.L. (1960). *Rev. Sci. Instr.* **31**, 1000–1001.

Silver, D.E.P. (1964). *Electron. Eng.* **36**, 374–377.

Spighel, M. (1957). *J. Phys. Radium* **18**, 108A–111A.

Spokas, O. E., and Danos, M. (1962). *Rev. Sci. Instr.* **33**, 613–617.

Stahlke, J. L. (1962). *Nucl. Instr. Methods* **17**, 157–160.

Symonds, J. L. (1955). *Rept. Progr. Phys.* **18**, 83–126.

Taieb, J., Guillon, H., Gabet, A., and Mey, J. (1955). *Onde Elec.* **35**, 1076–1083.

Tenzer, R. K. (1955). *Arch. Tech. Messen, Lfg.* **239**, 285–288.

Tickle, R. S. (1962). *Nucl. Instr. Methods* **18-19**, 98–101.

van der Walt, N. T. (1953). *Rev. Sci. Instr.* **24**, 413–416.

Vasilev, A. A., Batskikh, G. I., Vasina, Y.A., and Andryushchenko-Lutsenko, N. I. (1962). *Instr. Exptl. Tech.* (*USSR*) (*English Transl.*) **4**, 699–703.

Vasilevskaya, D. P., Vasiliev, L. V., and Denisov, Y.N. (1963). Rept. P-1475. Joint Institute for Nuclear Research, Dubna, USSR.

Vizir, V. A., Kuzmin, V. N., and Petrov, Y.K. (1963). *Instr. Exptl. Tech.* (*USSR*) (*English Transl.*) **2**, 319–321.

Voelker, P., and Leavitt, M. A. (1955). Rept. UCRL 3084. Radiation Laboratory, University of California, Berkeley, California.

Weiss, H. (1956). *Z. Naturforsch.* **11a**, 430–434.

Weiss, H., and Wilhelm, M. (1963). *Z. Physik* **176**, 399–408.

Welker, H., and Weiss, H. (1956). *Solid State Phys.* **3**, 1–78.

Williamson, K. I. (1947). *J. Sci. Instr.* **24**, 242–243.

Wills, M. S. (1952). *J. Sci. Instr.* **29**, 374–376.

Woodbridge, D. D., and Warner, W. R. (1958). *Am. J. Phys.* **26**, 490–492.

Zingery, W. L. (1961). *Rev. Sci. Instr.* **32**, 706–708.

2
Lenses

CHAPTER 2.1

HIGH BRIGHTNESS ELECTRON GUNS

M. E. Haine

A.E.I. RESEARCH LABORATORY
RUGBY, WARWICKSHIRE, ENGLAND

AND

D. Linder

A.E.I. INSTRUMENTATION DIVISION
HARLOW, ESSEX, ENGLAND

2.1.1. Introduction

The electron guns to be described herein are the type used in electron microscopes (Haine and Cosslett 1961), x-ray microanalyzers (Duncumb and Melford 1960), and other electron optical devices needing a low-current, high-brightness source. The most common type of gun is the triode gun that is used in most electron-optical applications, with the exception of certain sealed-in tubes that use a pentode gun. The basic design of the guns for the foregoing uses is similar but their appearances may be different as a result of the differing accelerating voltage used For electron microscopy the range of accelerating voltages may be from 30 to 1000 kV, with 100 kV being a common value. For x-ray microanalyzers the range is smaller, being from 10 to 80 kV. For both these applications, the high-voltage requirements play an important part in the physical design of the gun. For sealed-in tubes, however, the accelerating voltage may well be in the range 300 V–15 kV and here the high-voltage design is relatively unimportant.

In all the guns described the total gun current might well be less than 500 μA and the useful beam current less than 1 μA.

2.1.2. Electron-Gun Requirements and Limitations

In almost all electron-optical instruments the electron gun is required to give as high as possible a current density over a prescribed area in a plane with the additional requirement that the angular divergence should be limited to some prescribed maximum value. The angular limitation may derive from limitations met by lens aberration, or from the need to obtain electron-optical coherence, and so forth. Thus, the overall requirements is for the gun to give the maximum possible current density per unit solid angle or "brightness." It is convenient to separate out this parameter since, as will be explained, it is subject to a fundamental limitation. In addition, the gun must produce a maximum—or at least an adequate—brightness over a sufficient area and solid angle. The latter two auxiliary requirements are largely a matter of choice of emitting cathode size and the subsequent focusing system.

The theoretical limitations to brightness derives from the fact that the electrons emitted from the cathode have a Maxwellian-energy distribution with a most probable energy of kT and an average energy of $2\,kT$, where k is Boltzmann's constant ($1/11,600$ eV/°K) and T is the cathode temperature. Electrons emitted into a free-field space travel out in a Lambertian distribution in a solid angle (cosine law) with the foregoing distributions in energy. The electrons might now be accelerated in a parallel or uniform applied electrostatic field that would add to their velocity component that is parallel to the field but leave the perpendicular component unchanged. The semiangle of divergence would thus be reduced from π to the ratio of the parallel and perpendicular velocities. If the electrons have been accelerated through a potential drop of V_0, the new parallel velocity will be $(2eV_0/m)^{1/2}$ and hence the new semiangle will be $(kT/eV_0)^{1/2}$ giving a solid angle of divergence of $\pi\, kT/eV_0$, assuming all the electrons have the most probable energy of emission, namely, kT.

Now if the cathode is large in extent the current density will be the same at the accelerating anode as at the cathode; thus the average brightness (β) at the anode will be given by

$$\beta = \varrho_c eV_0/\pi kT$$

where ϱ_c is the cathode current density.

It will be noted that this limiting value of brightness is proportional to the emitted current density and the accelerating voltage and inversely proportional to the cathode temperature. However, for a thermionic source the cathode current density increases exponentially with the increase in temperature and hence the net effect on brightness is to give an increase in brightness with temperature. As will be explained later, the cathode life decreases with increase in temperature and hence there is an upper limit of current density consistent with reasonable life.

Equation (1), derived for the simple case of parallel acceleration, has in fact a much wider application; it represents the fundamental limitations to the brightness obtainable in the electron gun. Equation (1) was originally derived by Langmuir (1937) taking into account the Maxwellian energy distribution and considering an aberration-free imaging system. The expression can be applied irrespective of any focusing system that may follow the gun and, in fact, no focusing system can increase the brightness above that obtained at the gun itself. This is fairly evident when we consider that a focusing system of magnification M reduces the current density M^2 times and increases the angular aperture by $1/M^2$ times, thus leaving the brightness unchanged.

The brightness limit can also be derived from a statistical mechanical treatment involving Liouville's theorem, for a general electron–optical system. The application of Liouville's theorem assumes the electrons behave as noninteracting particles, which is a reasonable assumption for low-density beams but for high-density beams some interactions might be expected. This interaction has been observed and measured by Boersh (1954), Dietrich (1958), and Hartwig and Ulmer (1963) and it was found to result in a broadening of the velocity distribution in the beam to several times its Maxwellian width. This would cause a reduction of the limiting brightness below that predicted by Eq. (1), although no measurements of any reduction attributed to this cause have been reported.

A theoretical limit of this sort is only of value if in fact it is within the practical possibility of attainment. That this is so will later be explained. Knowing that it is so is of great value to the instrument designer or experimenter, since he can then use simple theoretical expressions in his designs or in calculating the limited possibilities of new instrumental applications. It is, for example, possible to calculate the required cathode current density to give sufficient light intensity for visual observation of the image in the electron microscope at a given magnification (Agar, 1957) and the limiting performance of a microprobe electron-beam machine as described elsewhere in this volume (cf. T. Mulvey 2.3).

Now that the brightness condition is established, we can consider the requirements for beam area, angle, and current density, and consider how we should proceed to design an electron gun. If we assume that the accelerating voltage is fixed by the instrumental requirement, then the spot current density and angle being defined, and thus the brightness, we can calculate the required cathode current density from Eq. (1).

It is seldom that calculations can be carried beyond this point as, in general, the source to be imaged will not be the cathode surface but an electron crossover. Although the size of the crossover is known, the area of emission from the cathode and hence the total gun current is undefined because it depends very much on the operating conditions of the gun. In general, purely practical matters tend to decide the general features of electron gun design and the focusing system is then designed to suit the gun performance. The practical design of the gun will be considered in two parts; first, the cathode design and, secondly, the accelerating electrode design.

2.1.3. Cathode Design

A. CATHODE MATERIALS

Since the requirement for the electron source is that it should have the highest possible brightness, we require a cathode to give the maximum current density while operating at minimum possible temperature. The choice of materials is decided mainly by the vacuum conditions in the gun. For a sealed system in which the vacuum is good ($< 10^{-5}$ mm Hg), oxide cathodes are used; these might typically be barium or strontium oxides having workfunctions of about 1.6 and 2.0 eV., respectively and giving emission current densities of up to 1 A/cm^2 for an operating temperature of 1000°K. However, for a demountable gun in which the operational pressure is higher than 10^{-5} mm, Hg oxide cathodes are not suitable because they are poisonous.

Thus refractory metal cathodes have to be used. Of the refractory metals, tungsten is the usual choice since it has a workfunction of 4.5 eV and a possible operating temperature up to 3000°K, giving emission current densities of up to 14 A/cm².

The most common form of the tungsten cathode is a hairpin filament. It consists of a 1-cm length of tungsten wire bent into the form of a "V." The emission occurs at the tip of the V, which presents an almost spherical surface. The choice of wire thickness for this cathode is governed by factors such as power input, filament life, and ease of handling; a common thickness value is 0.0125-cm diameter. The length of wire is chosen to minimize the power input; about a 1-cm length is found to be satisfactory for the 0.0125 cm diameter. Shorter wires than this produce excessive cooling because of heat conduction to the filament support with a consequent increase in the power requirements.

In most guns using this type of filament it is necessary to set the position of the filament with some accuracy and thus to ensure stability of the filaments, it is advisable to preheat the filaments for a few seconds before centering to remove any strain introduced during fabrication.

B. Point Cathodes and Ground Cathodes

In electron microscopy increasing use is being made of the so-called pointed and ground filaments. These are basically tungsten hairpin filaments that have either had a very fine tungsten point added at the tip or that have been ground to produce a point (Hibi, 1956a; Bradley, 1961).

It is claimed that these filaments have certain electron-optical advantages. The main difference between these and the hairpin cathode are their smaller source size and the maximum temperature of operation. The source sizes for the ground and point cathodes are, respectively, about two and three times smaller than that of the hairpin cathode and this results in a more coherent source (Thon, 1963).

This increase in operating temperature is important because, due to space charge, at temperatures above 2750°K the brightness of the hairpin cathode under normal operation falls below the Langmuir value as is shown later. It may be possible, however, to find an operating condition for the hairpin cathode wherein space-charge limitations only enter at correspondingly high temperatures. With the point and ground cathodes, however, operating under normal conditions, the space charge is reduced and hence the filament temperature can usefully be increased above 2750°K to give a corresponding increase in brightness.

It has been suggested that point cathodes will give brightnesses higher than the hairpin cathode due to an increased emission current density resulting from field assisted thermionic emission (the Shottkey effect). It is doubtful, however, whether the fields near the tip under normal operating conditions would be sufficient to increase the emission current density by the factor of two or three claimed. The increase in brightness is more likely to be due to an increase in the operating temperature which in general was not measured at the same time as the brightness.

The construction of pointed filaments has been described by Hibi (1956b). The main features are that the tip radius should be less than $1\,\mu$ and the tip length less than 0.3 m to avoid an excessive temperature drop along the spike. The production of such a point has been described by Niemick and Ruppin (1954) who use an electrolytic etch method. This method or a modification of it has been used successfully by many people since and it is possible to produce points of $0.1\,\mu$ diameter and less. It is doubtful whether points of this radius would be of much use because on heating to $2750°K$, surface migration of the tungsten takes place and the tip radius increases rapidly to near $0.5\,\mu$.

An increase in resolution from 30Å using a hairpin cathode to 10 Å using a $0.25\,\mu$ radius point cathode has been reported by Hibi (1962). Recent reports (Hibi, 1964) show that a lattice resolution of 3.4 Å is possible with a two-stage microscope using a single condenser lens and a point cathode.

The construction of ground cathodes for the electron microscope has been described by Bradley (1961) and their use by Thon (1963).

C. FILAMENT LIFE

Only the lives of hairpin cathodes will be described quantitatively since the life of point and ground filaments depend very much on the method of construction.

The life of the tungsten filament is determined mainly by thermal evaporation, unless a significant water vapor pressure is present (Bloomer, 1957). In normal operation in a continuously evacuated system, water vapor represents no serious limitation. However, in a system that is repeatedly let up to atmospheric pressure, water vapor can easily prove the limiting factor.

The shape of the hairpin filament makes the temperature slightly higher in the side legs just below the tip, and burnout normally occurs in this region. Evaporation is rapidly dependent on temperature and causes the wire to "neck" in this region. The necking causes the local increase in resistance and hence a local increase in heating and evaporation. Once the filament diameter has been reduced in this local region by about 6%, a thermal runaway condition is established and the remaining life is very short. The effects of a water vapor attack or of evaporation can be visually recognized in a burned out filament. In the former, a uniform reduction in diameter is observed with a final fairly gradual necking near the burnout. In the latter, no obvious diameter reduction is seen except in the immediate vicinity of the sharp neck at the burnout. In the more normal case of evaporation, burnout conditions depend fairly critically on the supply impedance as described subsequently.

From data of Reimann (1938), Bloomer (1957) computed the filament lives due to evaporation and obtained experimental verification of his results. His paper gives curves connecting life and brightness with filament diameter, heating current, and/or temperature. From these curves it can be shown that, to a practically significant approximation, the life is given by a simple expression that is only slightly dependent upon wire diameter:

$$t = 32/\varrho_c$$

where t is the life measured in hours and ϱ_c is the temperature-limited current density.

This expression immediately allows closer estimates to be made of the theoretical brightness obtainable at a given accelerating voltage for a given cathode life, and hence the available source image current, or current density for a given beam angle. Thus, we have all the necessary basic data for electron photometric calculations.

D. FILAMENT SUPPLIES

The filament heating may be either ac or dc. The use of ac is simpler as the filament is at a high voltage with respect to ground and an ac filament supply can easily be generated at this voltage using an isolating transformer. With ac heating, care must be taken to ensure that the emitting region is at the electrical center of the filament thus ensuring that its potential is fixed. Failure to ensure this will result in modulation of the electron beam due to variations in bias voltage. The problem is usually solved by connecting the HT to the wiper of a potentiometer connected across the filament.

The impedance of the filament supply can affect the filament life considerably in the case in which life is limited by evaporation. A high-impedance source, approximating to constant-current conditions, leads to a runaway conditions in which the temperature of the filament increases as the wire thins causing an increased evaporation rate and, hence, a shorter life. Under the conditions of constant voltage, that is, a low-impedance supply, the temperature of the wire drops as it thins and this gives a longer life but is of little use in a microscope in which we wish to maintain the filament tip temperature constant. For the special case in which the generator impedance is one third the filament impedance, the temperature of the wire remains substantially constant during life.

2.1.4. Gun Geometry and Performance

A. Gun Characteristics

The performance of the gun is not critically dependent upon the shapes of the electrodes. For a fairly wide variety of shapes, roughly the same performance can be obtained by a suitable choice of the cathode. The gun to be described is illustrated in Fig. 1.

It uses a tungsten hairpin cathode (1) of wire 0.0125 cm in diameter and

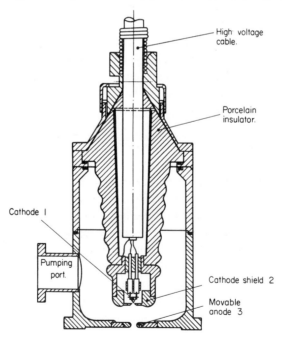

FIG. 1. Cross section of A.E.I.–EM6 electron gun.

approximately 14 mm long. The cathode is mounted so that the tip (which is approximately spherical) lies just behind the center of a negatively biased grid aperture (2), and opposite this is a parallel apertured anode electrode (3).

This basic electrode system can be defined through the dimension "h" (Fig. 2), the cathode height behind the grid aperture, "d" the grid-aperture diameter, and the cathode-anode distance. The anode is preferably flat

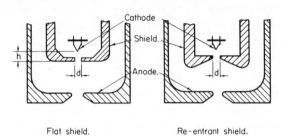

Flat shield.   Re-entrant shield.

FIG. 2. The flat and reentrant cathode shield geometries studied by Haine and Einstein.

but the cathode shield may be flat or reentrant. The exact shape of the cathode shield is not critical and if a reentrant shape is required, a conical shape is usefully adopted (see later). In operation, also the cathode temperature (T_c) and bias voltage (V_b) may be varied (it is assumed that the anode voltage is fixed by the beam requirement). The main beam properties that may be affected by varying these quantities are the brightness (β) and the beam semiangle (α), the current and the virtual source diameter, and axial position. By "virtual source" is meant the apparent source as it would appear looking back into the gun from beyond the anode. This, therefore, corresponds to the spot which would be focussed by a condenser lens into some plane beyond. This virtual source is not necessarily a cathode image; it may be a crossover or aberration caustic. All these quantities are of interest to the users of the electron gun.

Before discussing the established relationship between design and operating conditions—and the performance characteristics—a brief historical note will be introduced to emphasize an important aspect of electron-gun performance. In the early days of use of this type of electron gun it was observed that under conditions of increasing cathode temperature the current would at first increase until at some temperature a sharp saturation would appear, above which no further increase of current occurred. This saturation condition had exactly the appearance of the space-charge-limited current knee, familiar in certain types of thermionic diode. It was naturally assumed that this was indeed a space-charge effect, and for several years space charge was considered of great importance, not only in limiting emission, but in determining the beam and virtual cathode geometry. That this is not, in fact, the case was clearly shown by Haine and Einstein (1952). Space charge is virtually of no importance in this type of electron gun except at the highest usable current densities. The current saturation was shown to be due to the negative feedback effect of the series cathode resistor used to generate the bias. No such saturation occurs if the bias is derived from a battery, for example. The virtual cathode and beam geometry for a given gun configuration and bias was shown to be quite independent of cathode temperature and beam current, except at very high cathode temperatures. Thus was it demonstrated that space charge only affects the beam geometry at the highest cathode temperatures.

These points are stressed at this stage to avoid confusion in reading what follows and because subsequent literature does not indicate that the points have been fully appreciated. Haine and Einstein experimentally investigated the characteristics of this type of electron gun. For a given cathode shape, grid aperture diameter, and cathode-anode spacing, they obtain the characteristics shown in Fig. 3. The top curves show how the measured brightness

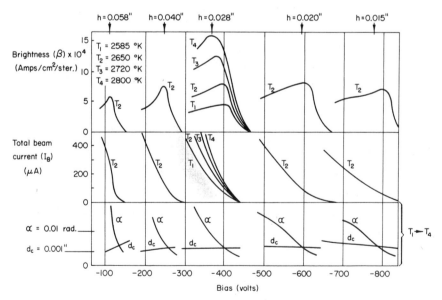

FIG. 3. Brightness (β), total beam current (I_B), semibeam angle (α), and virtual source diameter (d_c) plotted as a function of gun bias for five different values of filament height and for one height for four different cathode temperatures. Cathode-shield aperture 0.07 in diameter, accelerating voltage 50 kV.

was found to vary with bias voltage for a range of values at cathode temperature and cathode height; the lower curves show how the beam-current, beam
semiangle, and virtual source size vary with bias. The important characteristic is the rise in brightness from cutoff bias, as bias is reduced, to a maximum. This maximum was found to equal the theoretical value for the temperature concerned, except at the highest temperatures. The same characteristics were found for a wide range of cathode shapes, from flat to deeply
reentrant cones.

Thus, the conclusion is reached that over a wide range of geometric design
the theoretical brightness can be obtained, except at the highest temperatures, provided the gun is operated near the optimum bias condition. The
reason that the maximum brightness falls short of the theoretical value at
the highest cathode temperatures was shown to be due to space charge. At
these temperatures the life of a hairpin cathode is very short, the effect only
being just significant at 2750°K. Figure 4 shows the variation of the ratio
of observed-to-theoretical brightness as a function of evaporation-limited
filament life.

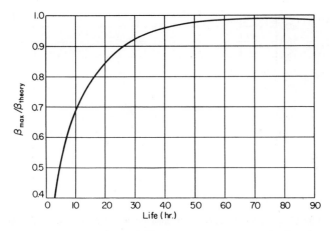

FIG. 4. The ratio of the maximum brightness obtained to the theoretical brightness
($\beta_{max}/\beta_{theor}$) as a function of cathode life.

Although theoretical brightness is obtained for a wide range of geometry,
other characteristics vary greatly. For example, both the current and the
beam angle at the optimum bias vary. This means that the current density
at the center of the divergent beam is independent of geometry, but that the
total angle of the beam, and hence the current varies. Since usually a limiting aperture is used to define the required beam angle, and any electron

current outside the aperture is wasted, control of geometry enables a minimum wastage by adjusting the total beam angle so that it just exceeds the angle subtended by the defining aperture. It is usual to allow some wastage: first, because the distribution of current density across the divergent beam is approximately Gaussian in shape, and it is desirable to use only the center peak to ensure maximum current through the aperture; and secondly because a larger beam angle is desirable to reduce the problem of maintaining alignment between beam and defining aperture.

Figure 5 shows how the total current (and beam angle) for the theoretical brightness condition varies with the filament height for a range of values of cathode-shield diameter for a flat cathode-shield geometry. Each value of filament height and cathode shield diameter requires a change in the bias condition to give the theoretical brightness value.

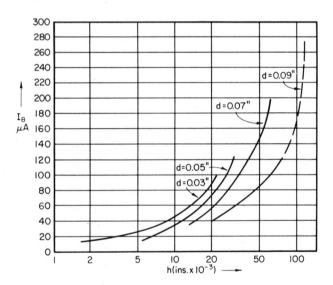

Fig. 5. Variation of beam current at maximum brightness ($\beta = 80,000 \text{ A}/\text{cm}^2/\text{sr}$) with filament height for four values of cathode-shield diameter.

In Fig. 6 it is shown how the axial position of the virtual cathode varies with the filament height and beam current. It will be noted how the axial virtual source moves between the axial cathode and the anode but never reaches the anode under the conditions covered. Attempts have been made to produce positive focus, or telefocal, guns giving a focused spot after the anode. Figure 7 shows the arrangements used by Steigerwald (1949) and by Bruck and Bricka (1948). The geometry is somewhat complicated and

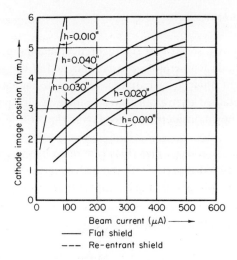

FIG. 6. The variation of cathode image position with beam current for three values of cathode height with flat shield and one value of height for reentrant shield (cathode-aperture diameter 0.05 in., accelerating voltage 50 kV).

not very suitable for very high-voltage operation. Somewhat misleading claims have been made suggesting that this type of gun gives a brighter image in the electron microscope, for example; in fact, they cannot give a brightness greater than the more conventional design. Their only advantage is the possible saving of a condenser lens in some systems. The characteristics of a gun operated with a point cathode are similar to those for a hairpin cathode and have been described by Swift and Nixon (1962).

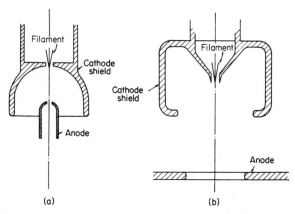

FIG. 7. Positive focus electron guns designed (a) by Bruck and Bricka, and (b) by Steigerwald.

B. Biasing Arrangements

It has been shown so far that the theoretical maximum brightness values for a given cathode temperature can be obtained at the same time as controlling the beam angle to a required value. Further, any required beam angle can be obtained over a fairly wide choice of dimensions and cathode shape. We can now consider a further factor that will lead to a limitation to this choice.

It is convenient to use self-bias (bias generated by the voltage drop across a resistance connected in series with the high-voltage supply to the cathode) (see Fig. 8). Not only is this much simpler than providing a separate bias

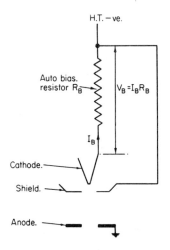

Fig. 8. Circuit for automatic bias.

supply (at the HT voltage above ground), but it also provides an important stabilizing action to the beam current. It has already been explained how, with self-bias, the current reaches a stable plateau at a certain temperature. In practice, it is required to so design the gun that not only is the maximum brightness obtained, but that this occurs for the desired operating temperature (for example, as limited by life), and that the current saturation point occurs at this temperature. Also, these conditions must occur for the desired beam divergence angle.

Haine *et al.* (1958) have investigated the variation of the conditions giving current saturation for three shapes of cathode shield (flat, 30° and 60° reentrant cone) for a range of values of filament height, bias resistor value, and cathode temperature. Their results are shown in Fig. 9. Each family of curves (full line) gives the variation of current with filament heating current and temperature, and (dotted lines) the variation of brightness,

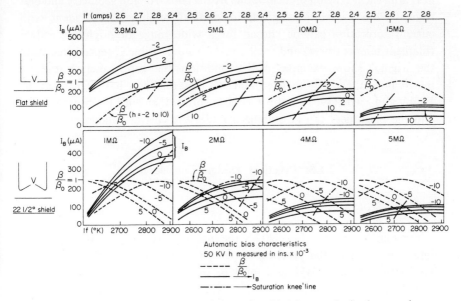

FIG. 9. Variation of beam current and the ratio of brightness obtained to maximum brightness with heating current and cathode temperature for various values of cathode-bias resistance and cathode height and three cathode-shield geometries.

TABLE I

OPTIMUM CONDITIONS FOR AN ELECTRON GUN UNDER SELF-BIAS

Shield	Filament temp. (°K)	Filament life (hr)	β_t theoretical (A/cm² sr)	Bias resistance (MΩ)	Filament height ($h \times 10^{-3}$in.)	Beam current (μA)
Flat	2615	80	$5 \cdot 8 \times 10^4$	$20 \rightarrow 15$	$-2 \rightarrow +10$	$80 \rightarrow 30$
	2680	40	$9 \cdot 6 \times 10^4$	$15 \rightarrow 20$	$0 \rightarrow +10$	$100 \rightarrow 50$
	2760	16	$1 \cdot 8 \times 10^5$	$10 \rightarrow 15$	$2 \rightarrow +10$	$150 \rightarrow 50$
	2840	8	$3 \cdot 1 \times 10^5$	$5 \rightarrow 10$	$5 \rightarrow +10$	$200 \rightarrow 80$
$22\frac{1}{2}°$ Re-entrant	2615	80	$5 \cdot 8 \times 10^4$	~ 10	~ -10	~ 50
	2680	40	$9 \cdot 6 \times 10^4$	$8 \rightarrow 10$	~ -10	~ 60
	2760	16	$1 \cdot 8 \times 10^5$	$6 \rightarrow 8$	~ -10	~ 80
	2840	8	$3 \cdot 1 \times 10^5$	$3 \rightarrow 4$	~ 10	~ 200

measured as a fraction of the theoretical value. The points of current saturation are shown by the chain-lines.

It is seen that the brightness curves have broad maxima where the brightness approaches the theoretical value. By the choice of gun geometry and bias
resistor value, conditions giving theoretical brightness coincident with
current saturation can be chosen for a wide range of brightness values
(filament temperatures) and total current values. Table I, reproduced from
the original, shows examples of the conditions that can be met. A study of
mixed bias, that is, both fixed and self-bias together, has been made by
Dolby and Swift (1961) and it has been shown that both the beam current
and the brightness can be stabilized under the correct conditions.

2.1.5. Practical Design Considerations

A. HIGH VOLTAGE

For guns operating above 20 kV the high-voltage considerations plays an
important part in the practical design. Both the filament and modulator
are at full negative EHT and must be insulated from the anode which is
usually at ground potential. The insulator must be capable of operation
in vacuums and at temperatures up to 100°C. These temperatures at the
insulator and the modulator junction arise from the heating of the modulator
by the hot cathode. These considerations limit the choice of materials to
glass, steatite, and porcelain—of which porcelain is to be preferred because
of the ease with which it can be moulded and ground to the correct shape
and size.

The shape and position of the insulator is governed by two factors:
first, it should not be near the direct path of the beam to avoid its charging
up; and secondly, it should be capable of standing the full EHT without
breaking down or tracking along its surface. The first condition is easily
met by using a shape similar to that shown in Fig. 1. The size of the insulator
needed to prevent breakdown through the bulk is easily determined, but the
shaping of the insulator to avoid tracking across its surface is more difficult.
The breakdown voltage for tracking across the surface depends on factors
such as the state of the surface and the geometry of the insulator metal
junctions (Kofoid, 1960; Gleichaudf, 1951). The insulator shown in Fig. 1
is capable of operating at voltages up to 100 kV; the corrugations on the
outside of the insulator increase the path length across the surface to reduce
the risk of tracking.

The design of the insulator usually allows the high-voltage cable to be sealed in thus avoiding any need for air insulation.

B. ALIGNMENT

For the most efficient operation of the gun, we require that the axis of the beam emerging from the gun be coincident with the axis of the rest of the system. Accurate spigotting of the various parts of the gun will ensure that the geometrical axis of the gun is correct but the beam axis will, in general, not be coincident with the geometrical axis unless the filament is perfectly central in the shield. Small misalignments of the filament can arise from two sources: first, inadequate initial centering of the filament; and secondly, distortion of the filament wire on heating. This distortion can be minimized as mentioned in Section 2.1.3. These small misalignments will result in the emerging beam being tilted with respect to the geometrical axis and will reduce the current passing through any defining aperture below —for example, the condenser aperture in an electron microscope.

There are two possibilities for aligning the beam. The first method is to provide controls for centering the filament in the shield while the gun is operating. This is theoretically the ideal method but the manipulation of the filament while it is at full high voltage presents many practical difficulties. The second method is to provide the gun with a traverse movement with respect to the rest of the column and to align the beam so that its axis passes through the defining aperture below. The beam axis will still be tilted but this will not be serious in any instrument except in the electron microscope wherein it can easily be corrected by other means. This method is practically possible, but there is a simpler method of achieving the same result—namely, to arrange for a section of the anode containing the anode hole to be mova-ble in a plane perpendicular to the gun axis. The anode hole acts as a diver-gent electrostatic lens and hence a lateral displacement of it will bend the beam and enable alignment to be carried out as just described. The focal length of such an aperture lens is independent of its diameter which is there-fore not critical.

REFERENCES

Agar, A. W. (1957). *Brit. J. Appl. Phys.* **8**, 410.
Bloomer, R. N. (1957). *Brit. J. Appl. Phys.* **8**, 83.
Boersch, H. (1954). *Z. Physik.* **139**, 115.
Bradley, D. (1961). *Nature* **189**, 298.
Bruck, H., and Bricka, M. (1948). *Ann. Radioelec. Compagn. Franc. Assoc. T.S.F.* **3**, 339.

Dietrich, W. (1958). *Z. Physik* **152**, 306.

Dolby, R. M., and Swift, D.W. (1961) *Proc. European Regional Conf. Electron Microscopy, Delft*, 1960 Vol. 1, p. 114. Nederlandse Vereniging voor Electronenemicroscopie, Delft, Netherlands.

Duncumb, P., and Melford, D.A. (1960). "X-ray Microscopy and X-ray Microanalysis." Elsevier, Amsterdam.

Gleichauf, P. H. (1951). *J. Appl. Phys.* **22**, 766.

Haine, M. E., and Cosslett, V. E. (1961) "The Electron Microscope." Spon, London.

Haine, M. E., and Einstein, P. A. (1952). *Brit. J. Appl. Phys.* **3**, 40.

Haine, M. E., Einstein, P. A., and Borchards, P. H. (1958). *Brit. J. Appl. Phys.* **9**, 482.

Hartwig, D., and Ulmer, K. (1963). *Z. Physik* **173**, 294.

Hibi, T. (1956a). *Proc. 3rd Intern. Conf. Electron Microscopy, London, 1954* p. 636 Roy. Microscop. Soc., London.

Hibi, T. (1956b). *J. Electronmicroscopy (Tokyo)* **4**, 10.

Hibi, T. (1962). *Proc. 5th Intern. Conf. Electron Microscopy, Philadelphia, 1962*, Vol. I, KK1. Academic Press, New York.

Hibi, T. (1964). *J. Electronmicroscopy (Tokyo)* **13**, 32.

Kofoid, M. J. (1960). *Trans. AIEE* **79**, Pt. III, 999.

Langmuir, D. B. (1937). *Proc. IRE* **25**, 977.

Niemeck, F. W., and Ruppin, D. (1954). *Z. angew. Phys.* **6**, 1.

Reimann, A.L. (1938). *Phil. Mag.* [7], **25**, 834.

Steigerwald, K. H. (1949). *Optik* **5**, 469.

Swift, D. W., and Nixon, W. C. (1962). *Brit. J. Appl. Phys.* **13**, 288.

Thon, F. (1963). *Zurich Cong.* p. 15. Abstract.

CHAPTER 2.2

ELECTROSTATIC LENSES

K.-J. Hanszen and R. Lauer

BRAUNSCHWEIG, GERMANY
PHYSIKALISCH-TECHNISCHE BUNDESANSTALT

2.2.1. General Properties of Image Formation

Any electrostatic field with rotational symmetry has imaging properties. Charged particles with equal energies, starting from points in the object plane, are focused into points in the (Gaussian) image plane if their paths are paraxial (that is, if they always run in a small distance from the axis and always have a small inclination to it). Rays such as this lead to a stigmatic (that is, a correct point-by-point) magnified or diminished image.

Magnification and position of the image depend on the energy of the particles: There is *energy dispersion*; the image formation is disturbed by *chromatic aberration*.

Off-axis rays are not focused into the same point as paraxial rays. The whole set of the intersection points of all adjacent rays starting from the same object point form a *caustic* shell: The image is affected by *geometrical aberrations*.

For *focusing* a beam with rotational symmetry, the area around the axis is utilized. With increasing demands on the imaging quality, the radius of the used lens area must be reduced. *Imaging with high resolution* (for example, in the electron microscope) requires an aperture, so small, that the diffraction at the edge of the diaphragm hole essentially influences the image quality. For *energy analyzing* (velocity spectrometry) one takes advantage of the strong chromatic aberration in the off-axis lens zones. Simultaneously, the particles with equal energy can be focused into a line focus on the caustic shell. Electrostatic lenses may also be used as *filter lenses*; with their aid it is possible to select the high energetical part of a beam.

A. OPTICAL LENS DATA

From the geometry of the lens electrodes, in principle it is possible (a) to determine the potential- and field-distribution in the lens and (b) from these, to determine all possible electron paths. In many cases, however, for the description of the paraxial lens data, it is sufficient to know only the paths of two paraxial rays entering the lens in parallel to the axis from the right hand side and the left hand side, respectively. Then the so-called "cardinal points"—which can be determined mostly by direct experiments —totally describe the image formation. Therefore, we shall concern ourselves foremost with these cardinal points. Without restriction we admit that the particles have different energies in front of and behind the lens. Consequently all equations are valid for immersion lenses too (see Section 2.2.2,B).

1. Cardinal Points

Particularly important for the description of the imaging conditions in the case of field-free object and image spaces are the axis coordinates of the following points:[1]

(a) The z coordinates z_{H*} and $z_{H'*}$ of the intersection points of the entrance and exit asymptotes of rays entering the lens parallel to the axis from the right- and left-hand side, respectively;

[1] In order to give f and f' the same sign, the values for the axis coordinates on the object and image side increase in the opposite direction (see Figs.). Always, when H, F, and f cannot be *strictly* regarded as cardinal elements, they will be marked by asterisks.

(b)   The intersection points z_{F*} and $z_{F'*}$ of the exit asymptotes of the same rays mentioned in (a) with the axis of the lens;

(c)   The coordinate differences $z_{F*} - z_{H*} =_{df} f^*$ and $z_{F'*} - z_{H'*} =_{df} f'^*$.

Applying these expressions to paraxial rays according to Fig. 1, we will omit the asterisks and notate the coordinates of the *principal planes* as z_H and $z_{H'}$; the coordinates of the *focal planes* as z_F and $z_{F'}$; and the corresponding focal lengths of the object and image side as f and f', respectively. We call the whole of these data as *(virtual) cardinal points* (Glaser, 1952, 1956).

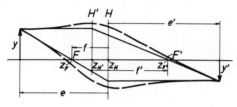

FIG. 1. Construction of the virtual cardinal points by the aid of the entrance and exit asymptotes. z_F and $z_{F'}$ are the axis coordinates of the virtual focal points F and F' on the object and image side, z_H and $z_{H'}$ the axis coordinates of the corresponding principal planes.

With the aid of these cardinal points, it is possible to determine *in field-free object and image spaces* the corresponding image distance e' and the lateral magnification M' (negative sign for inverted imaging!)

$$M' =_{df} \frac{y'}{y} \qquad (y = \text{object size}; \qquad y' = \text{image size}) \qquad (1)$$

of the Gaussian image from the equation

$$\frac{f}{e - f} = \frac{e' - f'}{f'} = -\frac{y'}{y} \quad \text{for any object distance } e. \qquad (2)$$

The *power K* of a lens is defined by

$$K =_{df} \pm 1/(ff')^{1/2} \text{ (positive sign for } f, f' > 0; \text{ negative sign for } f, f' < 0). \qquad (3)$$

Between the focal lengths f; f' and the kinetic energies W; W' of the particles in front of and behind the lens, the fundamental equation (Glaser, 1952, see p. 150)

$$\frac{f'}{f} = \left(\frac{W'(1 + W'/2\,W_0)}{W\,(1 + W\,/2\,W_0)}\right)^{1/2} =_{df} \left(\frac{W_r'}{W_r}\right)^{1/2} \tag{4}$$

with W; $W' =$ kinetic energies, see (44); and

$$W_0 = \text{rest energy } m_0 c_0^2$$

is valid. From this, we obtain for the nonrelativistic case (W; $W' \ll W_0$):

$$\frac{f'}{f} = \left(\frac{W'}{W}\right)^{1/2} \tag{4a}$$

Hence the focal lengths f, f' differ only for immersion lenses; they are equal for single lenses. For these the following relation holds:

$$K = 1/f = 1/f' \tag{5}$$

Usually the object and the image screen are located outside the field, because, otherwise, they would raise field deformations. Under these circumstances, (2) is always applicable. For image formation by means of several stages, however, the intermediate images may be located in the field of a subsequent lens.

In such cases, we can apply other cardinal points: the so-called *osculatory cardinal points* (Glaser, 1952, 1956); the scope of these points is limited however to a small interval of object displacements. Numerical information about the osculatory cardinal points of electrostatic lenses are not published.

2. Operating Ranges of a Lens

Except gauze lenses (Sections 2.2.2,A,3; 2.2.2,B,1), which raise particular problems, weak lenses always have the properties of converging lenses, but never those of diverging lenses. The principal planes coincide ("thin lenses"). With increasing excitation the principal planes withdraw from each other ("thick lenses"). Their position is always inverted, see Fig. 1. In strong lenses the point of intersection of parallel incident rays can move into the lens field, so as to produce a second or even a third crossover by subsequent focusing (see Fig. 2). According to the number of existing crossovers the operating ranges will be of first, second, or third order, respectively. Negative focal lengths occur in the operating range of second-order.

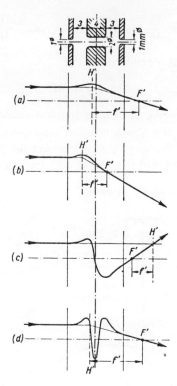

FIG. 2. Particle trajectory and positions of the image side focal and principal points of an electrostatic single lens according to Heise and Rang (1949). The electrical excitation increases from Fig. 2a to Fig. 2d (a, b first operating range, c second range, d third range). The whole focal and principal point characteristic of this lens is given in Fig. 9, the power characteristic in Fig. 20.

3. *Paraxial Chromatic Aberration*

Because of the chromatic lens aberration (see the beginning of this chapter), an image aberration disk of radius δ_{ch} is formed in the Gaussian image plane. Introducing the relative energy width $\Delta W/W$ of the particles and the angle of incidence α, we can establish the following relation:

$$\delta_{ch} = \alpha \, c_{ch} \frac{\Delta W}{W} \qquad (6)$$

c_{ch} is called the *chromatic aberration constant related to the image*. The radius of the image aberration disk corresponds to the interval ε_{ch} in the object plane

$$\varepsilon_{ch} = \frac{\delta_{ch}}{M'} = a\,\frac{c_{ch}}{M'}\,\frac{\Delta W}{W} = aC_{ch}\,\frac{\Delta W}{W} \qquad (7)$$

where M' is the lateral magnification and

$$C_{ch} =_{df} c_{ch}/M' \qquad (8)$$

C_{ch} is called the *chromatic aberration constant related to the object*. Both constants depend on the object position (and thus on the magnification too). Particles with high energy always intersect the axis at larger distances behind the lens (see Fig. 3b) than those with low energy.

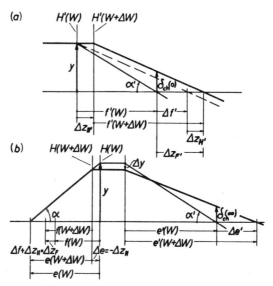

FIG. 3. Chromatic aberration disk δ_{ch} and chromatic focal point displacements Δz_F, $\Delta z_{F'}$ (a) for very strong diminution; and (b) for very strong magnification.

The chromatic aberration in the paraxial area may also be described by the energy dependence of the coordinates of the focal and principal points. With very strong diminution ($M' \to 0$) and with very strong magnification ($M' \to \infty$), very simple mathematical conditions emerge.

(a) $M' \to 0$ The incident beam can be regarded as a beam of parallel rays. As shown in Fig. 3a, δ_{ch} is only given by the displacement of the cardinal elements $\Delta z_{H'}$ and $\Delta f'$ of the image space. We find

$$\delta_{ch}(0) = \frac{y}{f'(W + \Delta W)} \cdot \Delta z_{F'} = \frac{y}{f'(W + \Delta W)}\,(\Delta f' + \Delta z_{H'}) \qquad (9)$$

with $f'(W + \Delta W) = f'(W) + \Delta f'$.

For $\Delta f' \ll f'$ and for small angles of emersion α', Eq. (9) becomes

$$\delta_{ch}(0) = \frac{y}{f'(W)} \cdot \Delta z_{F'} = \alpha' \, \Delta z_{F'}; \qquad \varepsilon_{ch}(0) = \frac{\alpha'}{M'} \Delta z_{F'} \qquad (10)$$

Considering the Helmholtz-Lagrange formula

$$M' = -\frac{\alpha}{\alpha'} \left(\frac{W_r}{W_r'} \right)^{1/2} \qquad (11)$$

and comparing (10) with (7), we shall get the required relation between the chromatic aberration constant $C_{ch}^{(0)}$ and the chromatic aberration $\Delta z_{F'}$ the rear focal point

$$C_{ch}(0) = -\frac{\Delta z_{F'}(W_r/W_r')^{1/2}}{M'^2 \, \Delta W/W} \qquad (12)$$

(b) $M' \to \infty$ Since the object distance e does not differ from f practically, the emerging beam may be interpreted as a beam of parallel rays. Now the image distance e' is so large that we can equate $e' - f' \approx e'$ and the displacement of the coordinates of the principal and focal points on the image side will be of no importance. As shown in Fig. 3b, for the radius of the chromatic-aberration disk, we may write

$$\delta_{ch}(\infty) = \frac{y + \Delta y}{e' + \Delta e'} \Delta e' \approx \frac{y}{e'} \Delta e' = \alpha' \, \Delta e' \qquad (13)$$

if $\Delta y \ll y$, $\Delta e' \ll e'$. e' is regarded here as a function of $e, f,$ and f'. In this manner we obtain with (1) and (2) the total differential of e':

$$\Delta e' = M' \left[\frac{f'}{e-f} (\Delta e - \Delta f) - \Delta f' \right] = -M' \left[\frac{f'}{e-f} \Delta z_F + \Delta f' \right]$$

$$\approx -M' \Delta z_F \, \alpha/\alpha' = -M' \Delta z_F \, e'/f \qquad (14)$$

considering (1) - (4); (11) and

$$\Delta e = -\Delta z_H; \quad \Delta f - \Delta e = \Delta z_F; \quad \Delta f' = \Delta f (W_r'/W_r)^{1/2} \qquad (15)$$

Substituting (14) into (13) and converting it to $\varepsilon_{ch}(\infty)$, we get

$$\delta_{ch}(\infty) = -(y/f)M' \Delta z_F = -\alpha M' \Delta z_F; \qquad \varepsilon_{ch}(\infty) = -\alpha \Delta z_F \qquad (16)$$

Comparing this result with (7), we find the required relation between the chromatic-aberration constant $C_{ch}(\infty)$ and the chromatic aberration Δz_F of the focal point on the object side:

$$C_{ch}(\infty) = - \frac{\Delta z_F}{\Delta W / W} \qquad (17)$$

c. *Inversed Ray Tracing.* In opposition to our former procedure the rays now shall enter from the right-hand side. In this case we must convert all the symbols with primes into the symbols without primes. The symbols of the corresponding chromatic-aberration disks will be indicated by the letter u. With

$$M \underset{df}{=} \frac{1}{M'} = - \frac{a'}{a} \left(\frac{W_r'}{W_r} \right)^{1/2} \qquad (18)$$

we obtain

$$\delta_{ch}^u(0) = \quad a\, \Delta z_F \quad ; \quad \varepsilon_{ch}^u(0) = \quad \frac{a}{M}\, \Delta z_F \qquad (19)$$

$$\delta_{ch}^u(\infty) = - a' M\, \Delta z_{F'} \quad ; \quad \varepsilon_{ch}^u(\infty) = - a'\, \Delta z_{F'} \qquad (20)$$

Especially the combination of (19), (16), (17) leads to the result

$$\varepsilon_{ch}(\infty) = - \delta_{ch}^u(0) = C_{ch}(\infty)\, a \cdot \Delta W / W \qquad (21)$$

Hence, the chromatic aberration constant for *strong magnification* can be determined by using the inversed ray tracing with strong diminution. This method is analogous to the method for determining the spherical aberration (see Section 2.2.1,A,5c).

4. *General Information about Off-Axis Data*

In discussing the data about the off-axis lens zones, again we confine ourselves to field-free object and image spaces. It is sufficient to know the entrance asymptotes and the corresponding exit asymptotes of each ray. The entrance asymptote is characterized (see Fig. 4a) by its *intersection point p with the axis* and by its *slope* $\tan \alpha = y/(p - q)$; the exit asymptote (see Fig. 4a) is characterized by its *intersection point p' with the axis* and its *slope* $\tan \alpha' = y'/(p' - q')$. The asymptotes of rays incident parallel to the axis are characterized by their *heights of incidence y* instead of by their slopes $\tan \alpha$ (Hanszen, 1958b).

Only in the paraxial area, the data of all rays in a beam emitted by a point source in any finite distance from the lens can be evaluated by the data of an incident and an emerging ray both parallel to the axis (see Fig. 1). As to the off-axis lens zones this is not possible. In order to know the lens properties completely, it is indispensable to measure or calculate the entrance and exit asymptotes for all possible rays. Moreover, on account of the chromatic

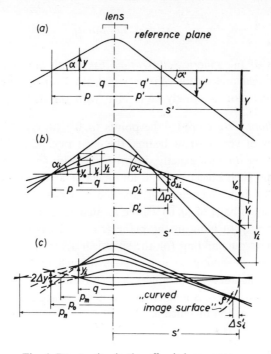

Fig. 4. Ray tracing in the off-axis lens zones.

errors, these measurements or calculations have to be done for particles of different energies. Mostly, however, it is sufficient to determine only the data of a beam that emerges from a *fixed* point source (necessarily not located on the axis) and to evaluate from these the interesting *geometrical aberrations*. The data derived in this manner are not confined to an approximate theory of aberrations.

a. *Spherical Aberration.* The intersection point with the axis p_i' on the image side and the angle of emersion α_i', belonging to the rays originating from an axis point with the fixed coordinate p, depend only on the specific angle of incidence α_i (and for parallel rays on the heights of incidence y_i), see Fig. 4b. If we take the axis point p to be an object point, then p_i' is the corresponding image point, and, especially, for $a \to 0$, $p_i' \to p_0'$ it is the corresponding (paraxial) Gaussian image point. Then

$$\Delta p_i' =_{df} p_i' - p_0' \tag{22}$$

is the *spherical longitudinal aberration* (see Fig. 25) and

$$\delta_{s_i} =_{df} \Delta p_i' \tan \alpha_i' \tag{23}$$

is the *radius of the spherical-aberration disk*. Both are defining quantities for the spherical aberration. As we will learn from the next section, the spherical aberration is closely connected with the distortion.

b. *Distortion.* We consider the points $(q, 0)$; (q, y_0); . . . ; (q, y_i) to be point sources of very narrow beams, the chief rays of which are drawn in Fig. 4b; two beams are exactly drawn up in Fig. 4c. Beams such as this originate from each object point, if an object at a distance q is illuminated by a source of finite diameter $2 \Delta y$ located at a distance p.

The Gaussian image of $(q, 0)$ will be located at $(s', 0)$. Then, the image of the plane $(z = q, y)$ lying in the plane $(z = s', Y)$ is distorted. The same error, which is outstanding for the spherical aberration if an object point at the distance p—acting as a pupil—is imaged in the Gaussian image plane at the distance p', produces a distortion in another plane at the distance s', if the object is located at another distance q and if the entrance pupil is located at the distance p.

c. *Field Curvature, Coma, Off-Axis Astigmatism.* Figure 4c shows how to derive the field curvature (curvature of the "image surface," marked by $\Delta s_i'$) and the *coma* (marked by ϱ) from the hitherto drawn rays.

Rays originating from the same object point, which do not proceed in the drawing plane, are not imaged in the same surface. The deviation of this surface from the one drawn describes the *off-axis astigmatism*.

d. *Summary.* Except for the off-axis astigmatism, which is not only determined by the rays proceeding in the plane of incidence, we can characterize all geometrical aberrations according to Fig. 4a by the intersection points and the slopes of the entrance and exit asymptotes of all rays in the system. Hence it is, for example, sufficient to know the cardinal points and the spherical aberrations in the form of $\Delta p'(a)$ and $\delta_s(a)$ for the rays with all possible axis intersections p_m, p_n, p_0, . . . in order to determine distortion, field curvature, and coma.

5. *Third-Order Aberrations*

In the lens zones immediately annexed to the paraxial area (in the so-called Seidel area), we are allowed to make further statements about the geometrical aberrations, if we expand the deviation of the off-axis image points from the paraxial image points into a series and consider only the first term. In this article we will do this only for the spherical aberration and the distortion. But first, we will point out the relation between spherical aberration and distortion that holds in this area.

a. *Relations between the Spherical-Aberration Disk for Very High Diminution and the Distortion of a Shadow Image at a Large Distance.* According to Fig. 4b, we project the shadow image of an object located at the distance q, by a point source at the distance $p \to \infty$. The image screen will arbitrarily be placed at a large distance s'. According to the particulars of Fig. 5, the exit

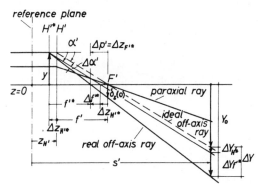

FIG. 5. Spherical aberration disk $\delta_s(0)$ for strong diminution and the distortion $\Delta Y/Y_0$.

asymptote of a ray, entering parallely with a great height y of incidence into an aberration-free lens, intersects the axis at the focal point F' of the image side with the slope $\tan \alpha' = y/f'$. The ray emerging from an imperfect lens[2] differs from this ray by a longitudinal displacement $\Delta z_{H'*}$ and by a variation $\Delta \alpha'$ of the angle of emersion α'. This variation leads to a displacement $\Delta f'^*$ of the intersection point with the axis. Thus the total displacement of this point, that is, the longitudinal spherical aberration, amounts to

$$\Delta p' = \Delta z_{H'*} + \Delta f'^* = \Delta z_{F'*} . \tag{24}$$

[2] As to the following notations, see Section 2.2.1,A,1. The asterisked quantities are derived in the same manner as the cardinal points from the entrance and exit asymptotes of an parallel incident beam, but they do not establish an imaging in the sense of Eq. (2).

The radius of the spherical aberration disk has the value

$$\delta_s(0) = \Delta z_{H'*} \tan \alpha' + \Delta f'^* \tan(\alpha' + \Delta \alpha') \approx$$
$$(\Delta z_{H'*} + \Delta f'^*) \tan \alpha' = \Delta z_{F'*} \cdot \tan \alpha'. \tag{25}$$

The second line of equation (25) holds as long as $\Delta \alpha' \ll \alpha'$. Similarly, we can write for the distortion with reference to Fig. 5

$$\frac{\Delta Y}{Y_0} = \frac{\Delta Y_{H'*}}{Y_0} + \frac{\Delta Y_{f'*}}{Y_0} \approx \frac{\Delta z_{H'*} + (s' - z_{H'}) \Delta f'^*/f'}{Y_0} \cdot \tan \alpha', \tag{26}$$

whereas $Y_0 = -(s' - z_{H'} - f') \tan \alpha'$. If $(s' - z_{H'}) \gg f'$, the first term of the numerator of Eq. (26) can be neglected with respect to the second one. Then, we obtain for the distortion

$$\Delta Y/Y_0 \approx - \Delta f'^*/f'. \tag{27}$$

The image-aberration disk is characterized by the *total* longitudinal aberration $\Delta z_{F'*}$ according to (25), whereas, the distortion is marked mainly by a part of the longitudinal aberration, namely, by $\Delta f'^* = \Delta z_{F'*} - \Delta z_{H'*}$. This is to say, the size of the spherical-aberration disk is essentially given by the displacement of the axis intersection $\Delta z_{F'*}$, whereas the distortion is chiefly expressed by the variation $\Delta \alpha'$ of the ray's slope and therefore by $\Delta f'^*$. From this we recognize the great difficulty in the *exact* determination of the spherical aberration from the distortion. Moreover, it is difficult to fix $z_{F'*}$ accurately. The published values, however, are mostly due to such measurements.

b. *The Spherical-Aberration Disk for Very Strong Magnification.* For $M' \to \infty$ the intersection of an incident paraxial ray with the axis (signed by terms with the subscript 0) coincides with the first focal point. The corresponding ray of emersion is nearly parallel to the axis. With the analogous approximations in reference to the image side, as made in 2.2.1,A,3b, we can write for the radius of the image-aberration disk (see Fig. 6)

$$\frac{\delta_s(\infty)}{\Delta e'} = \frac{y_1}{e'(\alpha_0) + \Delta e'} \approx \frac{y_1}{e'(\alpha_0)} \tag{28}$$

A ray emerging from a large distance y_1 parallel to the axis (represented in Fig. 6 with sufficient approximation by the line —·—) may intersect the axis on the object side in F^*. We assume its distance from the focal point to be Δz_{F*}. In spite of the impossibility in describing the imaging properties of the off-axial lens zones by the aid of the virtual cardinal points, we assume that between the rays $\overgroup{FH^*ED_1}$ and $\overgroup{F^*H^*ED_0}$ there is an, at least, osculatory dependence in accordance with the lens equation. Then, in the

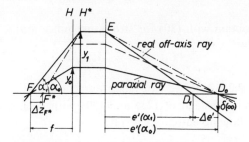

FIG. 6. Spherical aberration disk $\delta_s(\infty)$ for strong magnification. (In order to get a clear drawing the object distance e is equated with the focal length f.)

same procedure and with the same restrictions as in 2.2.1,A,3b, we can write analogously to (14)

$$\Delta e' = - M' \Delta z_{F^*} e'/f: \tag{29}$$

Applying (28), we obtain in this manner for the radius of the aberration disk

$$\delta_s(\infty) = - y_1 M' \Delta z_{F^*}/f = - \alpha M' \Delta z_{F^*} \tag{30}$$

c. *The Constants of Spherical Aberration.* For the radius of the spherical-aberration disk, the following expansion is valid:

$$\delta_s = c_s \alpha^3 + \cdots \tag{31}$$

where c_s is the *spherical aberration constant related to the image.* The radius of the image-aberration disk corresponds to the interval on the object side:

$$\varepsilon_s = \frac{\delta_s}{M'} = \frac{c_s \alpha^3}{M'} + \cdots \qquad =_{df} C_s \alpha^3 + \cdots \tag{32}$$

Here

$$C_s = c_s/M' \tag{33}$$

is the *spherical-aberration constant, related to the object.*

The values of these two aberration constants depend on the object position and thus on the magnification. For *strong diminution* we obtain with (11), (25), (31), and for very small α, the following connection between the aberration constant related to the object $C_s(0)$ and the longitudinal aberration $\Delta z_{F'*}$ the following relation:

$$C_s(0) = - \frac{\Delta z_{F'*}}{M'^2 a^2} \cdot \left(\frac{W_r}{W_r'}\right)^{1/2} \tag{34}$$

Similarly, we obtain with (30) for strong magnification:

$$C_s(\infty) = - \Delta z_{F*}/a^2 \tag{35}$$

As in Section 2.2.1,A,3c we can relate our considerations to the inversed ray tracing again. Instead of (25) and (30) with (32), and (18), for very small a, the following equations result:

$$\delta_s^u(0) = a \, \Delta z_{F*} \quad ; \quad \varepsilon_s^u(0) = \frac{a}{M} \Delta z_{F*} \tag{36}$$

$$\delta_s^u(\infty) = - a'M \, \Delta z_{F'*} \quad ; \quad \varepsilon_s^u(\infty) = - a' \, \Delta z_{F'*} \tag{37}$$

$$C_s^u(\infty) = - \Delta z_{F'*}/a'^2 \tag{38}$$

Of special interest is the comparison between (30) and (36). Considering (32), we have for very small a

$$\varepsilon_s(\infty) = - \delta_s^u(0) = - a \, \Delta z_{F*} = C_s(\infty)a^3 \tag{39}$$

Hence it is possible to determine the "spherical-aberration constant related to the object $C_s(\infty)$ for infinite magnification and object position on the unprimed lens side" from the displacement of the axis intersection Δz_{F*} belonging to this side. There are two different ways to determine Δz_{F*}: either from $\delta_s(\infty) = \varepsilon_s(\infty)M'$ in the direct ray tracing with very strong magnification, or from $\delta_s^u(0)$ in the inversed ray tracing (that is, incidence on the primed side) with very strong diminution.

On the same way we can derive from (25) and (37) two methods for the determination of the spherical-aberration constant $C_s^u(\infty)$ belonging to the other lens side. For these we obtain

$$\varepsilon_s^u(\infty) = - \delta_s(0) = - a'\Delta z_{F'*} = C_s^u(\infty)a'^3 \tag{40}$$

The given relations that do not depend on alterations of the particles' energies, are *valid for immersion lenses* too.

Trajectories with greater angle of incidence intersect the axis always in shorter distances behind the lens than trajectories with smaller angle of incidence (Exception: gauze lenses, see 2.2.2,A,3).

d. *The Constant for Distortion.* For the distortion we can give the following expansion:

$$\frac{\Delta Y}{Y_0} = c_d \, a'^2 + \cdots \tag{41}$$

From this, in the special case of (27) we have for the distortion constant c_d

$$c_d = -\frac{\Delta f'^*}{f' a'^2}. \tag{42}$$

Measured values for this constant, will be given in 2.2.2,A,2d.

6. *Experimental Methods*

Fortunately, we are able to rely on extensive direct experimental material for the data of electrostatic lenses. We will consult these measurements as much as possible and comment on the performed comparison between theory and experiment in suitable passages herein. Next, we will give a short outline of the experimental measuring methods in use. More details about measuring arrangements may be drawn from the original literature.

a. *General Methods.* As mentioned in 2.2.1,A,4, the most general method of measuring the lens data is based on determining all correlated couples of entrance and exit asymptotes in a beam. In the further description we use the notations of Fig. 4a,b. The source of the beam is placed at the distance p in front of the lens. (We are mostly interested in $p \rightarrow \infty$; see 2.2.1,A,5a,c). The most exact method in determining the functions $\tan a'$ ($\tan a$) and $p'(\tan a)$, which fix the required asymptotes, is demonstrated in Fig. 7. In the planes with the axis coordinates q and q' we successively displace a diaphragm with a fine hole along the coordinates y and y', respectively, and register each shadow image of the hole on a photographic plate, the translation of which in x direction is coupled linearly with the diaphragm movement (see Hanszen, 1964a;c). The curves, plotted in this way, give information about $Y(y)$ and $Y(y')$, their mutual correlation gives the wanted connections directly (see Figs. 23–26). Measurements such as these enable us to evaluate, particularly, the paraxial and the Seidel data.

b. *Determination of the Cardinal Points.* The former method is also suitable for determining the positions of the principal and focal points, because the coordinate $z_{F'}$ of the focus on the image side can be found from an incident parallel beam ($p \rightarrow \infty$) by $z_{F'} = p'(y \rightarrow 0)$, and similarly the corresponding focal length f', by

$$f' = \lim_{y \to 0} \frac{y}{\tan \alpha'} \; ; \quad \text{with} \quad \alpha' = \alpha'(y) \tag{43}$$

In order to determine the positions of the principal and focal points on the object side, we have to use the inversed ray tracing (in the case of symmetrical single lenses, we need of course with $f = f'$ and $z_H = z_{H'}$ only one measuring procedure).

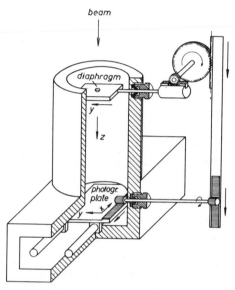

Fig. 7. Measurement of the rays' exit asymptotes by coupling the movement of a fine hole along the y coordinate with the translation in x direction of a photographic plate, which registrates the shadow image of the hole (Hanszen, 1964a).

If our intention is to determine only the paraxial lens data, it is sufficient for us to use a more simple arrangement in which the shadow magnifications of two small objects (such as thin wires) are measured; the first of them is placed before the lens and the second behind it in the distances q and q', respectively (Klemperer and Wright, 1939; Spangenberg and Field, 1942; Heise and Rang, 1949; Liebmann, 1949; Everitt and Hanszen, 1956; Septier, 1960; Fink and Kessler, 1963; Hamisch and Oldenburg, 1964).

For asymmetrical lenses, *two measuring procedures* are required again. In this case there is no need to place the source at a *large* distance from the lens. As shown by Everitt and Hanszen (1956), simply, the lens must be turned around the intersection point of the reference plane and the optical axis between the two measurements. In doing this, the electrodes of immersion lenses must keep their original potentials. Very simple equations result if $q = q'$ and $p = s'$ are chosen.

c. *Determination of Distortion and Spherical Aberration.* In order to determine the distortion by the above shadow method, we have to introduce extended objects into a practically parallel incident beam in front of the lens [with advantages: nets (Heise, 1949; Jacob and Schah, 1953); rows of teeth (Hanszen, 1958b); rows of holes, movable edge (Liebmann, 1949)]. The shadow image of these objects at very large distances s' behind the lens delivers the information (41) about *the distortion.* With the general method 2.2.1,A,6a, the relative distortion for large s' can be read off immediately from the deviation of the registrated curves from a straight line.

By employing the exact formula (24) to (26), (39), (40), the data for the *spherical-aberration disk* (and thus for the longitudinal spherical aberration) can be determined in the same way from the distortion (considerations concerning the exactness of this method are pointed out in Section 2.2.1,A,5a). Particularly, the spherical aberration constant for strong magnification $C_s(\infty)$ can be determined by the inversed ray tracing. But the accuracy of these measured values is not high at all. It must be emphasized that this method allows us to measure $C_s(\infty)$ for a (virtual) object position in the lens field too. But this case is hardly of any practical advantage since for objects in the lens field the osculatory cardinal elements are decisive instead of the virtual ones.

In the *direct methods* of determining the spherical-aberration constant for strong magnification, the exit asymptote is measured in direct dependence on the angle of incidence using the normal ray tracing. According to Seeliger (1948), Shipley (1952), and Septier (1960), a fine hole, placed at the intersection point of the axis with the first focal plane, is irradiated under various inclinations α. Thus, the spherical aberration constant $C_s(\infty)$ can be calculated from the positions of the corresponding exit asymptotes. This method is only valid if the focal points are located out of the lens field.

Mahl and Recknagel (1944) employed as "electron pencils" the diffraction intensities emitted into discrete glancing angles by a microcrystalline object, positioned on the axis. With this method, the spherical aberration can be determined from the distortion of the Debye-Scherrer-diagram in the focal plane of the image side.

B. SIMILARITY LAWS

For nonrelativistic velocities of the particles, the properties of an electrostatic lens are only determined by the mutual *ratio R* of the electrode-potentials referred to the particles' rest potential U_0, but not by the absolute values of the electrode potentials. Under these circumstances, particularly the particles' trajectories are independent of the ratio: particle charge to particle mass. Hence, we shall introduce in the following passages a suitable *voltage ratio R* as an electric parameter. We shall call the dependences of the optical quantities on R, the *"characteristics"*; for example, $K(R) = 1/[ff'(R)]^{1/2}$ will be called the lens power characteristic, and so on. For relativistic velocities, however, the trajectories of the particles depend on the ratio of kinetic energy to rest energy. Then, the trajectories are not only functions of the voltage ratio, but also of the absolute values of the individual voltages. Data for a special lens are given in 2.2.2,A,2b.

In lenses with *similar geometry*, operating at the same ratio R, all particles describe similar paths. In this case, we are able to give the geometrical lens data, in a reduced scale; that is, the required data will be divided by a significant length (for example, electrode distance, bore diameter or focallength). Since in the relativistic range, only particles with the same ratio of kinetic energy to rest energy describe similar paths, the reduced geometrical lens data are functions of this ratio.

C. CONFRONTATION OF ELECTROSTATIC AND MAGNETIC LENSES

1. *Consequences Taken from the Ray Equations*

In electrostatic lenses the trajectories of particles with nonrelativistic velocities and with the same ratio of kinetic energy to electric charge, that is, particles which have passed the same accelerating voltage, are independent of the particles' mass. In magnetic fields, however, under the same conditions the particle trajectories depend on the particles' mass: Heavy particles in particular are less focused than light ones. On that account, electrostatic lenses are favored to focus *heavy particles* (*ions*).

2. *Consequences Taken from the Equation of Motion*

Whereas the kinetic energy

$$W = e(U_0 - U) \tag{44}$$

(where e is the charge of the particles, U_0 the equilibrium rest potential of the particles, and U the electrostatic potential at the considered point) of a particle does not alter along the whole path in the field of a magnetic lens, in the electrostatic lens field considerable energy alterations are possible. By an opposing field of sufficient strength the particles may even be reflected instead of being transmitted. In a beam with particles of equal energy this is done in the off-axis lens zones by a lower opposing electrode potential than in the paraxial zone. This effect is relevant for the action of the grid in electron guns as "iris diaphragm." In beams with particles of different energies, the particles with the lower energy likewise are reflected by a lower opposing potential. In this case, the lens has filtering properties (see Section 2.2.2,A,5d).

If the particles endure energy alterations only in the lens field (then their energy is the same on both lens sides, and object and image space have the same electrostatic potential), we call these lenses "*single-lenses.*" For these we must substitute $W = W'$ in all equations. Especially the two focal lengths f and f' are equal, see Eq. (4). If the electrostatic potentials in the object and image space are different, we have *accelerating* or *decelerating lenses.* For both, we use the notion "*immersion lenses.*" In the following section the data of single and immersion lenses are referred to separately.

3. Consequences of the Object Position

In order to avoid deformations of the imaging field, the objects are not allowed, particularly for strong magnifications, to immerse into the imaging field of electrostatic lenses. With the exception of ferromagnetic materials, however, the objects are generally allowed to immerse into the imaging field of magnetic lenses. This leads to the following consequences:

a. *Applicability of the Virtual Cardinal Points to Electrostatic and Magnetic Lenses.* The virtual cardinal points, the applicability of which is confined to field-free object and image spaces, are of main importance for electrostatic lenses, but of secondary importance for magnetic lenses. Since the virtual cardinal points can be easily derived from the course of the asymptotes by experiments, the experimental methods for determining the lens data—in opposition to the analytical methods—are more relevant for electrostatic than for magnetic lenses.

b. *Applicability of Electrostatic and Magnetic Lenses as Microscope Objectives.* As pointed out in 2.2.2,A,5a, on account of the demand for field-free object position and the requirement of high-voltage technology (2.2.3,A,B), only a minimum focal length about 6 mm of electrostatic objective lenses can be reached, whereas magnetic lenses with focal lengths 2 mm or less are commonly in use. Moreover, it is not possible to reduce the spherical aberration of the objective, which is decisive for the resolving power of transmission microscopes, to its minimum value in electrostatic objectives. Therefore these lenses are not adequate for high-resolution microscopes (particulars are reported in 2.2.2,A,2c and 2.2.2,A,5a).

4. Consequences Taken from Designing

From the examples of Section 2.2.3,C, it follows, that electrostatic lenses mostly can be more simply manufactured than magnetic lenses. On that account, for more simple experiments in laboratories the electrostatic lenses are often favored above magnetic lenses. But their limited high-tension rigidity is a great disadvantage. It is unpromising to operate high-voltage lenses with short focal lengths by much more than 60 kV.

2.2.2. Data of Particular Lenses

In most cases electrostatic lenses are realized by a sequence of coaxially arranged tubes or diaphragms on different electrostatic potentials. Complicated geometries only exceptionally occur. If one or several lens electrodes are equipped with gauzes, which must be crossed by the beam, we call these lenses *gauze-lenses*. We shall deal with them in connection with the belonging diaphragm and tube lenses.

A. SINGLE LENSES

Single lenses may be realized by three-diaphragm or three-tube systems, whose external electrodes have the same potential U_a as the field-free external spaces. We call the potential of the intermediate electrode U_i.

According to the performance of Section 2.2.1,B, the electrical excitation will be designated by the characteristic voltage ratio

$$R_{sl} = \frac{U_i - U_0}{U_a - U_0} \tag{45}$$

Mostly, the potential of the intermediate electrode has such a sign that the particles are decelerated in the field between entrance and intermediate electrode. Then the operating voltage ratio is $R_{sl} < 1$. In this case, the potential of the intermediate electrode can be drawn from the high-tension generator, which supplies the accelerating voltage. Owing to the low energy of the particles near the intermediate electrode, lenses operating at a voltage ratio smaller than 1 by the amount x have considerably stronger power than lenses operating at a voltage ratio greater than 1 by the amount x.

Moreover, for the realization of voltage ratios $R_{sl} > 1$, an additional high-voltage supply is needed.

1. An Illustrative Model of the Single Lens

With voltage-ratios $R_{sl} < 1$, the potential distribution around the intermediate electrode of an electrostatic single lens acts as converging lens; the potential distributions around the external electrodes act as diverging lenses. With $R_{sl} > 1$, the conditions are just opposite. For details see 2.2.2,B,2. But always the converging parts of the potential field are predominant; thus the whole lens always acts as a converging lens. For qualitative considerations the diverging part can be thoroughly neglected.

Real diverging lenses can be realized only by suppressing the converging part of the potential distribution. This can be done for $R_{sl} < 1$ by replacing the intermediate aperture electrode by a conducting, but transparent foil (for practical purposes represented by fine gauzes). Since all parts of the potential distribution for $R_{sl} > 1$ act inversed, the lens with the above gauze electrode operates at this voltage ratio as a converging lens. All other lenses in the first operating range are always converging lenses (see Section 2.2.1, A,2), irrespective of $R_{sl} \lessgtr 1$ (that is, $U_i \lessgtr U_a$). Both focal lengths of these lenses are equal. In the operating ranges of higher order the lens power may

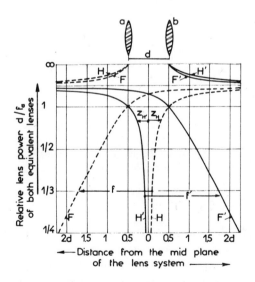

FIG. 8. Entire course of the focal and principal-point coordinates of a lens system consisting of two equal weak converging lenses a and b with the fixed mutual distance d in dependence of the lens power $1/f_a = 1/f_b$ belonging to each lens (Hanszen, 1958a).

increase in such a degree that several successive focusings occur (see Fig. 2). An illustrative model for the influence of the increasing electrical excitation on the cardinal points in the first and second operating range represents a system of two thin "equivalent lenses" a and b, placed at the fixed mutual distance d, whose lens powers $1/f_a$ and $1/f_b$ increase in a constant ratio (Hanszen, 1958a; Felici, 1959; there are given further mathematical evaluations). Symmetrical single lenses, that is, lenses whose intermediate electrode lies in a symmetry plane, may be represented by two "equivalent lenses" with equal lens powers $1/f_a = 1/f_b$ thusly (see Fig. 8; compare a real characteristic of cardinal points, for example, Fig. 9).

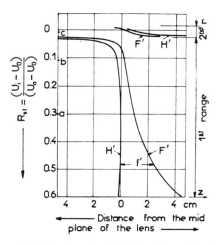

FIG. 9. Entire course of the focal and principal point coordinates (focal and principal point characteristics) of the lens shown in Fig. 2 in dependence of the voltage ratio R_{sl} for the first and second operating range, according to Heise and Rang (1949) (U_i = potential of the intermediate electrode, U_a = potential of the outer electrodes, U_0 = rest potential of the particles, z = axis coordinate). The voltage ratios which refer to the trajectories a and b of Fig. 2 are marked on the ordinate scale. The cardinal points on the object side and the cardinal point on the image side lie symmetrically to the midplane of the lens. The power characteristic of this lens is represented in Fig. 20.

2. Three-Diaphragm Single Lenses

As to three-diaphragm single lenses, extensive theoretical calculations have been done (Regenstreif, 1951; Glaser and Schiske, 1954, 1955; Kanaya et al., 1966). The ray tracing in these lenses also has been investigated with the aid of analogical systems (electrolytic trough, resistor network, and so on). But there exist such a lot of direct measurements that we exclusively can relate on these in the following sections. Comparison with the other methods may be drawn from Lippert and Pohlit (1954), Archard (1956), and Vine (1960).

a. *Characteristics of Lens Power. Focal Points and Principal Points.* The following geometrical parameters influence the cardinal points decisively:

(a)   The distances a and a' between the intermediate and the external electrodes;

(b)   The thickness d of the intermediate electrode;

(c)   The diameter b (and possibly the geometrical shape) of the aperture in the intermediate electrode.

The dependence of the cardinal points on the voltage $R_{sl} < 1$ is demonstrated for various lenses in Figs. 9–16. Special interest requires the general

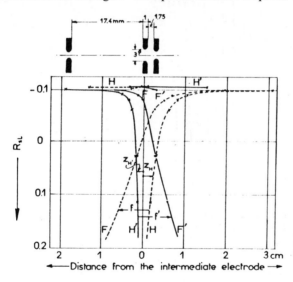

FIG. 10. Focal and principal point characteristics of the drawn asymmetrical lens (Everitt and Hanszen, 1956).

behavior of the lens power characteristics $1/f(R_{sl})$. With increasing electric excitation, that is, with decreasing voltage ratio downward from $R_{sl} = 1$, we find in the Figs. 11 to 16 the first operating range with $1/f > 0$, which is followed by the second operating range with $1/f < 0$. For some lenses even operating ranges of higher order with alternating sign of the lens power have been observed. Between each two consecutive operating ranges there is a *telescopic operating point* with $1/f = 0$. Figures 11–13 (similar as Fig. 9) refer on *symmetrical single lenses.* They show the influence of several geometrical shape parameters. The lenses with the first maximum of the lens power at $R_{sl} = 0$ are particularly interesting. The geometrical data of these lenses may be taken from Fig. 14.

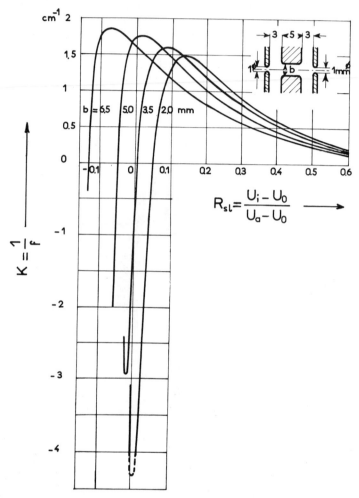

FIG. 11. Power characteristics of symmetrical single lenses. Parameter is the bore diameter b of the intermediate electrode (Heise and Rang, 1949).

Asymmetrical single lenses are expediently discussed in connection with symmetrical single lenses; their properties can be derived from the properties of the latter by varying a single parameter (for example, the ratio of the electrode distances a'/a) *under maintenance of the first telescopic operating point* (Everitt and Hanszen, 1956). Both focal lengths of asymmetrical single lenses are equal too. Figures 15 and 16 show some lens power characteristics. The power maximum of the asymmetrical lens is reached almost at the same voltage ratio as the maximum of the corresponding symmetrical lens and is somewhat higher than the latter. The position of the principal planes of these lenses is no longer symmetrical with respect to the intermediate electrode (see Fig. 10). The "midplane of the lens" appears displaced towards the side with the higher potential gradient.

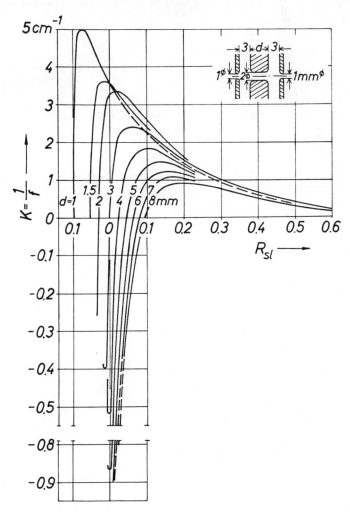

FIG. 12. Power characteristics of symmetrical single lenses. Parameter is the thickness *d* of the intermediate electrode (Heise and Rang, 1949).

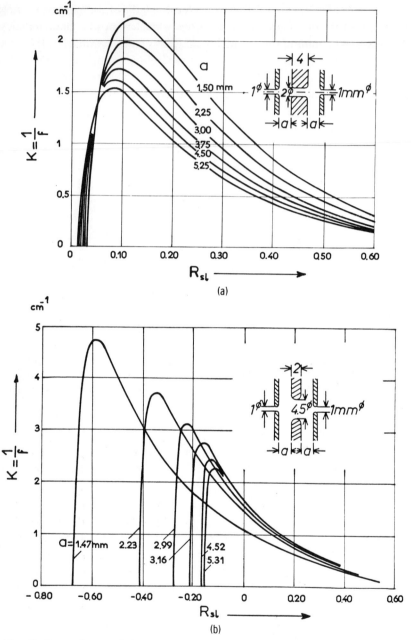

FIG. 13. Power characteristics of symmetrical single lenses (a) with a narrow bore in a thick intermediate electrode and (b) with a wide bore in a thin intermediate electrode. Parameter is the electrode distance a (Heise and Rang, 1949).

b. *Cardinal Points of Unipotential Lenses.* Very simple relations result, if the potential of intermediate electrode is equal to the rest potential of the particles, for in this case R_{sl} has the constant value 0 for all accelerating voltages $U_a - U_0$. Therefore, the cardinal points of these lenses, for particles with nonrelativistic velocities, do not depend on the amount of the accelerating voltage. We call these lenses *unipotential lenses.*

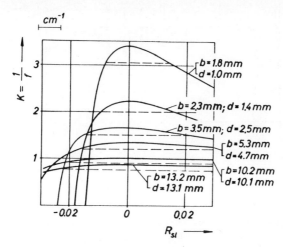

FIG. 14. Lens power of symmetrical single lenses with the first power maximum at $R_{sl} = 0$ (that is, operating as unipotential lens). The R_{sl} interval in which the power decreases by 10% is plotted with dotted lines (Lippert and Pohlit, 1952); b is the bore diameter; d the thickness of the intermediate electrode; and a the distance of the electrodes. The lens length $l = 2a + d$ of all treated lenses amounts to 20 mm.

The manner in which the lens power $K = 1/f$ of symmetrical unipotential lenses depends on the geometrical data is shown in Fig. 17; the manner in which the focal point coordinates z_F depend on the same data is given in Fig. 18. The solid curves refer to the first, the dotted curves to the second operating range. The comparison of both figures leads to the following statement: For distances $> 0.5l$ between the focal points and the midplane of the lens (that is, focal points beyond the lens field) and for a fixed position of the focal points, the focal length—and the positions of the principal planes too—are nearly independent of the geometrical data b/l and d/l of the intermediate electrode. It is easy to understand that this behavior occurs in the first operating range in as much as the principal planes are located close to the midplane and the conditions of weak lenses exist. As calculated by Laplume (1947), in the relativistic range the focal length of unipotential lenses first increases approximately 8% with increasing beam

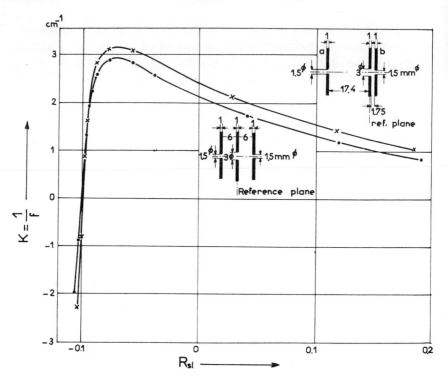

FIG. 15. Lens power characteristics of an asymmetrical single lens and of a symmetrical single lens with the same telescopic operating point (Everitt and Hanszen, 1956).

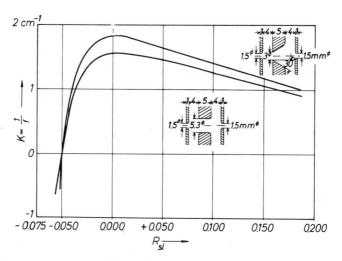

FIG. 16. Lens power characteristics of an asymmetrical and a symmetrical single lens with equal telescopic operating points (Hanszen, 1958a).

energy until a maximum is reached for electrons with the kinetic energy of 1 MeV. With still higher energies, the focal length decreases again and with energies of 5 MeV it falls even below the nonrelativistic value of the focal length; for still higher energies a finite boundary value is reached. The distance of the·principal planes from the midplane changes in the opposite sense as the focal length. Their displacements however are, by far, smaller.

c. *Chromatic Aberration of Single Lenses.* The particles within a beam have a finite energy width $\Delta W = e\,\Delta U_0$ because of their different rest energies. Since U_0 is connected with R_{sl} according to (45), the cardinal points for particles with different energies of incidence can be determined from the given information about the lens characteristics, respecting the validity of

$$\Delta R_{sl} = \Delta W/W\,(1 - R_{sl}) \qquad (46)$$

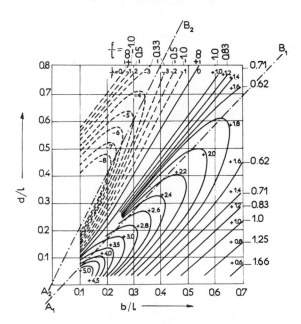

FIG. 17. Dependence of the relative lens power l/f and of the relative focal length f/l of symmetrical unipotential lenses ($l = 2a + d$ = lens length) on the relative geometrical lens data (see Figs. 12 to 14) according to Lippert and Pohlit (1952) (b is the bore diameter of the intermediate electrode; d the thickness of the intermediate electrode; and a the distance of the electrodes). Solid curves = 1st operating range; dotted curves = 2nd operating range. The straight line A_1B_1 marks the first lens power maximum; the line A_2B_2 marks the second power maximum. Both maxima are characterized by focal length achromatism. Curves with the parameter value $l/f = 0$ (that is $f/l = \pm \infty$) mark the telescopic operating points.

and particularly for unipotential lenses with $R_{sl} = 0$ of

$$\Delta R_{sl} = \Delta W / W \qquad (47)$$

Since the chromatic-aberration constants are determined by the chromatic displacements Δz_F and $\Delta z_{F'}$ of the focal points [compare, (12), (17)], the constants of chromatic aberration can be evaluated directly from the focal point characteristics (for example, Figs. 9 and 10). In this manner, the "chromatic-aberration constant related to the object for strong magnification" $C_{ch}(\infty)$ for single lenses (that is, for $W = W'$) can be expressed by the equation

$$C_{ch}(\infty) = \frac{\Delta z_F}{\Delta W / W} = \frac{\Delta z_F}{\Delta R_{sl}} (1 - R_{sl}) \qquad (48)$$

and particularly for unipotential lenses by

$$C_{ch}(\infty) = \frac{\Delta z_F}{\Delta R_{sl}} \qquad (49)$$

With this we directly learn from Fig. 10, that the chromatic aberration of asymmetrical single lenses is unequal on both lens sides.

Because $\Delta z_F / \Delta R_{sl}$ does not vanish anywhere, the chromatic aberration does not vanish at any operating point too. According to Fig. 3, we can

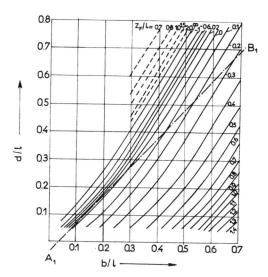

FIG. 18. Dependence of the relative focal point coordinates z_F/l on the geometrical lens data (Lippert and Pohlit, 1952). Notations as in Fig. 17; the origin of the coordinate system is located in the lens center.

see that the displacements Δz_F and $\Delta z_{F'}$ of the focal points are the sums of the displacement of the principal points Δz_H and $\Delta z_{H'}$ and the alteration of the focal lengths Δf and $\Delta f'$, respectively. The latter vanish at the maximum of the lens power (at this points we have *focal-length* achromatism). For this reason, the chromatic error reaches its minimum value near this point—or, strictly speaking, at an operating point just below the power maximum. The flatter the maximum, the wider the range around the maximum with low chromatic aberration (see Fig. 14). Another chromatic error is the *chromatic error of magnification*. It depends essentially on the position of the exit pupil (Wendt, 1940; Hanszen and Lauer, 1965). If the pupil is located at the second focal plane, for strong magnifications, this error is given by $\Delta f/f$, that is, by the lens power characteristic. The chromatic error of magnification vanishes for an appropriate position of the pupil. For initially decelerating lenses, the ratio $C_{ch}(\infty)/f$ decreases if the electrical excitation decreases and runs to 2 for $R_{sl} \to 1$. For $R_{sl} > 1$ (that is, for initially accelerating lenses) $C_{ch}(\infty)/f$ further drops (see Glaser, 1956).

The chromatic-aberration constant of unipotential lenses in dependence on their geometrical data is demonstrated in Fig. 19. The minimum value of this aberration constant amounts to about three times the length of the

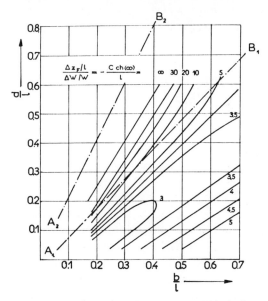

FIG. 19. The ratio of the chromatic aberration constant $C_{ch}(\infty)$ of symmetrical unipotential lenses to the lens length l in dependence of the geometrical lens data according to Lippert and Pohlit (1953). Δz_F is the focal point displacement; $\Delta W/W$ the relative energy shift; the other notations are as in Fig. 17.

lenses. For a lens with minimum aberration, the dislocation of the focal
point amounts therefore to 3% of the lens length, if the energy shift of the
particles amounts to 1%. At the telescopic operating point, the ratio of
focal-point displacement to energy shifting becomes infinite. In weak lenses
(that is, first operating range and focal points beyond the lens field) that
have the same focal-point position, the chromatic aberration of the focal
point is nearly independent of the lens geometries. Relativistic effects scarcely influence the chromatic aberration (Laplume, 1947).

d. *Distortion.*   According to (27) and to Fig. 5, the distortion $\Delta Y/Y_0$ at a
large distance behind the lens depends only on $\Delta f^*/f$, at which $\Delta f^* = f^* - f$,
and f^* is equal to the focal length f for $\alpha' \to 0$ or $y \to 0$. Figure 20 demon-

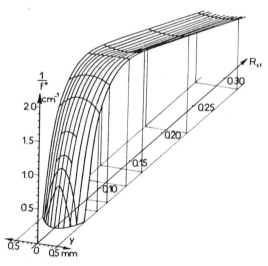

Fig. 20. Characteristic of the function $1/f^*(R_{sl})$ for the lens represented at the top of
Fig. 2 (Heise, 1949). The principal point characteristic of this lens is shown in Fig. 9.
y is the height of incidence. For $y = 0$, the characteristic is identical with the lens power
characteristic.

strates the dependence of $1/f^*$ on the angle of incidence α or on the height
of incidence y; the $1/f^*$ characteristic can be obtained in the first approximation by displacing the lens power characteristic (compare Figs. 11–16)
towards increasing values of R_{sl}. These displacements are very small and
augment with increasing α or y. (The lens focuses "stronger" in the off-axis
zones.) From this we recognize that within the direct vicinity of the power
maximum, $1/f^*$ remains equal to the lens power $1/f$ for finite but small α or
y. Hence the distortion vanishes at this point. (An example for application:

the distortion-free projective lens, see Section 2.2.2,A,5b). For operating points with more positive values of R_{sl}, cushion-shaped distortion occurs and with more negative R_{sl}, barrel-shaped distortion occurs; for the second operating range the corresponding statements are valid.

As to the distortion of asymmetrical lenses, see 2.2.2,A,2f.

Numerical values for the distortion constant $c_d = -\Delta f^*/f\,a'^2$ (see Eq. (42)] of *unipotential lenses* may be drawn from Fig. 21. Since the represented curves can be approximated mostly by straight lines, the distortion depends essentially on the ratio: thickness of the intermediate electrode to its bore diameter. We have to note that the distortion of weak lenses (that is, focal points beyond the lens field, cf. Figs. 17 and 18), which have the same z_F and f, severely depends on the lens geometry.

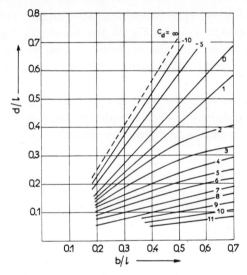

FIG. 21. The distortion of unipotential lenses in dependence of the geometrical data according to Lippert and Pohlit (1953). Notations are as in Fig. 17. Parameter is $c_d = \Delta Y/Y_0\,a'^2$. $\Delta Y/Y_0$ is the relative distortion; a' the angle of emersion; positive sign of c_d indicates cushion-shaped distortion, negative sign of c_d barrel-shaped distortion.

e. *Spherical Aberration.* Mostly the values for spherical aberration are reported in the form $C_s(\infty)/f$. [$C_s(\infty)$ is the spherical-aberration constant related to the object for strong magnification (compare 2.2.1,A,5c)]. For the lens, known from the diagrams 2, 9, 20, Heise (1949) determined the minimum value $C_s(\infty)/f = 16$ at the maximum of the lens power ($R_{sl} = R_{slmax}$). If the excitation of this lens is decreased ($R_{sl} > R_{slmax}$) until the half-maximum power is reached, $C_s(\infty)/f$ raises to its double value.

If the lens excitation is increased ($R_{sl} < R_{sl\max}$) until the half-maximum power is reached, $C_s(\infty)/f$ raises to its 30-fold value.

As it was pointed out in Section 2.2.1,A,5c, the spherical aberration is completely determined by $\Delta z_{F'*}$, consequently, not by Δf^* alone. A graphical representation of the $z_{F'*}$ characteristic, from which the spherical aberration could be drawn in a similar manner like the distortion from Fig. 20, is not known. The spherical aberration has always the same sign, whereas the distortion is able to assume positive or negative sign (see Fig. 21). Rays, emerging from the object in greater angles, intersect the axis always at smaller distances behind the lens than the paraxial ones.

According to a rough rule, $C_s(\infty)/f$ has the order of 10 for initially decelerating single lenses, the focal points of which lie beyond the field.

Initially accelerating lenses have lower spherical aberrations. The spherical-aberration constants of *unipotential lenses* can be drawn from Fig. 22, which shows, in comparison with Fig. 19, that the spherical-aberration minimum nearly coincides with the minimum of the chromatic aberration and appears at operating points with voltage ratios something more positive than $R_{sl\max}$.

The spherical aberration of weak lenses (focal points beyond the lens field; compare Fig. 22 with Figs. 17 and 18) with the same focal point position depends only slightly on the lens geometry, and the spherical aberration of

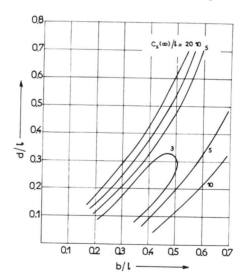

FIG. 22. Ratio of the "spherical-aberration constant $C_s(\infty)$ for strong magnification" to the "length l of the lens" in dependence of the geometrical lens data. Notations are as in Fig. 17 (Lippert and Pohlit, 1953).

weak lenses with the same focal length depends but scarcely on the lens geometry. According to Laplume (1947) the influence of relativistic effects on the spherical aberration of unipotential lenses is but inferior. For strong magnification, asymmetrical lenses have smaller or larger spherical aberrations than the corresponding symmetrical lenses, if the beam incides on the lens side with the higher or lower potential gradient, respectively (Hanszen, 1958b); compare to this, Section 2.2.2,A,2f and Fig. 26 (see furthermore, Seeliger, 1948). Septier (1959, 1960) utilized an asymmetrical lens with $C_s(\infty)/f = 2$ behind the ion source of a linear accelerator for protons.

Whereas it is mostly sufficient to know the relative spherical-aberration constant, the knowledge of the absolute value of the spherical-aberration constant is necessary for the determination of the electron-microscopical resolving power. The smallest distance ε, which can be resolved by transmission microscopes, depends on the wavelength λ of the electrons in the rays and on the spherical-aberration constant according to the equation, $\varepsilon \approx \approx (C_s(\infty)\lambda^3)^{1/4}$. Since, for electrostatic lenses used as objectives, the focal point always has to be located out of the lens field (see 2.2.2,A,5a), the spherical-aberration constant of usual electrostatic objectives (symmetrical unipotential lenses) has the large value of $C_s(\infty) = 60$ mm (see Seeliger, 1948). Only by very asymmetrical shaping of the intermediate electrode, Seeliger 1948 succeeded in reducing $C_s(\infty)$ to 23.5 mm. Compared with this, the spherical-aberration constants of the objectives currently used in the high-resolution magnetic microscopes amount to only 4 mm or less.

f. *Trajectories in the Off-Axis Lens Zones.* Figure 23 states, in which manner the slope of the exit asymptotes of a nearly parallel incident beam depends on the angle of incidence (or on the height of incidence); Fig. 24 shows the dependence of the axis intersection point on the same parameters. Under these circumstances $\Delta p' = \Delta z_{F'*}$. From the course of the represented curves for $\tan \alpha \to 0$ or $y \to 0$, we can determine the cardinal elements (see Section 2.2.1,A,6b), and for small α or y the geometrical aberrations of the Seidel area (see Section 2.2.1,A,6c). The main advantage of these diagrams is to give the aberrations up to the outermost lens zones, without being tied to an approximate aberration theory.

Of particular interest is the knowledge of that lens zone which focuses parallel incident rays at a given distance behind the lens (that is we ask for details about the caustic). If we are concerned with an observation plane at the far distance s' behind the lens, in this plane the coordinate Y_i of a ray with the incident height y_i is only determined by the dependence of the emersion angle a_i' on y_i (see Fig. 4b). This holds because $\Delta p_i' \ll s'$.

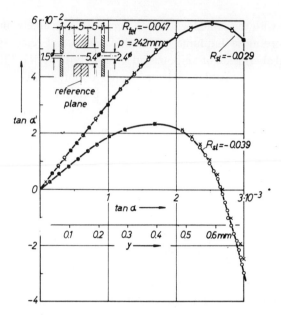

FIG. 23. Slope tan α' of the exit ray, emerging from an electrostatical single lens (similar to the lens of Fig. 16) in dependence of the slope tan α of the incident ray, or the height of incidence y, for different voltage ratios R_{sl} (Hanszen, 1964a).

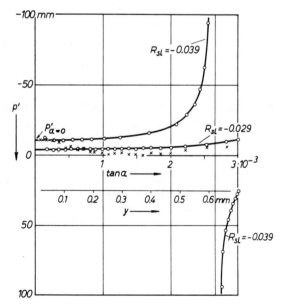

FIG. 24. Axis intersection points p' of the same lens as shown in Fig. 23 for two different voltage ratios. Notations as in Fig. 23.

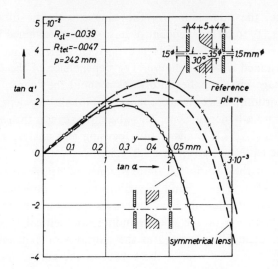

FIG. 25. Confrontation of the slope tan α' of the exit rays emerging from symmetrical and asymmetrical single lenses. The beam enters the lens nearly parallely to the axis from the *left-hand* side. Notations are as in Fig. 23 (Hanszen, 1958b).

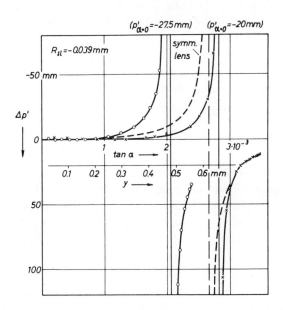

FIG. 26. Confrontation of the longitudinal spherical aberration of the symmetrical and asymmetrical single lenses shown in Fig. 25. Notations are as in Fig. 23 (Hanszen, 1958b).

Therefore, the intersection of the caustic with the observation plane (this means, Y_i to be independent of a_i or y_i) is determined by the maxima of the curves shown in Fig. 23. Hence, the focusing lens zone can immediately be determined from this figure. The dependence of the curve tan $a'(y)$ on the voltage ratio supplies information about the "chromatic aberration of the distortion," since in accordance to (46) the R_{sl} dependence can be converted into a W dependence. This aberration causes the *energy dispersion* of energy analyzers. Details about focusing and analyzing properties of electrostatic lenses are given in 2.2.2,A,5c.

The data of the rays in the off-axis zones of *asymmetrical single lenses* can be drawn from Figs. 25 and 26. According to the direction of the parallel incident beam, the longitudinal *spherical aberration* is larger or smaller than the aberration of the corresponding symmetrical lenses, the higher potential gradient being located at the entrance or exit side of the lens, respectively. For small a or y, the course of the curves shown in Fig. 26 gives according to (38) and (40) the "spherical aberration constant for strong magnification" in the inversed ray tracing. (These measurements are, however, of low accuracy.)

For small a or y, the course of the curve shown in Fig. 25 determines the focal length of the lens [see (43)]. Since both focal lengths of asymmetrical single lenses are equal, the curves belonging to both lens positions must have the same slope at the origin of the coordinate system. Because of the result given in Fig. 16, however, this slope is larger than that of the curve belonging to the corresponding symmetrical lens. This slope and further details about the trajectories in the external zones of the symmetrical lens can be drawn from Fig. 23.

3. *Gauze Lenses as Single Lenses*

The general properties of these lenses have already been reported in Section 2.2.2,A,1. Of particular interest are the voltage ratios $R_{sl} > 1$ (the incident particles are initially accelerated) at which the gauze lenses work as converging lenses. According to Fig. 27, the lens power $1/f$ equals to 0 for $R_{sl} = 1$ and increases linearly with ($R_{sl} - 1$). For this reason, with R_{sl} somewhat above 1, the lens power is considerably stronger than the power of an ordinary three-diaphragm lens, the power of which increases proportional to $(R_{sl} - 1)^2$. Hence gauze lenses may always be utilized where particles with very high energies have to be focused by the aid of limited lens voltages (for example, in linear accelerators for ions. For data on gauze lenses as accelerating lenses see Section 2.2.2,B,1).

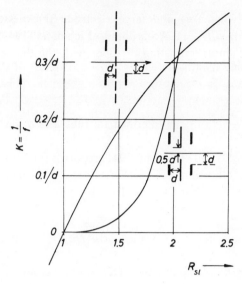

FIG. 27. Lens-power characteristic of a gauze lens compared to the characteristic of the corresponding three-diaphragm single lens with very thin electrodes and for an initially accelerating field between entrance and intermediate electrode ($R_{sl} > 1$) (Bernard, 1953a).

The geometrical aberration of gauze lenses are smaller than the aberrations of the corresponding three diaphragm lenses. According to Bernard (1952), the spherical aberration of diverging gauze lenses can even get negative values. Also the other aberrations of these lenses are so small that they appear suited for the imaging of extended object fields (Verster, 1963).

Mostly electroplated gauzes are utilized as gauze electrodes. At present they can be manufactured with a minimum lattice constant of 12.5 μm (and with a transparency of about 70%). Each mesh of the gauze has a weak diverging effect in converging lenses and a weak converging effect in diverging lenses. These effects limit the resolving power of the gauze lenses more than the spherical aberration. The smallest resolved distance is somewhat smaller than the grating constant of the used gauze (Bernard, 1953b).

4. Three-Tube Lenses

Three tube lenses are scarcely used for imaging purposes but frequently for the collecting of beams with high divergence. These lenses mostly consist of three coaxial tubes of equal diameter with short mutual distance apart. Even tubes with diameters up to some decimeters are in use. Their lens properties can be reduced from those of three diaphragm lenses with small electrode distances and large electrode thicknesses (for example, from Fig. 17 with $d/l \to 1$). According to Liebmann (1949) the focal length of unipotential lenses consisting of tubes with the diameter $b \geq d + a$ is given by the rule of thumb,

$$f = b^2/(d + a) \qquad (50)$$

where d is the length of the intermediate tube and a the distance between two tubes.

Liebmann (1949) and Gobrecht (1941) reported further details about these systems for the Gaussian and the Seidel area. Lens-power maxima occur with these lenses too. For the spherical-aberration constant of unipotential lenses it yields approximately:

$$C_s(\infty)/f = 10(f/b)^2 \qquad (51)$$

5. Examples for the Application of Electrostatic Lenses

a. *Electrostatic Lens as the Objective of a Microscope.* Because a very strong magnification is required in the microscope, the object must be located close to the first focal point of the objective. Therefore, we are allowed to use only lenses with focal points outside the field because the object is not permitted to disturb the lens field. Since for objective lenses, customary symmetrical unipotential lenses are in use, according to Fig. 22 in connection with Figs. 17 and 18, we have to reckon with spherical-aberration constants $C_s(\infty)$ not remarkably below 10 lenses lengths. Since for an operating point with minimum spherical aberration the focal point is always located within the lens field, it is not possible to take any advantage of this operating point for objective lenses. For the operating point with minimum chromatic aberration, the same statement is valid. These are serious disadvantages of the electrostatic objectives contrary to magnetic objectives, since for the latter the object is generally allowed to dip into the magnetic field. To assure a satisfactory breakdown rigidity (see Section 2.2.3,B), the electrodes have to be arranged with sufficient interspaces. This requirement prohibits production by electrostatic lenses of focal lengths as small as those produced by magnetic lenses. According to Seeliger (1948), the focal lengths of the common electrostatic objectives are 6–7 mm. Magnetic objectives however have focal lengths of 2 mm or less.

b. *The Electrostatic Projective Lens.* The image generated by the objective lens acts as object for the projective lens. This image of course is allowed to dip into the projective field. Therefore, this lens can be operated at its power maximum and it is possible to generate a distortion free image see Section 2.2.2,A,2d. (Strictly speaking the osculatory cardinal points are competent instead of the virtual ones for this imaging purpose.)

   c. *Electrostatic Lens as Energy Analyzer.* In the energy analyzer, the particles of an parallel incident beam with finite energy width will be separated according to their energies. Simultaneously, the particles with equal energy will be focused in the recording plane. The separation is done by the "chromatic dependence of the distortion" in the external zones of the lens. This dependence can be derived from the family of curves, drawn in Figs. 23 and 28, considering the fact, that the relative energy deviation $\Delta W/W$ of the particles is coupled with the voltage ratio according to (46). The focusing is realized according to Möllenstedt and Dietrich (1955) and Dietrich (1958) in the caustic ray tracing, that is, by applying the lens zones, for which the curves of Fig. 23 have their maximum values. Unfortu-

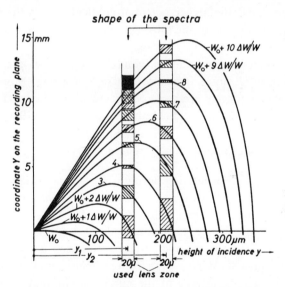

FIG. 28. Design of an energy spectrum $W = W_0 + n\,\Delta W/W$ (equidistant spectral lines) if two lens zones with different heights of incidence y_1 and y_2 are utilized (schematic)

nately, this condition is only fulfilled for one single energy, so that we have exact focusing only for one single spectral line. These facts are demonstrated in Fig. 28.

   The slope, and especially the radius of curvature at the extrema of the curves are decisive for the admissible latitude of the lens zone which can be utilized and on that account also for the *transmission* of the spectrometer. The rise between the maxima of two adjacent curves is decisive for the dispersion of the spectrometer. The run between the maxima of two adjacent curves is decisive for the unsharpness of the spectral lines not exactly focused; hence it determines the *defocusing*. The properties of each lens are described by a family of curves like Figs. 23 and 28. The families belonging to every possible lens approximately can be deduced from one another by suitable linear but generally *independent* transformations *of both coordinates* (see Lippert, 1955b). Lenses with thin intermediate electrodes generally have stronger dispersion, but also larger defocusing than lenses with thick intermediate electrodes.

To either side of an asymmetrical lens belongs a different diagram, similar to Fig. 28. One diagram results from the other with sufficient accuracy by *transformation of both coordinates* in an *equal scale*. [For example, proved for the asymmetrical lens of Fig. 15 by Hanszen (1956) published only as Congress report.] If the higher potential gradient lies on the exit side, the dispersion—and the defocusing too—rises in an equal ratio above the values belonging to the opposite ray tracing.

In order to be independent of voltage fluctuations, there are nearly always unipotential lenses in use for energy analyzing. Moreover, lenses with rotational symmetry are replaced by cylindrical lenses, for which the above statements are also valid (Metherell and Whelan, 1966).

There is a lack of detailed experimental material about cylindrical lenses; theoretical data are published by Archard (1954).

d. *Filter Lenses.* For filtering purposes we take advantage of the strong retarding effects in electrostatic lenses (Möllenstedt and Rang, 1951; Boersch, 1954; Lippert, 1955a; Hahn, 1959, 1961, 1964): Filter lenses for the electron miscroscopes are required to exclude all electrons with higher energy losses than 6 eV from the beam and to focus the electrons with energy losses of only some tenths of eV in order to form an image with the demanded resolution. On one hand, because of the minimum chromatic- and spherical-aberration values at the lens power's maximum, filter lenses should be operated at this maximum in order to have good imaging properties for the electrons without energy losses. On the other hand, to fulfill the filtering condition, it is indispensable for the saddle potential to approach the original rest potential within 6 V. Unfortunately this does not happen in the lenses at the first power maximum. Hence filter lenses give an example for the practical importance of higher order operating ranges. Since it is not possible for objectives to operate at a lens power's maximum, only projectives or intermediate lenses may be used as filter lenses; according to Section 2.2.2,A,5b and under the above conditions, simultaneously the image is distortion free too. Filter lenses for microscopes generally have a very thin intermediate electrode with a very small bore.

Lenses with a narrow bore in a thick intermediate electrode are suitable for filter lenses in beams with big cross section, if only less demands on focusing quality and energy resolution are wanted (Deichsel and Keck, 1960). Long-focus filter lenses (practically operating with telescopic ray tracing) with large aperture can be realized by five–electrode systems (Simpson and Marton, 1961; Kessler and Lindner, 1964). An interesting filter lens with coupled electric and magnetic fields has been designed by Brack (1962) and Hartl (1966).

e. *Electron Mirror.* With increasing retarding potential at the intermediate electrode, first the off-axis lens zones and finally the axis area exceed the rest potential of the particles. Therefore, first the off-axis zones and then by degrees the whole lens acts as a mirror for particles with equal energy. The power of the mirror—like the power of lenses—has oscillatory sign in dependence on the potential of the intermediate electrode. For very strong reflecting potentials diverging mirrors result. Details and the description of specific systems can be drawn from Hottenroth (1937) and for two-tube mirrors from Nicoll (1938).

B. IMMERSION LENSES

"Immersion lenses" is the collective name for accelerating and decelerating lenses. Mostly these lenses consist of two electrodes. As significant voltage ratio we introduce the quantity

$$R_{il} = \frac{U' - U_0}{U - U_0} \tag{52}$$

where U_0 is the rest potential of the particles, U the potential of the electrode on the object side, and U' the potential of the electrode on the image side.

If the object and image spaces are field free, that is to say, if they have uniform potentials U and U', respectively, for the immersion lenses, we can introduce cardinal elements again. For nonrelativistic particle velocities the ratio of the focal lengths f and f' on the object and image side is given by

$$f/f' = + [(U - U_0)/(U' - U_0)]^{1/2} ; \tag{53}$$

the corresponding relation for the relativistic range can be easily deduced from (4) and (44).

The foregoing statements hold particularly for two-diaphragm and two-tube lenses, the properties of which shall be discussed in this article only in reference to gauze immersion lenses; details about two-electrode lenses will be described in Chapter 5.1 in connection with accelerators.

1. *Gauze Immersion Lenses*

Verster (1963) investigated two gauze immersion lenses, consisting of two coaxial tubes of equal diameter arranged with short distance apart, wherein the tube on the image side was equipped with a flat or convex gauze. The results of these measurements are represented in Figs. 29 and 30. The

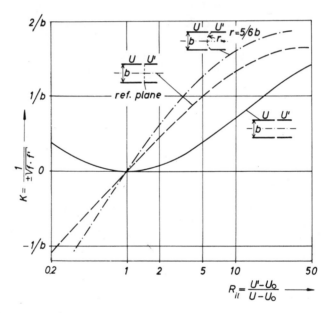

FIG. 29. Confrontation of the lens power characteristics of two gauze immersion lenses with the characteristic of the correspondent gauzeless two-tube immersion lens (according to Verster, 1963).

lens power characteristics of these lenses can be compared in Fig. 29 with the characteristic of a two-tube lens without gauze. Similarly as the gauze single lens described in Section 2.2.2,A,3 and Fig. 27, the gauze immersion lenses diverge the beam in the operating range with $R_{il} < 1$ and collect it with $R_{il} > 1$. For $R_{il} \approx 1$ their power increases proportional to $(\log R_{il})$:

therefore, the absolute value of the power in this range greatly exceeds the power of a lens without gauze, which rises proportional to $(\log R_{il})^2$. Thus, gauze immersion lenses can be utilized for the same purpose as gauze single lenses (see Section 2.2.2,A,3).

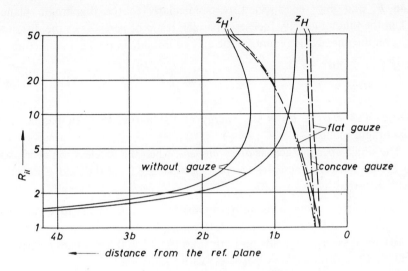

FIG. 30. Principal point characteristics of the lenses shown in Fig. 29 (Verster, 1963).

Figure 30 represents the principal point characteristics of both gauze immersion lenses. Whereas the principal planes of normal immersion lenses move toward $z = \infty$ on the object side for $R_{il} \to 1$, the principal points of gauze lenses remain at a finite distance.

Chromatic-aberration constant C_s, focal lengths f, f', and tube diameter b are connected by the following equation:

Object on the side with the potential U: ... $C_s (\infty)/f = Aff'/b^2$ (54)

Object on the side with the potential U': ... $C_s'(\infty)/f' = A'ff'/b^2$ (55)

Herein, the quantities A and A' are nearly independent of R_{il} and scarcely differ from one another. For accelerating lenses without gauzes and for not too large R_{il}, $A \approx A' \approx 2.4$; for both gauzes lenses, shown in Fig. 29, under the same conditions, $A \approx A' \approx 1.9$. About further geometrical aberrations see Verster (1963). About the influence of the gauze meshes on the image quality see 2.2.2,A,3.

2. The Single Aperture

A *small* bore of the diameter b in a very thin diaphragm at the potential U leads to an inhomogeneity in the electrical *fields* in the halfspaces on both sides of the diaphragm. The original fields may have the field strengths E and E' and may be directed perpendicularly to the diaphragm plane. Then the inhomogeneity focuses a parallel beam of particles at the distance $z_{F'*}$ on the image side (counting the axis coordinate in the beam direction!):

$$z_{F'*} = \frac{4\,(U - U_0)}{E - E'} \qquad (56)$$

where $W = e(U_0 - U)$ is the energy of the particles, when passing the diaphragm plane.

The single aperture acts as *diverging lens*, when the acceleration decreases or the deceleration increases for particles moving from the field E into the field E'; that is:

(a) for accelerating fields on both sides (E; $E' < 0$ for negative particles and E; $E' > 0$ for positive particles) with $|E| > |E'|$;

(b) for decelerating fields on both sides (E; $E' > 0$ for negative particles and E; $E' < 0$ for positive particles) with $|E| < |E'|$;

(c) for an accelerating field on the incidence side and decelerating field on the exit side ($E < 0$; $E' > 0$ for negative particles and $E > 0$; $E' < 0$ for positive particles) with arbitrary amounts of E; E'.

In all other cases the aperture acts as *converging lens*.

Particularities about the influence of the diaphragm thickness are published by MacNaughton (1952), about the influence of the bore diameter by Hoeft (1959).

If the fields have opposite directions in both halfspaces and the field on the incidence side is a decelerating one, the saddle potential in the bore of the diaphragm may assume such a height, that the particles no longer are allowed to pass the aperture. If the decelerating field decreases, the potential barrier opens for the beam in the manner of an iris. (Application: grid-electrode in electron guns).

3. Immersion Objective Lens for Electron Microscopes

The immersion objective consists

(a) of an emitting plane cathode surface, whose potential can be identified, without a large error, with the rest potential U_0 of the particles;

(b) of a single aperture at the potential U as grid electrode; and

(c) of a single aperture at the potential U' as anode.

Consequently, the object is a part of the lens, and the lens properties depend strongly on its position. Because the object space is not field free, it is not possible to determine image position and magnification by means of cardinal elements according to (2).

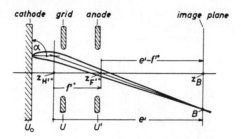

FIG. 31. Ray path of an elementary bundle in an immersion objective (schematic).

In most cases, we are allowed to assume the field just in front of the cathode to be homogeneous and axially directed. The electrons emitted from each element of the cathode into the angle $2\alpha = \pi$, in this case, are bundled into a narrow elementary beam with a chief ray initially parallel to the axis (see Fig. 31). The field, generated by grid electrode and anode, focuses the elementary beam into the image element B. The tangent line to the chief ray in the image point B intersects

(a) the axis at the point $z_{F'*}$ and

(b) the tangent line to the chief ray in the object point, at the point with the z coordinate $z_{H'*}$. With the quantities f'^* and e' (the meaning of which can be seen on the figure) the magnification can be expressed by

$$M' = -\,(e' - f'^*)/f'^* \qquad (57)$$

Hence, image position and magnification are known, if $z_{H'*}$, $z_{F'*}$ and the coordinate z_B of the image plane are known as a function of the geometrical parameters (essentially electrode distances, thickness and bore diameter of the grid electrode) and the voltage ratio. The manner in which M' and f'^* depend on the geometrical data, if the image plane has a fixed position at a large distance from the lens, is examined in great detail by Septier (1954a). Beyond that, Soa (1959) published systematic investigations for other image distances. The example, given in Fig. 32, permits to compare the course of $z_{H'*}$, $z_{F'*}$, z_B with the course of the cardinal elements of single lenses (see for example, Fig. 9); but it must be emphasized that for immersion lenses an object displacement (that is, a cathode displacement) will involve an alteration of the whole characteristic. Soa (1959) proved the

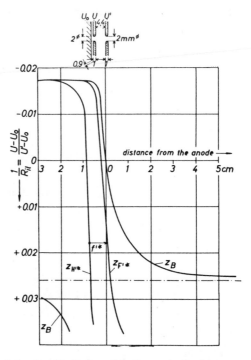

FIG. 32. Characteristic of an immersion objective. z_B is the image coordinate; $z_{F'*}$ the axis intersection of the chief ray's asymptote on the image side; and $z_{H'*}$ the axis coordinate of the intersection point between the two asymptotes of the chief rays on the object and image side (Soa, 1959).

quantity $1/f'*$ to pass a maximum value in dependence on the voltage ratio R_{il} (see Fig. 33) as it is known for the lens power of single lenses. His supposition that the imaging qualities of immersion objectives improve if the operating point approaches to this maximum could be confirmed by experiments. According to the published figures, especially the field curvature and the off-axial astigmatism [more about these errors can be read in Septier (1954a)] decrease when approaching to the maximum. But unfortunately it is not possible to reach this maximum with strong magnifications (that is, with large image distances). In order to raise $1/f'*$ toward its maximum value, we are compelled to make the grid electrode thin, its

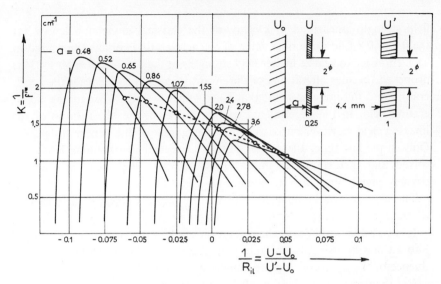

FIG. 33. The $1/f'^*$ characteristics of an immersion objective in dependence on the distance between cathode and grid electrode (Soa, 1959). The dotted line refers to image positions $z_B \to \infty$.

bore as large as possible and the distance between grid electrode and anode small (with this compare Bas and Preuss, 1964).

But under these circumstances no extreme resolution power is attainable. According to the geometric–optical consideration of Recknagel (1941); [compare the synopsis given by Septier (1954b)], the minimum resolved point distance (gained without applying an aperture stop) approximately amounts to $2kT/eE$ independent of the specific field distribution. (kT is the half-energy width of the emitted electrons and E the field strength in front of the cathode.) This expression, which still calls for a wave mechanical correction (Recknagel, 1943), in the most cases has been experimentally proved to be sufficient for high field strengths. According to these results, the resolving power increases with the cathode field strength. The relatively low resolution of the previously discussed lens therefore is caused by the weak cathode field strength. While maintaining a fixed *large image distance*, the field-strength cannot be raised by diminishing the distance between cathode and grid, because the required variation of the grid potential counteracts this raise. Only the bore diameters of the two diaphragms have influence on the field strength. Satisfactory values of the field strengths and the possibility to operate at the maximum of $1/f'^*$ only can be expected, according to Soa (1959), for small image distances, that is, for weak magnifications. According to the statements of Septier (1954a) the value for the cathode field strength (measured in kV/cm) of the classical immersion lenses has the 0.5 to 0.8-fold numerical value of the accelerating voltage (measured in kV). In order to realize higher field strengths at the cathode, Septier (1954a) proposed (besides alterations of the electrode shapes) to separate the accelerating and the imaging field; that is, to annex a *single lens* to the accelerating field. In the experiments to this proposal, the aberrations of the imaging field however surpassed the aberrations of the accelerating field. Only in the last time, Illenberger (1964) succeeded in obtaining highly resolved and distortion free images by utilizing a single lens with an extremely small bore in a thin entrance electrode.

The electrostatic imaging field of this immersion objective also can be replaced by a magnetic lens.

It is possible, to improve the resolution of all immersion objectives about 3 to 4 times by using an aperture stop. Instructive micrographs for the influence of the aperture stop are shown by Bayh (1958).

For theoretical investigation of the immersion objective see Hahn (1958).

4. *The Electron Gun as Immersion Lens*[3]

In electron guns too, the electrons emitted from the cathode, must be accelerated and focused. But contrary to the purpose to be fulfilled with immersion objectives, electron guns are not intended to image the cathode surface, but instead to transfer as many electrons as possible through a pupil cross section (crossover) as small as possible, located at the fixed distance $z_{F'*}$ (see Fig. 31).

Generally, the cathode surface is curved. In the simplest case of rotational symmetry the cathode has the shape of a spherical cap. Such a system has strong geometrical aberrations. The chief rays of the elementary bundles, starting from all elements of the cathode surface, intersect by no means at one single point, but they form an intricate caustic (Hanszen, 1962) whose shape can be exactly determined by the aid of the shadow method (Hanszen, 1964a). In this way, complicated intensity distributions over the beam cross section, especially in the so-called "hollow beams," are surveyed easily. In the course of these investigations it could be shown (Hanszen, 1964b,c), that even at the normal operating point of electron guns, a unitary crossover may not be present. The conditions for a point-focus or a cone-focus at a large distance from the cathode can be determined also by this shadow method. A special contribution in this book, see Chapter 2.1, is dedicated to the specific properties of electron guns.

---

[3] Compare also the recent publications of Lauer and Hanszen (1966) and Lauer (1967).

C. Electrostatic Stigmators (Correction of Axial Astigmatism)

Because of the mechanical tolerances, the shape of all electrostatic lenses departs, more or less, from an exact rotational symmetry; *axial astigmatism* results from this. External disturbances (such as electrical charges on the glass windows and insulators, the magnetic-earth field, contaminations on the electrodes, and so on) act in a similar manner. As far as the astigmatism originates from the single parts of the lens, it is possible to compensate the astigmatism components of the different lens parts by turning them around the axis against each other (see for example, Hahn, 1954).

Over and above this, it is possible to compensate the residual astigmatism —it is nearly always binary—by a cylindrical lens, that is, a so-called stigmator. In the simplest case, it can be realized by a single lens with a slot-shaped aperture in the intermediate electrode. This lens has to be located at an appropriate place of the ray path, and its strength has to be adjusted by varying the potential of the intermediate electrode.

A quadrupole lens serves for the same purposes. Since an azimuthal mechanical turning of the stigmator lens under vacuum conditions cannot easily be realized, an electrical turning is often preferred. To enable this, according to Rang (1949), two quadrupole lenses, which are composed of a system with "octopole-like" geometry, are used. By proper regulation of the excitation of both quadrupoles, it is even possible to choose direction and amount of the compensating astigmatism independent from one another.

2.2.3. Practical Aspects for the Construction of Electrostatic Lenses

A. Mechanical Tolerances, Shielding of Stray Fields, Highest Admitted Voltages

During the construction of electron lenses, exact rotational symmetry of all field influencing components must be observed. For example, in electron microscopes the noncircularity of each electrode bore and the deviation of their centers from a concentric position should be less than 1/100 mm. The lens chamber should be made of soft iron in order to shield external —particularly alternating—magnetic fields. For extraordinary cases, an additional "Mu-Metal" shielding may be used. Good materials for the insulator between high-voltage supply line and lens-chamber are nonporous ceramics or synthetics like "acryl-glass" and "epoxy-resin," which can be worked easily on a lathe. Size and shape of the insulator are guided by the applied voltage (see Figs. 36, 37). Mostly, the point of the question is to obtain strong lens powers. Therefore, we need a very high field strength between the lens electrodes. For this reason the designing of the electrode system including the insulators between the electrodes is closely referred to the problem of breakdown rigidity. Particular details in the design of these lens elements are given in the following chapter in connection with the discussion of the breakdown rigidity. As known by experience, it is possible to maintain the following field strength between the lens electrodes:

approx. 100 kV/cm without greater difficulty;

approx. 150 kV/cm with careful design of the components;

approx. 200 kV/cm only with best design and meticulous cleanness.

These facts will be illustrated by an example: The focal lengths, needed in the electron microscopy, are smaller than 1 cm. According to the statements of 2.2.2,A,5a and Figs. 17, 18, the realization of the foregoing objective focal length requires an electrode distance smaller than the focal length itself. For this reason, and with the fore-mentioned breakdown field strengths, we can hardly maintain higher voltages than 60 kV between the lens electrodes. For the same reason, it is not possible to design electrostatic lenses with focal lengths remarkably below 1 cm for particle energies higher than 60 keV.

B. BREAKDOWN RIGIDITY OF THE SYSTEM "LENS-ELECTRODES AND INSULATOR"

Just before reaching the breakdown-voltage, the so-called "microdischarges" occur (Arnal, 1955). They are induced by cascades of positive and negative ions and electrons that release each other by impact from the contamination layers on the electrodes (Boersch et al., 1961; 1966; Arnal and Bouvier, 1965). They are extinguished if the voltage is lessened by approximately 1 kV; (facility: resistance in the feed line). If a protective resistance of 50 MΩ is used, the voltage dip endures only 10^{-4} sec. But even then, the microdischarges disturb the operation of electron-optical apparatus severely (see Boersch et al., 1961). Some causes for microdischarges and some methods on how to avoid them are discussed in the following sections.

When the discharge current increases, electrode material is evaporated and sparkovers with complete voltage breakdowns occur. The evaporation of electrode material can be reduced with the aid of the above-mentioned protective resistance; for in that case, the charge transported by one spark is only determined by the capacitance of the lens system which, for this reason, should be as small as possible (Rabinowitz, 1965).

1. Microdischarges between Two Electrodes

Microdischarges are released by electrons which are emitted from very small protrusions on the negative electrode by field emission (Chatterton, 1966). Provided that the electrodes are covered with a thin insulating layer of hydrocarbons (such as oil from vacuum pumps), field emissions occurs already at a field strength of 10^4 V/cm, independent of the cathode material, if the contamination layer is polarized by the impinging of positive ions ("Malter-Effect", see for example, von Ardenne, 1956; Boersch et al., 1964). The way, in which the released discharges run off, however, depends on the anode material and the quantity of adsorbed hydrogen (Arnal, 1955). The rush of the microdischarge current leads to sudden evaporation of the cathode protrusions and thus to sparkover formation (Davies and Biondi, 1966).

A convenient anode material is stainless steel; for laboratory experiments duralumin often suffices (Gölz, 1940). The breakdown rigidity is raised by rounding off the edges and polishing the surfaces of electrodes. Moreover, it is advantageous to confine the area of highest field strength to these parts of the field which are important in leading the beam. Curved electrode surfaces are of advantage (Miller, 1966). The electrodes have to be cleaned by polishing; in critical cases the hydrocarbons must be cracked by heating or by glow-discharge at a pressure of 10^{-1} to 1 Torr. Residual emissions centers may be destroyed by arbitrary produced flashes with limited current. The breakdown rigidity is raised by a very small leak in the vacuum chamber near the critical electrodes (Schumacher, 1966). As shown by Arnal (1955), the geometry of the electrodes should be designed in such a way that flashes are not allowed to strike the insulator and to form contamination layers on its surface.

2. Microdischarges on the Insulator Surface

As is known from experience, the junction between negative metal and dielectric is a source of electrons and therefore a starting point of flashovers. The electron emission increases with the permittivity of the insulator material, and is nearly independent of the electrode materials (Kofoid, 1960a). Further details are reported in Grivet's textbook (1958). The electron emission can be suppressed by reducing the field strength at the junction.

This can be done by suitable geometrical design (see Fig. 34) and by using an insulator material with low permittivity (Kofoid, 1960b).

Beyond this, a positive charge on the insulator surface, caused by secondary emission of electrons, must be avoided. According to Boersch *et al.*

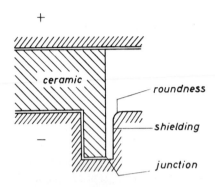

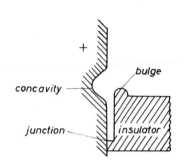

FIG. 34. Design of the junction between insulator and negative electrode. The electron emission is suppressed by lowering the field strength at this place. This can be done by "electrostatic shielding" of the junction and by rounding off the electrode edges (Kofoid, 1960b).

FIG. 35. Design of the junction between the insulator and the positive metal chamber of an electron lens (Gribi *et al.*, 1959). The concavity in the chamber wall reduces the field strength. By this shaping and the insulators' bulge, it is avoided that the electrons are stripped off from the negatively charged insulator.

(1963), by choosing a proper angle between insulator surface and field direction, it is principally possible to gain an uncharged or only negatively charged insulator surface near the junction. If such an angle cannot be realized, the distance between negative electrode and point of impact of the electrons on the insulator should be so large that a sufficiently high-po-

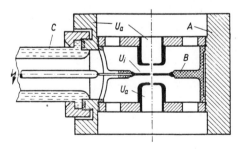

FIG. 36. Electron single lens for the electron-optical bench designed by Boersch (1951); see also Boersch (1940) and Mahl (1940). *A* is the lens chamber, consisting of iron or aluminium; *B* the high-voltage insulator, consisting, for example, of hard rubber (ebonite); and *C* the insulator for the high-voltage supply line, consisting of acrylglass.

tential difference occurs. Then, the electrons striking the insulator have energies high enough so that the secondary emission factor falls below 1. Under these circumstances, the insulator surface is only enabled to keep negative charge or none. Thus, the noxious field strength between insulator and cathode is diminished. A negative charge of the insulator raises however the potential difference near the positive electrode and may possibly lead to a stripping off of electrons from the insulator's surface. This effect can also be avoided by suitable geometrical design of the insulator and the positive electrode, see Fig. 35.

C. EXAMPLES FOR DESIGN

Figure 36 represents a single lens, designed for laboratory devices; Fig. 37 shows an objective-lens, which especially is flashproof and which is used in commercial electron microscopes. Further proposals for designing may be taken out of the tabular compilation by von Ardenne (1964).

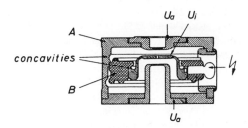

FIG. 37. Objective lens of a 70-kV microscope (Gribi *et al.*, 1959). *B* is the insulator, consisting of a mixture of epoxy-resin and quartz. (For less demanding purposes an epoxy-resin and talcum mixture is sufficient). All junctions metal-to-insulator are located at places with low field strength. According to the junction insulator B to lens chamber A compare to Fig. 34.

REFERENCES

Archard, G. D. (1954). *Brit. J. Appl. Phys.* **5**, 179 and 395.
Archard, G. D. (1956). *Brit. J. Appl. Phys.* **7**, 330.
Arnal, R. (1955). *Ann. Phys. (Paris)* [12] **10**, 830.
Arnal, R., and Bouvier, P. (1965). *Compt. Rend.* **260**, 4944.
Bass, E. B., and Preuss, L. (1964). *Optik* **21**, 261.
Bayh, W. (1958). *Z. Physik* **151**, 281.
Bernard, M. Y. (1952). *Compt. Rend.* **235**, 1115.
Bernard, M. Y. (1953a). *J. Phys. Radium* [8] **14**, 381.
Bernard, M. Y. (1953b). *J. Phys. Radium* [8] **14**, 451.
Boersch, H. (1940). *Jahrb. AEG Forsch.* **7**, 34.
Boersch, H. (1951). *Z. Physik* **130**, 517.

Boersch, H. (1953). Z. Physik **134**, 156.

Boersch, H., Hamisch, H., and Wiesner, S. (1961). Z. Angew. Phys. **13**, 450.

Boersch, H., Hamisch, H., and Ehrlich, W. (1963). Z. Angew. Phys. **15**, 518.

Boersch, H., Hamisch, H., and Markmann, G. (1964). Proc. Int. Symp. on Insulation of High Voltages in Vacuum, Cambridge (Mass.), p. 113.

Boersch, H., Hamisch, H., and Schirrmeister, H. (1966). Z. Angew. Phys. **22**, 5.

Brack, K. (1962). Z. Naturforsch. **17a**, 1066.

Chatterton, P. A. (1966). Proc. Phys. Soc. (London) **88**, 231.

Davies, D. K., and Biondi, M. A. (1966). J. Appl. Phys. **37**, 2969.

Deichsel, H., and Keck, K. (1960). Optik **17**, 401.

Dietrich, W. (1958). Z. Physik **151**, 519.

Everitt, C. W., and Hanszen, K.-J. (1956). Optik **13**, 385.

Felici, N. J. (1959). J. Phys. Radium [8] **20**, 97 A.

Fink, M., and Kessler, J. (1963). Z. Physik *174*, 197.

Glaser, W. (1952). " Grundlagen der Elektronenoptik," pp. 147 and 221. Springer, Vienna.

Glaser, W. (1956). In "Handbuch der Physik" (S. Flügge, ed.), Vol. 33, pp. 188 and 254. Springer, Berlin.

Glaser, W., and Schiske, P. (1954). Optik **11**, 422.

Glaser, W., and Schiske, P. (1955). Optik **12**, 233.

Gobrecht, R. (1941). Arch. Elektrotech. **35**, 672.

Gölz, E. (1940). Jahrb. AEG Forsch. **7**, 57.

Gribi, M., Thürkauf, M., Villiger, W., and Wegmann, L. (1959). Optik **16**, 65.

Grivet, P. (1958). "Optique Electronique," Vol. II, p. 170ff. Bordas, Paris; (engl. Edition: Electron Optics, Pergamon, Oxford, 1965).

Hahn, E. (1954). Jenaer Jahrb. **1954**, part 1, 63.

Hahn, E. (1958). Jenaer Jahrb. **1958**, part 1, 184.

Hahn, E. (1959). Exptl. Tech. Physik **7**, 258.

Hahn, E. (1961). Jenaer Jahrb. **1961**, part 2, 325.

Hahn, E. (1964). Jenaer Jahrb. 1964, 217.

Hamisch, H., and Oldenburg, K. (1964). Proc. 3rd Regional Conf. Electron Microscopy, Prague, 1964. Vol. A, p. 41. Publishing House of Czech. Acad. Sci., Prague.

Hanszen, K.-J. (1958a). Optik **15**, 304.

Hanszen, K.-J. (1958b). Z. Naturforsch. **13a**, 409.

Hanszen, K.-J. (1962). Proc. 5th Intern. Conf., Electron Microscopy, Philadelphia, 1962. Vol. I, KK 11. Academic Press, New York.

Hanszen, K.-J. (1964a). Z. Naturforsch. **19a**, 896.

Hanszen, K.-J. (1964b). Naturwissenschaften **51**, 379.

Hanszen, K.-J. (1964c). Proc. 3rd Regional Conf. Electron Microscopy, Prague 1964. Vol. A, p. 47. Publishing House of Czech. Acad. Sci., Prague.

Hanszen, K.-J., and Lauer, R. (1965). Optik **20**, *23*, 478.

Hartl, W. A. M. (1966). Z. Physik. **191**, 487.

Heise, F. (1949). Optik **5**, 479.

Heise, F., and Rang O., (1949). Optik **5**, 201.

Hoeft, J. (1959). Z. Angew. Phys. **11**, 380.

Hottenroth, G. (1937). Ann. Physik [5] **30**, 689.

Illenberger, A. (1964). Mikroskopie **19**, 316.

Jacob, L., and Shah, J. R. (1953). J. Appl. Phys. **24**, 1261.

Kanaya, K., Kawakatsu, H., Yamazaki, H., and Sibata, S. (1966). *J. Sci. Instr.* **43**, 416.

Kessler, J., and Lindner, H. (1964). *Z. Angew. Phys.* **18**, 7.

Klemperer, O., and Wright, W. D. (1939). *Proc. Phys. Soc. (London)* **51**, 296.

Kofoid, M. J. (1960a). *Power App. Systems* No. 51, 991.

Kofoid, M. J. (1960b). *Power App. Systems* No. 51, 999.

Laplume, J. (1947). *Cahiers Phys.* [5] **29-30**, 55.

Lauer, R., and Hanszen, K.-J. (1966). *6th Intern. Congress for Electron Microscopy, Kyoto*, Vol. I, p. 129.

Lauer, R. (1967). *Z. Naturforsch.* **22a**, in print.

Liebmann, G. (1949). *Proc. Phys. Soc. (London)* **B62**, 213.

Lippert, W. (1955a). *Optik* **12**, 173.

Lippert, W. (1955b). *Optik* **12**, 467.

Lippert, W., and Pohlit, W. (1952). *Optik* **9**, 456.

Lippert, W., and Pohlit, W. (1953). *Optik* **10**, 447.

Lippert, W., and Pohlit, W. (1954). *Optik* **11**, 181.

MacNaughton, M. M. (1952). *Proc. Phys. Soc. (London)* **B65**, 590.

Mahl, H. (1940). *Jahrb. AEG Forsch.* **7**, 43.

Mahl, H., and Recknagel, A. (1944). *Z. Physik* **122**, 660.

Metherell, A. J. F., and Whelan, M. J. (1966). *J. Appl. Phys.* **37**, 1737.

Miller, H. C. (1966). *J. Appl. Phys.* **37**, 784.

Möllenstedt, G., and Dietrich, W. (1955). *Optik* **12**, 246.

Möllenstedt, G., and Rang, O. (1951). *Z. Angew. Phys.* **3**, 187.

Nicoll, F. H. (1938). *Proc. Phys. Soc. (London)* **50**, 888.

Rabinowitz, M. (1965). *Vacuum* **15**, 59.

Rang, O. (1949). *Optik* **5**, 518.

Recknagel, A. (1941). *Z. Physik* **117**, 689.

Recknagel, A. (1943). *Z. Physik* **120**, 331.

Regenstreif, E. (1951). *Ann. Radioelec. Compagn. Franc. Assoc. T.S.F.* **6**, 51, 114, 245, and 299.

Schumacher, B. W. (1966). *Rev. Sci. Instr.* **8**, 1092.

Seeliger, R. (1948). *Optik* **4**, 258.

Septier, A. (1954a). *Ann. Radioelec. Compagn. Franc. Assoc. T.S.F.* **9**, 374.

Septier, A. (1954b). *J. Phys. Radium* [8] **15**, 573.

Septier, A. and Ruytoor, M. (1959). *Compt. Rend.* **249**, 2175.

Septier, A. (1960). *CERN Rept.* **60-39**.

Shipley, D. W. (1952). *J. Appl. Phys.* **23**, 1310.

Simpson, J. A., and Marton L. (1961). *Rev. Sci. Instr.* **32**, 802.

Soa, E. A. (1959). *Jenaer Jahrb.* **1959**, part 1, 115.

Spangenberg, K. L., and Field, L. M. (1942). *Proc. IRE* **30**, 138.

Verster, J. L. (1963). *Philips. Res. Rept.* **18**, 465.

Vine, J. (1960). *Brit. J. Appl. Phys.* **11**, 408.

von Ardenne, M. (1956). "Tabellen der Elektronenphysik, Ionenphysik und Übermikroskopie," Vol. I, p. 118. VEB Deutscher Verlag der Wissenschaften, Berlin.

von Ardenne, M. (1964). "Tabellen zur angewandten Physik," Vol. II. VEB Deutscher Verlag der Wissenschaften, Berlin.

Wendt, G. (1940). *Z. Physik* **116**, 436.

CHAPTER 2.3

MAGNETIC ELECTRON LENSES

C. Fert and P. Durandeau

FACULTÉ DES SCIENCES
TOULOUSE, FRANCE

2.3.1. Introduction

A coil of axial symmetry around the axis $z'z$ (Fig. 1a), when it conducts a current, gives a *field of magnetic induction $B(z)$*, whose distribution is the same in all meridian planes, such as the plane xOz of Fig. 1. The effect of this field on a paraxial beam of charged particles is comparable to that of a centered system on a beam of light rays: the induction field created by the coil constitutes a *magnetic lens*.

In a wider sense, the coil itself is referred to as a magnetic lens, but it is advisable to bear in mind that the lens is in fact the induction field.

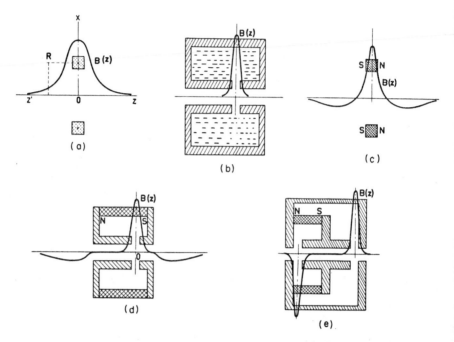

FIG. 1. Some examples of magnetic lenses. (a) Ironless lens; (b) lens of the electromagnet type; (c) lens with a permanent simple magnet; (d) lens having a permanent magnet with magnetic circuit and one gap; (e) lens having a permanent magnet with magnetic circuit and two gaps.

In order to obtain a more intense or better localized field, the coil may be surrounded by a *magnetic circuit* of ferromagnetic material. The field is localized in the neighborhood of the air gap; a hole bored in the pole-faces and in the central core allows passage of the beam of particles (Fig. 1b). The lens is then an electromagnet.

In order to avoid the need of current supply, magnetic lenses with permanent magnets may be used. Figure 1c shows a very simple model: a ring of ferrite that is magnetized parallel to its axis; Fig. 1d,e shows a more complex model, which involves a magnetic circuit.

Magnetic lenses are used mainly, if not exclusively, for the focalization of electron beams.[1] The most widely used apparatus with magnetic lenses is undoubtedly the electron microscope, but instruments with electronic spots (such as microanalyzers, scanning microscopes), electron diffractographs, and some velocity spectrometers must be mentioned as well.[2] In addition, less powerful magnetic lenses are sometimes utilized for the focalization of electron beams on tubes, such as image tubes or television tubes.

2.3.2. General Properties of Magnetic Lenses

These properties are recalled very briefly here, in as much as the object is to develop, in Sections 2.3.3 and 2.3.4., practical rules for the construction of magnetic lenses.

A. MAGNETIC INDUCTION ON THE AXIS

From the scalar function $B(z)$, which represents the intensity of induction on the axis of the lens, calculation of the induction on all neighboring points of the axis can be made by a power series; the latter is obtained easily because of the fact that induction in vacuum comes from a potential that satisfies the Laplace equation.[3]

Hence, it is sufficient to know the distribution $B(z)$ along the axis in order to calculate all the properties of a lens. The function $B(z)$ is shown in Fig. 1 for different lenses.

The function $B(z)$ can sometimes be calculated; it can always be measured. Calculation is sometimes difficult in the case of an eventual saturation of the magnetic circuit, but the measurement can always be made without ambiguity.[4]

A simple *rule of similitude* may be given which will prove valid for a magnetic lens without iron, or for a magnetic lens with a coil, the iron of which

---

[1] The trajectory Eq. (6) makes the factor $qB^2/8\,m_0V^*$ appear, where m_0 is the mass of the particle, q is its charge. For ions, the value of the term m_0 imposes very intense or very extended fields, that leads to the preference of electrostatic lenses for the focalization of ions.

[2] The question of Beta-spectrometers in which the lens behaves as a separator will not be discussed here.

[3] See a standard handbook on electromagnetism.

[4] Among the articles concerning the calculation or measurement of $B(z)$ in the case of magnetic lenses, consult the following: Dosse (1941b), Ments and Le Poole (1947), Liebmann (1951), Durandeau *et al.* (1959b), and Gautier (1957). A representation of $B(z)$ for an extended series of magnetic lenses will be found in Durandeau *et al.* (1959b).

presents no saturation. For these lenses, $B(z)$ is proportional to the number of ampere-turns NI of the magnetizing coil (N number of turns; I, intensity of current). If we consider different lenses of *like geometry*, but that are built according to different scales (for example, coil and magnetic circuit), and designate by S a characteristic length of the scale (S represents, for example, the gap), the function $B(z)$ is proportional to NI/S. This rule of similitude permits us to utilize in a large interval of excitation or dimensions the distributions $B(z)$ given in the articles referred to.[4]

Note on Ampere's Theorem. Ampere's theorem permits, along the axis of revolution of the lens, the notation

$$\int_{-\infty}^{+\infty} B(z)\, dz = \mu_0 \cdot NI \qquad (\mu_0 = 4\pi \cdot 10^{-7}) \tag{1}$$

NI being the number of ampere-turns of the magnetizing coil.

For a lens with permanent magnet, $NI = 0$, and the preceding integral is always null.

It is often useful to calculate the integral $\int_{z_0}^{z_1} B(z)\, dz$ between two abscissas z_0, z_1 and to define conventionally a number of ampere-turns nI by the relation

$$\int_{z_0}^{z_1} B(z)\, dz = \mu_0 \cdot nI \tag{2}$$

For purposes of convenience, we shall say that there exists nI ampere-turns distributed between the abscissas z_0 and z_1.

B. Gaussian Approximation

1. *Relativistic Correction*

In a magnetic lens, induction B is independent of time, and the force $\mathbf{f} = q[\mathbf{v} \wedge \mathbf{B}]$ is normal to the velocity of the particle. It follows that the velocity v of the particle remains constant the full length of the trajectory; the same holds true of the mass $m = m_0/(1 - \beta^2)^{1/2}$ (m_0, mass at rest; $\beta = v/c$).

The symbol V is used to designate the acceleration potential of the particle; if q is the charge of the particle, its energy in joules will be qV, and its momentum

$$m v = \left[2\, m_0\, |\, q\, V\,| \left(1 + \frac{|\, q\, V\,|}{2\, m_0\, c^2} \right) \right]^{1/2} \tag{3}$$

It is convenient to introduce the expression

$$V^* = V \left(1 + \frac{|\, q\, V\,|}{2\, m_0 c^2} \right) \tag{4}$$

In the case of magnetic lenses, it happens that the same expressions (equation of trajectories, focal length, aberration coefficients, etc.) are obtained for relativistic particles on condition that V^* is written in place of V, and m_0 (mass at rest) in place of m. These conventions will be employed here; consequently, the results given below are valid for relativistic particles.

Magnetic lenses are used almost exclusively for focalizing electrons; for that reason, e is used in place of q for the charge of the particle.

For electrons, $e/m_0 = 1.76 \cdot 10^{11}$ C/kg; $V^* = V(1 + 0.98 \cdot 10^{-6} V)$. The corrective term is weaker for ions of like energy, of which the mass M_0 is much higher.

2. Equation of Trajectories

Usually, the cylindrical variables r, φ, z are introduced. It is preferable (Durand, 1954) to use, in place of r, φ, z, the complex variable $u = x + jy$ $= r \exp(j\varphi)$; x and y are the coordinates of point M of the trajectory in

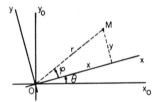

FIG. 2. Reference axes.

the abscissa plane z, in relation to the axes x, y. The equation of trajectories is particularly simple if the axes (x, y) have an orientation varying with z. If (x_0, y_0) represent the axes in the abscissa plane z_0, and (x, y) in the abscissa plane z (Fig. 2), we can postulate

$$\theta = \left(\frac{e}{8 \, m_0 V^*}\right)^{1/2} \int_{z_0}^{z_1} B(z) \, dz \tag{5}$$

With these conventions, the equation of trajectories takes the simple form

$$\frac{d^2 u}{dz^2} + \frac{e B^2}{8 \, m_0 V^*} \, u = 0 \tag{6}$$

regardless of the initial conditions.

3. Optical Properties

a. *Stigmatism.* If the rotation of axes defined by (5) is not taken into account, it follows from the equation of trajectories that a lens is stigmatic for a beam of particles of the same nature and energy. This means, as Fig. 3 shows, that particles passing through the same object point B_0, in the frontal plane z_0, converge on the same point image B_1, in the frontal plane z_1. The figure represents only trajectories in a plane, without taking the rotation into account, but the form of Eq. (6) shows that this result remains true for all trajectories.

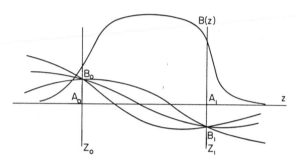

FIG. 3. Stigmatism of magnetic lenses.

Image Rotation. A change in the orientation of the axes defined by (5) does not modify this conclusion, because the change of orientation has a value θ, characteristic of z_0 and z_1.

Correspondence and Magnification Equations: Real Object, Real Image. Different correspondence conditions may be distinguished between an "object" and its "image" for the same magnetic lens, in the material sense of the term.

Since a real nonferromagnetic object can be immersed in an induction field without perceptible perturbation, a first correspondence may be defined between a real object $\overline{A_0B_0}$ immersed, and a real image $\overline{A_1B_1}$ (Fig. 4a). Suppose $\sigma(z)$, $\varrho(z)$ are two particular and independent trajectories. It can easily be shown that, for an object in abscissa z_0, the abscissa of the image plane and the enlargement are given by the correspondence and magnification equation:

$$\frac{\varrho(z_1)}{\varrho(z_0)} = \frac{\sigma(z_1)}{\sigma(z_0)} = M \tag{7}$$

In general, *this relationship does not define a homographic correspondence between z_0 and z_1.*

It should be noted that the field zone useful for the formation of the image is the one that is contained between abscissa plane z_0 and z_1. It varies with the position of the object.

Correspondence and Magnification Equation: Asymptotic Correspondence.
Figure 4b, defines another type of correspondence: In this case, the object
A_0 is a virtual object (A_0 is the point of convergence of a beam given by a
first lens, the objective of an electron microscope, for example). In this
case, the entire field of the lens contributes to the formation of image A_1,
presumed to be distant. Relationship (7) remains valid; but the asymptotes
of $\sigma(z)$ and $\varrho(z)$ should be taken in order to find the correspondence equation, which takes the form

$$\frac{p_1 z + q_1}{p_0 z + q_0} = \frac{r_1 z + s_1}{r_0 z + s_0} \tag{8}$$

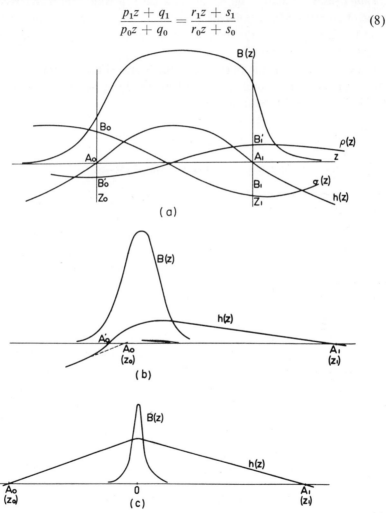

FIG. 4. Different types of correspondence. (a) Real object, real image; (b) asymptotic
correspondence; (c) thin lens.

where $p_1 z + q_1$ is an asymptote of $\varrho(z)$ when z tends toward $+\infty$; $p_0 z + q_0$ is an asymptote of $\varrho(z)$ when z tends toward $-\infty$; $r_1 z + s_1$ and $r_0 z + s_0$, asymptotes of $\sigma(z)$ when z tends toward $\pm\infty$. This correspondence is frequently referred to as *asymptotic correspondence*. It will be noted that the relation (8) that links z_0 and z_1 is a *homographic relation*, as in optics. In this case, and only in this case, two focuses, two principal planes, and a focal distance characteristic of the lens can be defined as in optics.

Thin Lenses.   Lastly, Fig. 4c shows a case in which the two preceding types of correspondence coincide: A_0 and A_1 are both far enough removed from the useful zone of the field for the trajectories to be rectilinear in the neighborhood of A_0 and A_1.

In extreme cases, if the thickness of the lens is slight in comparison with the distances from A_0 and A_1 to the lens, the term *thin lens* may be used; its focal distance f is expressed by

$$\frac{1}{f} = \frac{e}{8\,m_0 V^*} \int_{-\infty}^{+\infty} B^2 \, dz \tag{9}$$

and the correspondence equation takes the standard form

$$\frac{1}{p} + \frac{1}{p'} = \frac{1}{f} \quad (p = \overline{A_0 O},\ p' = \overline{OA_1}) \tag{10}$$

Osculatory Correspondence.   To return to the case of a real object and a real image immersed in the field (Fig. 4a) (Section 2.3.2,B,3c), the correspondence equation (7) is not, except in very rare cases, a homographic relation.

Consider, however, a lens used in such a way that the real object is always located in the neighborhood of a point A_0 on abscissa z_0 and the image in the neighborhood of a point A_1 on abscissa z_1, conjugated from A_0. Such is the case, for example, with the objective lens of an electron microscope: the image is always at a great distance; the object is in the neighborhood of F_0 conjugated from an image at infinite distance.

Let $z_0 + dz_0$, $z_1 + dz_1$ represent the abscissa of two conjugated points A_0', A_1' near A_0 and A_1, respectively. Let us develop $\sigma(z)$ and $\varrho(z)$ and use only the terms of first and second order in dz_0 and dz_1; the following correspondence equation is obtained:

$$\frac{\sigma(z_1) + \sigma'(z_1) \cdot dz_1}{\sigma(z_0) + \sigma'(z_0) \cdot dz_0} = \frac{\varrho(z_1) + \varrho'(z_1)\, dz_1}{\varrho(z_0) + \varrho'(z_0)\, dz_0} = M \tag{11}$$

In this relation, $\sigma(z_1)$, $\sigma(z_0)$, $\sigma'(z_1)$, $\sigma'(z_0)$, $\varrho(z_1)$, $\varrho(z_0)$, $\varrho'(z_0)$, $\varrho'(z_1)$, are constants, and the correspondence relation is a homographic relation. It is valid to the third order, since the terms of the second order have been eliminated. This relation is called an *osculatory homographic relation* near to A_0, A_1 (Glaser, 1952). For this zone, focal points, principal planes and focal distance can be defined as in optics (Fig. 5).

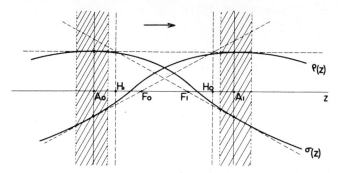

FIG. 5. Construction of the elements of the osculatory homographic correspondence in the neighborhood of (A_0, A_1); $\sigma(z)$ was chosen such that the tangent to $\sigma(z)$ is parallel to t the axis for abscissa z_0; $\varrho(z)$ was chosen so that the tangent to $\varrho(z)$ is parallel to the axis for abscissa z_1. F_0, F_1, focal points; H_0, H_1, principal planes; $F_0H_0 = H_1F_1 = f$, focal distance. These elements are valid when the object and the image have abscissas in the neighborhood of z_0 and z_1; for example, when the object and the image are in the zones shown in cross-hatching.

But it must be borne in mind that, generally speaking, the elements of this homographic correspondence are not to be used except for an object near to A_0, and an image near to A_1. These elements are different in the neighborhood of another pair of conjugated points.

C. ABERRATIONS

Aberration coefficients have been calculated by various authors. The reader is referred, in particular, to the general study by Durand (1954), in which a relatively simple method is used, but the various works mentioned in the bibliography may also be consulted.[5] Anisotropic aberrations relative to the rotation of images should be added to the standard optical aberrations.

D. MATHEMATICAL MODELS

Analytical calculation of trajectories supposes that the expression $B(z)$ is known. At the beginning of the development of electron optics, different

[5] For example, Glaser (1956, p. 219).

mathematical "models" which would permit a simple calculation were proposed to represent the function $B(z)$. The most frequently used models are

(a) The homogeneous field: $B(z) = B_0$ between abscissas $z = \pm L/2$ which limit the field (Fig. 6a);

(b) The "Glaser" field (1941), (Fig. 6b).

$$B(z) = B_M/[1 + (z^2/a^2)] \qquad (12)$$

(c) The Grivet-Lenz field (Grivet, 1952)

$$B(z) = B_M/ch\,\frac{z}{b}$$

The homogeneous field is naturally the simpler. It constitutes a convenient model and leads to expressions which represent rather well the real properties of shielded lenses in the Gaussian approximation (see Section 2.3.3).

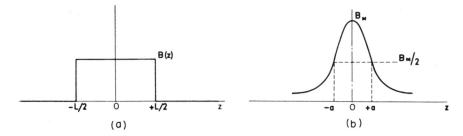

FIG. 6. Models. (a) constant field; (b) Glaser field.

The Glaser field was, and continues to be, much used. It is to be noted, however, that, in this formulation, $B(z)$ decreases much more slowly than in a real lens. It is sometimes said that the Glaser field effectively represents the case of a lens of which the polar faces are saturated. This is not exact: If saturation appears on the polar faces, it is accompanied by a widening of the distribution which is not correctly represented by the Glaser field. If saturation appears far from the pole separation, it does so in an entirely different way.[6]

---

[6] It should also be noted that the Glaser field has the following exceptional property: The osculatory relation defined by (11) is represented by the same cardinal elements, regardless of the conjugated points. In other words, the relation (7) proves to be homographic (Glaser, 1941). But this is only casual coincidence, and is not true for real lenses. We shall consider other examples of differences between experimental values and the calculated ones from the Glaser field.

At present, it appears preferable to us to use the sufficiently complete results obtained on real lenses by calculation and measurement. We refer the reader particularly to the extended research done by Liebmann (1951, 1952a,b,c, 1955a,b,c) who utilized an analogous method for the calculation of the field, and to that of Durandeau, (1956; Durandeau and Fert, 1957; Durandeau *et al.*, 1959a,b) whose results were obtained from measurements of induction on real lenses. The older study of Ments and Le Poole (1947) and the publications of Lenz (1950a,b) will also prove useful.

2.3.3. Ironless Magnetic Lenses

The study of ironless magnetic lenses is simple. We review here the results concerning:

(a) A coil assimilable to a circular turn of ray R,
(b) A long solenoid

A. THIN COIL

Supposing that the coil gives a field that can be perceptibly reduced to that of a thin coil of radius R, let

$$B(z) = \frac{B_M}{[1 + (z^2/R^2)]^{3/2}} \quad \text{with} \quad B_M = \mu_0 \frac{NI}{2R} \tag{13}$$

This lens will generally behave in the same way as a thin lens; its focal length, according to (9) is expressed by the following equation:

$$\frac{R}{f} = \frac{3\pi}{8} \frac{e B_M^2 R^2}{8 m_0 V^*} = \frac{3\pi}{32} \frac{e}{8 m_0} \mu_0^2 \frac{(NI)^2}{V^*} \tag{14}$$

and the rotation of the image will as a matter of course have the value

$$\theta = \mu_0 \left(\frac{e}{8 m_0}\right)^{1/2} \frac{NI}{(V^*)^{1/2}} \tag{15}$$

Lastly, the chromatic aberration coefficient C_c and the C_s spherical aberration coefficient have the respective values

$$C_c \quad = f \tag{16}$$

$$C_s R^2 / f^3 = 0.47 \tag{17}$$

For electrons, (14) and (15) become

$$\frac{R}{f} = 1.02 \cdot 10^{-2} \frac{(NI)^2}{V^*}$$ (18)

$$\theta = 0.186 \frac{NI}{(V^*)^{1/2}}$$ (19)

N.B.1. Recently, Le Poole (1964) built miniature and powerful magnetic lenses without iron. The winding is built so that, without overheating, the current density can be greatly increased, for example, up to 60–80 A/mm² as compared with 2–3 A/mm² in conventional lenses. As a result of the high current density, a small coil, 25–50 mm long, having a 3–4-mm i.d. and a 7–9-mm o.d. suffices for some 2.000–2.400 At. These lenses are proposed particularly to produce fine electron probes.

N.B.2. It has been proposed to use magnetic lenses without iron, including a supraconducting coil. In that case, one can obtain a very powerful lens without iron, which can be no more assimilated to a thin lens. But the distribution $B(z)$ on the axis can always be calculated, and hence the electron-optical properties.

B. LONG SOLENOID

The field is uniform, with $B = \mu_0 nI$ (n, number of turns per unit of length).

The trajectory is a helix the step of which, in the Gaussian approximation, is

$$z_1 - z_0 = \pi \left(\frac{8 \, m_0 V^*}{B^2} \right)^{1/2}$$ (20)

The distance between the two successive images is $\Delta z = z_1 - z_0$. The magnification is 1, but the rotation is equal to π.

It is noted that, between the frontal planes of the two successive images,

$$B \cdot \Delta z = \mu_0 \cdot nI \cdot \Delta z = \pi \left(\frac{8 \, m_0 V^*}{e} \right)^{1/2}$$ (21)

For electrons,

$$nI \cdot \Delta z = 16.9(V^*)$$ (22)

The calculation being well known and easy, even when the Gaussian approximation conditions are not respected, we will not develop in further detail the study of this case.

2.3.4. Magnetic Lenses of the Electromagnet Type

Figure 7 shows schematically a lens of the electromagnetic type and recalls the names of its different elements. Figures 8a and 8b show the distribution $B(z)$ on the axis adjacent to the pole separation in two specific cases.

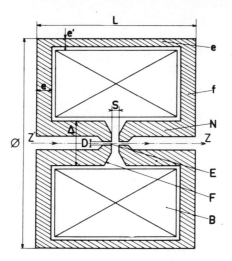

FIG. 7. General structure of a lens of the electromagnet type: S, gap; F, polar faces; N, polar cores; f, endplates; e, outer envelope; b, coil.

A. MAGNETIC CIRCUIT

The rules for construction of a magnetic circuit are the same as those for the construction of an electromagnet. One difference must be pointed out, however.

In an electromagnet, the field which eventually exists in the channel that was bored in the polar cores outside the pole separation brings no particular difficulty; its presence necessitates only supplementary ampere-turns of the coil in the meaning indicated in Section 2.3.2,A.

For an electron lens, on the contrary, it is necessary to avoid the presence of an intense field in the channel, since the electronic beam undergoes its action; the latter takes the form of a modification, usually in an unfavorable way, of the electron-optical properties of the lens.

For a magnetic lens, the curve $B(z)$ must be reduced to that expressed in Figs. 8a or 8b. The role of the magnetic circuit is to obtain this result. For that reason, any saturation of the magnetic circuit outside the polar faces limiting the gap must be avoided.

The common soft iron used in the construction of the magnetic circuit of the lens possesses great permeability, higher than 1000, as long as induction in the iron does not exceed 1.25–1.4 Wb/m². In order to render negligible the field of induction in the channel, a section must be given to the different parts of the magnetic circuit such that induction in the iron remains lower than that value.

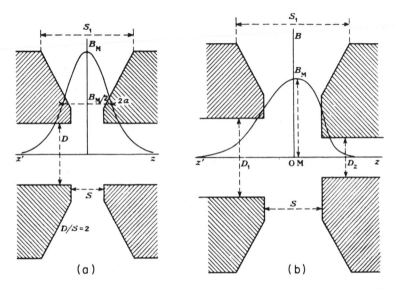

FIG. 8. Distribution of the field along the axis. (a) Symmetrical lens $(D/S) = 2$; (b) dissymmetrical lens $(D_1 + D_2)/2S = 1$, $(D_1/D_2) = 2$.

1. Polar Cores

In the polar core the flux increases from the gap into the endplate (Mulvey, 1952; Durandeau, 1956); consequently, a value which increases from the gap toward the endplate must be given to the area of the section of the polar cores, in such a way that, in soft iron, the average density φ/σ of flux φ through the area σ remains weak enough to avoid incipient saturation.

Mulvey (1952) proposes to attain this result by reducing the plane portion of the polar faces (Fig. 9), and prolonging it by a long portion in the shape of a truncated cone NPQ of an angle θ_1 varying from 60 to 80°, then by a cylindrical core QQ'. Under these conditions, φ/σ diminishes appreciably between sections σ_0 and σ_2, then increases between σ_2 and σ_3, but without exceeding the saturation threshold if the cone NPQ has been sufficiently extended.

Durandeau (1956) prefers to reduce the conical portion of angle θ_1 and to adopt beyond it the conic core of angle θ_2 (10° to 15°). Under these conditions, φ/σ varies first, as in the preceding case, between the section σ_0 and σ_1, then, if θ_2 is well chosen, remains constant between σ_1 and σ_3.

For Figure 9, it is supposed that, at the level of σ_0, φ/σ_0 corresponded to an incipient saturation of the polar faces, whereas at the level of σ_1, φ/σ_1 was lower than the saturation threshold.

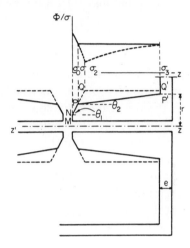

FIG. 9. Magnetic circuit of a lens of the electromagnet type, and variation of the average density φ/σ of the flux of induction in the polar core. Broken line: polar faces, truncated cone, cylindrical cores (Mulvey, 1952). Continuous line: polar faces and cores in the form of truncated cones (Durandeau, 1956).

The solution of the core in the shape of a truncated cone has already been advocated for electromagnets. It should be retained whenever a powerful lens is desired, for which φ/σ at the level of the polar faces is large. It will be useful to consult the lens objective presented by Ruska (1964) or that for lenses of a high-tension microscope (Dupouy and Perrier, 1962).

2. Endplate and Outer Envelope

The flux of induction from the base of the core (Fig. 9, section σ_3) spreads out in the endplate. The latter should be given a thickness such that there is no diminution in the section of the magnetic circuit. In moving away from the axis, the flux finds an increasingly greater area, which avoids all possibility of saturation.

For the outer envelope, a section somewhat higher than the area of the core is suitable.

Experience shows that, with a lens constructed to respect these rules, induction in the channel is negligible for the strongest excitation used; $B(z)$ is reduced to the distribution shown in Fig. 8; this distribution brings together all the ampere-turns of excitation.

If the lens is always to work at weak excitation, the rules set forth do not have an imperative character and may, of course, be relaxed.

B. POLAR FACES

Suppose we have a lens whose magnetic circuit satisfies the rules set forth above: There is no saturation of the magnetic lens except at the level of the polar faces on both sides of the gap; induction in the channel is negligible for all excitations used; the distribution $B(z)$ is reduced to a bell curve the maximum of which is in the gap (Figs. 8a and 8b); this bell curve brings together all the ampere-turns of excitation.

It remains to be specified the manner of drawing the opposite polar faces, on both sides of the gap, for the given values of S, D_1, D_2, in order to obtain the most favorable distribution, that is, in general, distribution that has the lowest spread for a given excitation.

1. Distribution of the Induction on the Axis at Increasing Excitation

Lay out the curve $B(z)/B_M$ at increasing excitation, for the given polar faces.

If the excitation is weak, this curve remains invariable when NI is augmented; the maximum value B_M of B augments proportionally to NI.

Figure 10 shows the variation of B_M and of width a for a symmetrical lens in function of NI/S; the excitation may be considered weak as long as NI/S remains lower than 1100 or 1200, if S is expressed in millimeters. Under these conditions, the rules of similitude set forth in Section 2.3.2,A are valid.

But Figure 10 shows that, if the excitation, expressed by NI/S, exceeds a certain critical value NI_c/S, the induction B increases less rapidly than the number of ampere-turns, and the width a of the distribution increases.

The phenomena just described can be easily explained.

For a weak excitation, the permeability of the iron is very high at all points of the magnetic circuit, including the polar faces. The proportionality of B to NI and the independence of $B(z)/B_M$ and of NI result from this hypothesis, which has permitted different authors to make the numerical calculation of induction in a magnetic lens (Liebmann and Grad, 1951; Durand, 1955).

But the permeability of iron varies rapidly from a certain induction value, near 1.4 Wb/m² for a common soft iron. After this point, B_M is no longer proportional to NI and the distribution a widens.

A simple chain of reasoning makes it possible to find the order of magnitude of the zone of transition illustrated in Fig. 10.

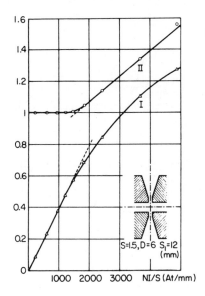

FIG. 10. Variation of the maximum induction B and of the width "a" of the distribution in function of NI/S; polar pieces of soft iron; symmetrical lens. Curve I: B (tesla); curve II: a/a_0.

Suppose the polar faces are plane and parallel. It may be assumed that, in the gap, at some distance from the zone where the holes bored along the axis of the lens are situated, induction is uniform and equal to $B = B_1 = \mu_0 \, NI/S$, in the hypothesis of a high permeability. As a result of the conservation of the normal component of induction in passing through the surface of separation between the iron and the vacuum, it may also be stated that B_1 represents the induction in the iron on the polar faces. The effects of saturation will become perceptible when, for a common soft iron, reaches 1.4 Wb/m², that is to say for

$$NI/S = B_1/\mu_0 = 1.1 \cdot 10^6 \text{ At/m}, \quad \text{or} \quad 1100 \text{ At/mm} \tag{23}$$

This is shown in Fig. 10. In practice and following the desired approximation, $NI_c/S \sim$ 1100–1300 At/mm may be stated as the value for polar faces
of soft iron.

It should be noted that, for $NI/S > 1100$, saturation appears only on the
polar faces and not in other parts of the magnetic circuit, if the latter has
been well designed. Under these conditions, when NI/S exceeds the critical
value, the distribution widens, but no field appears in the channel outside
the zone adjacent to the gap.

2. Influence of the Profile of the Polar Faces

a *Weak Excitation.* If $NI/S < 1100$, the distribution $B(z)$ depends very
little on the profile of the polar faces for the given values of S, D_1, D_2
(Fig. 11). The function $B(z)$ is almost the same as that obtained in the
hypothesis of infinite permeability, with polar pieces having plane and paral-

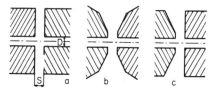

FIG. 11. For the polar faces of this figure, S and D have the same value; is $NI < 1100\,S$
(soft iron, S in mm), the distribution $B(z)$, in a first approximation, is the same for the
three profiles.

lel faces (Fig. 11a). The larger the plane part around the hole bored in the
polar faces, the better is the approximation.

This result, immediately comprehensible, entails the extensive validity of
the results obtained by the authors who adopted these hypotheses for the
calculation of the field on the axis, and, in particular, the important research
of Liebmann.

The rules of similitude take, at this point, a particularly simple form.
On taking the gap S as unit of length and drawing the distribution $B(z)$
in reduced coordinates, postulating $\beta(z/s) = B(z/s)/B_M$, a unique curve
is obtained for all lenses which have the same value of $(D_1 + D_2)/2S$ and
D_1/D_2.[7]

These curves are given, as well as those of the first and second derivatives,
for an entire series of lenses, by Durandeau *et al.* (1959b).

---

[7] For the value $(D_1 + D_2)/2S < 0.5$, see the observation p. 90A in the work of Durandeau *et al.* (1959b).

The maximum induction B_M is suitably represented by the empirical formula,

$$B_M = \frac{\mu_0 NI}{\left[S^2 + 0.45\left(\dfrac{D_1 + D_2}{2}\right)^2\right]^{1/2}} = \mu_0 \frac{NI}{L} \qquad (24)$$

The difference between the experimental value and the deduced value of (24) is given in Fig. 12: It is always lower than 5% for the most usual lenses.

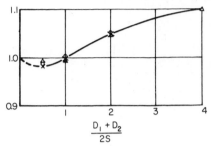

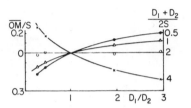

FIG. 12. Maximum intensity of induction on axis in function of $(D_1 + D_2)/2S$: comparison of the given values for the formula (24) and the experimental values. The curve is drawn for $D_1/D_2 = 1$. The relation between the maximum induction calculated according to (2) was noted as ordinated and the maximum induction measured. 0: $D_1/D_2 = 2$ or $\frac{1}{2}$; Δ: $D_1/D_2 = 3$ or $\frac{1}{3}$.

FIG. 13. Abscissa of the maximum of induction, origin taken in the middle of the gap (OM is defined in Fig. 8b). $(D_1 + D_2)/2S = 0.5$: ● ; $1 : \Delta$; $2 : \bigcirc$; $4 : x$.

The distance OM (Fig. 13) from point M of the axis, where induction is maximal, to the central point O of the gap is given in Fig. 13 (Durandeau et al., 1959b).

Lastly, for symmetrical lenses ($D_1 = D_2$), the width $2a$ (Fig. 8a) is suitably represented by[8]

$$2a = 0.97(S^2 + 0.45D^2)^{1/2} \qquad (25)$$

b. *Strong Excitation.* If NI/S exceeds the critical value noted in the foregoing, the distribution $B(z)$ widens, whereas B_M no longer increases proportionally to NI. The phenomena then depend on the form of the polar faces.

---

[8] See Durandeau and Fert (1957). The difference between the experimental value and the deduced value of (25) is lower than 3% when D/S varies from 0.5 to 4.

In order to obtain a minimal widening, polar faces presenting a plane area adjacent to the hole should be chosen, followed by a portion in the shape of a truncated cone, the angle of which is $\theta_1 = 60°{-}80°$, until the gap reaches a value S_1 (Fig. 14) such that $NI_M/S \sim 500$, approximately (NI_M

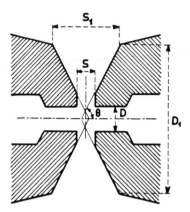

FIG. 14. Profile of polar faces, symmetrical lenses.

maximum number of ampere-turns that can be utilized). Mechanical reasons may, of course, necessitate plottings such as those of Fig. 15; the essential thing is assuring that there is no longer any saturation when the periphery of the polar faces is reached, as Fig. 9 indicates.

3. Influence of the Channel Bored in the Polar Pieces

In order for the above results to remain valid, it is necessary that the channel bored in the polar pieces remain cylindrical for a length comparable to the diameter of the holes. Beyond that, it is possible to increase its diameter without noticeable effect on the distribution $B(z)$.

4. Observation

(a) The above results, completed by the distributions $B(z)$ of Liebmann and Grad (1951) or of Durandeau et al. (1959b), permit plotting a profile of polar pieces and providing for the essential given conditions of the distribution for a lower or slightly higher excitation than NI_c. In actual practice, it is rare that anything is to be gained from greatly exceeding NI_c.

(b) It is well known that a very serious defect of the magnetic lenses that are used in electron microscopes is the astigmatism of the objective lens;

the slighter this astigmatism is, initially, the more easily it can be corrected. It seems, according to numerous results presented by various authors, that it is advantageous, in order to reduce the astigmatism of a lens before correction, to choose a value for the gap such that there is incipient saturation of the polar faces. This saturation probably attenuates the defect of homogeneity of the metal [see, for example, results from Morito *et al.* (1960)].

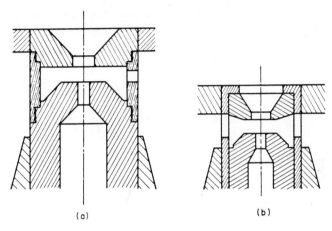

(a)   (b)

FIG. 15. Two examples of polar pieces.

(c) By a comparison of the value of B_M measured by certain authors with that provided for in the formula (24), certain differences are observed, even when NI/S is lower than 1000 or 1200. This is not due to a defect in validity of the formula (24) but to the fact that the measurements were made on a lens whose magnetic circuit was partly saturated; it is evident that if ampere-turns are lost in order to create a field in the channel far from the gap, they cannot be recovered in the gap.

C. ELECTRON-OPTICAL PROPERTIES

1. *Form Parameter and Excitation Coefficient*

The electron-optical magnitudes of a magnetic lens are often given as a function of the parameter $k^2 = eB_M^2 a^2/8m_0 V^*$ (B_M maximal induction; a halfwidth for the Glaser model, or radius of the hole bored in the polar faces, for Liebmann).

The study of a lens design is very much simplified if the electron-optical parameters are represented directly as functions of geometric (S, D_1, D_2)

and electric (NI, V^*) parameters that intervene directly in the construction of the lens. We shall give preference to this representation.

For the lenses that we are considering at present (electromagnet type), the rules of similitude set forth in Section 2.3.2,A show that, for a number of ampere-turns lower than $1100 S$ or $1300 S$ (S gap in mm), the distribution $B(z/s)/B_M$, in an first approximation, depends only on the two parameters of form $(D_1 + D_2)/2S$ and D_1/D_2 whereas B_M is proportional to NI.

The *most general* trajectory equation[9] then shows that, if the electron-optical parameters are compared to a standard length (S, for example), and if the variation of one of these parameters is represented as a function of $NI/(V^*)^{1/2}$, a unique curve is obtained for a lens for which the form parameters $(D_1 + D_2)/2S$ and D_1/D_2 are given, at whatever scale the lens is built.

If the maker has at his disposal a set of these curves for different values of $(D_1 + D_2)/2S$ and D_1/D_2, he can easily set up the design for a lens.

Figure 17 shows this graph for the case in which $D_1/D_2 = 1$, and $D/S = 1$ (f_1, f_0, z_1, z_0 are defined subsequently).

If NI/S exceeds the critical value corresponding to incipient saturation of the polar faces, it will suffice to substitute corrected values for the real values of S, D_1, D_2 (see Paragraph 2.3.4,B,4). For various reasons, strong excitations much in excess of the critical value are to be avoided; the correction then remains slight enough to be known with sufficient accuracy.

2. Choice of Electron-optical Parameters

In Section 2.3.2 we recalled several results concerning the properties of magnetic lenses.

In the following paragraphs, we give the value of the most important properties in the case which corresponds to the most frequent use of lenses of the electromagnet type, namely, for electron microscopes and related instruments.

For the lens projective of an electron microscope or for the intermediary lens, the asymptotic correspondence (Section 2.3.2,3) defines the relationships between object and image. The most important magnitude is the focal distance f_1; the position of the focal points (or that of the principal planes) plays a secondary role. The most important aberrations are field aberrations.

For the lens objective of an electron microscope, the object is habitually

---

[9] See, for example, Dugas *et al.* (1961).

immersed and conjugated with a point image at a considerable distance. It is then necessary to know:

(a) Abscissa z_0 from the real object point conjugated with an image at infinite distance;

(b) The focal distance f_0 of the osculatory homographic correspondence which defines the correspondence in the vicinity of these operating conditions (that is, as long as the object does not move too far away from abscissa z_0).

For the last lens of a microanalyzer with electron spot, z_0 represents the abscissa of the target where the real image of the remote source is formed and f_1 represents the focal length of the osculatory homographic correspondence that defines the correspondence in the vicinity of these functioning conditions.

Figure 16 recalls the corresponding definitions.

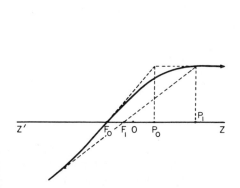

FIG. 16. Definition of the electron-optical magnitudes studied herein. Asymptotic correspondence: $f_i = \overline{F_1 P_1}$, $z_1 = \overline{F_1 O}$. Osculatory correspondence for a real object, immersed conjugated with an image at great distance: $f_0 = \overline{F_0 P_0}$; $z_0 = \overline{F_0 O}$.

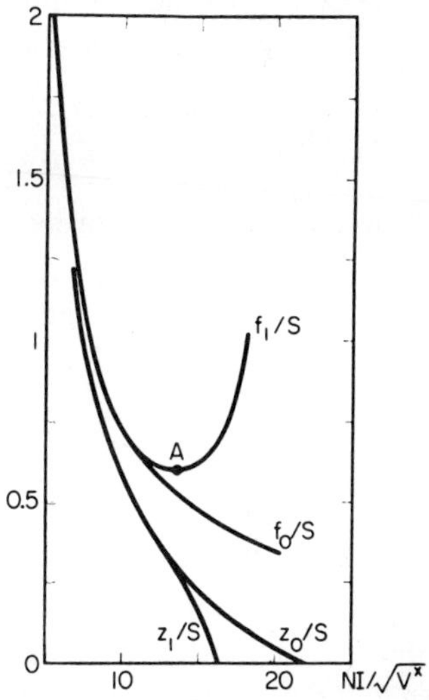

FIG. 17. Electron-optical magnitudes of a lens in function of the coefficient $NI/(V^*)^{1/2}$; $D/S = 1$.

In this case, the important aberrations are the chromatic aberration and the spherical aberration for an axial point object.

It is understood, of course, that, for thin lenses, when $NI/(V^*)^{1/2}$ is small the characteristic magnitudes of a lens objective and a lens projective tend toward the same values.

3. *Reduced Coordinates*

Figure 17 shows the example of a symmetrical lens for $D/S = 1$, in the Gaussian approximation.

Theoretically, we should draw a group of curves for each form of the lens for each value of $(D_1 + D_2)/2S$ and D_1/D_2. These groups of curves all have the same general appearance.

Various authors have tried to obtain universal curves from a suitable choice of coordinates (Liebmann, 1955c; Durandeau and Fert, 1957). The system of reduced coordinates, which seems the simplest, is adopted here.

Let NI_0 represent the number of ampere-turns corresponding (for a given value of V^*) to the minimum f_{1m} of f_1. The plotted curves will be referred to as "reduced curve" and noted:

(a)   On abscissas, NI/NI_0 instead of $NI/(V^*)^{1/2}$;
(b)   On ordinates, f_1/f_{1m}, f_0/f_{1m}, and so on, instead of f_1/S, f_0/S.

In actual practice, in order to use a curve plotted in reduced coordinates, the values of NI_0 and of f_{1m} must be known, in function of the parameters of form $(D_1 + D_2)/2S$ and D_1/D_2. These values are given for a whole series of lenses, by the graph of Fig. 18 with the convention

$$L = \left[S^2 + 0.45 \left(\frac{D_1 + D_2}{2} \right)^2 \right]^{1/2} \tag{26}$$

which has already been done [see Eq. (24)].

This figure calls for some commentary. It is first noted that symmetrical and dissymmetrical lenses give a unique curve for f_{1m}/L and for $NI/(V^*)^{1/2}$. In a first approximation the value of f_{1m}/L is little different from 0.5 regardless of $(D_1 + D_2)$ and D_1/D_2; $NI_0/(V^*)^{1/2}$ is little different from 13.5 when $(D_1 + D_2)/2S$ is higher than 1. Consequently, for a rough calculation, the simple postulate

$$f_{1m} = 0.5 \left[S^2 + 0.45 \left(\frac{D_1 + D_2}{2} \right)^2 \right]^{1/2} \tag{27}$$

$$NI_0 = 13.5(V^*)^{1/2} \tag{28}$$

may be made. These two simple expressions are often sufficient. Their use is extremely practical. If necessary, Fig. 18 may be referred to for more precise values.[10]

It seems worthwhile to cite here the values usually retained for different parameters: $(D_1 + D_2)/2S$ and D_1/D_2 are frequently contained between 0.5 and 2.

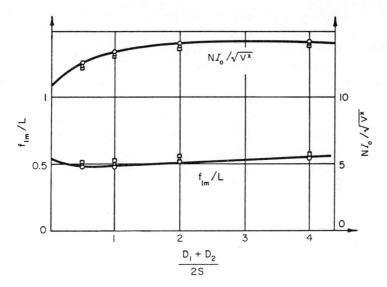

FIG. 18. Minimum focal length f_{1m} and corresponding value of $NI/(V^*)^{1/2}$ (the curve is drawn for $D_1/D_2 = 1$; The points marked correspond to $D_1/D_2 = 2$ and $\frac{1}{2}$, 3 and $\frac{1}{3}$

For the lens projective of an electron microscope, NI/NI_0 is usually near 1 ($f_1 = f_{1m}$). For a condenser, an intermediate lens, a lens for a diffractograph, NI/NI_0 varies within very wide limits; for the lens objective of an electron microscope, a reducing lens for a microanalyzer or scanning microscope, NI/NI_0 varies from 0.4 to 0.8. Recently, the use of lens objectives for which NI/NI_0 is near 1.5 (object in the middle of the field) has been advocated (Ruska, 1964).[11]

The presentation adopted below conveniently shows the magnitudes for

[10] The numerical values of $NI_0/(V^*)^{1/2}$ are given for electrons. The trajectory equation shows that, for ions, one need to put q/M instead of e/m. Thereby, numerical values of $NI_0/(V^*)^{1/2}$ for ions are equal to $NI_0/(V^*)^{1/2}$ for electrons multiplied by the term $[(M/m)/ \cdot (q/e)]^{1/2}$.

[11] At present, there is no practical case of application corresponding to $NI/NI_0 > 1.5$. Then "multiple focuses" and "intermediary images" appear; but in practice, it is not necessary to take them into consideration.

NI/NI_0 between 0.3 (approximately) and 1.5. For values of NI/NI_0 lower than 0.3, the approximation for thin lenses may be used; the corresponding simple formulas will be given here.

4. *Extreme Case*: D/S *Decreases toward Zero*; *Gaussian Approximation*

It is interesting to compare the properties of lenses to those that can be easily calculated: for example, in the case in which the diameter D of the holes bored in the polar pieces is small in front of the gap S ($D/S \ll 1$). The polar pieces being plane and parallel to a sufficient extent, induction is uniform in the gap and equal to $B = \mu_0 NI/S$ in the interval $- S/2, + S/2$.

The integration of paraxial trajectories is quite simple and the calculation of the electron-optical parameters easy. If their variation is represented as a function of $NI/(V^*)^{1/2}$, a graph is obtained which closely resembles that of Fig. 17. Figure 18 recalls that if D/S tends toward zero, $NI_0 = 10.9(V^*)^{1/2}$ and $f_{1m} = 0.55\ S$. By introducing the reduced coordinates set forth in Section 2.3.3, the following expressions are obtained (Durandeau and Fert, 1957):

lens projective

$$\frac{f_1}{f_{1m}} = \frac{0.897}{(NI/NI_0) \sin 2.029(NI/NI_0)} \tag{29}$$

$$\frac{z_1}{f_{1m}} = 0.91 + \frac{0.897}{(NI/NI_0) \tan 2.029(NI/NI_0)} \tag{30}$$

lens objective

$$\frac{NI}{NI_0} < 0.774 \qquad f_0 = f_1 \qquad z_0 = z_1 \tag{31}$$

$$\frac{NI}{NI_0} > 0.774 \qquad \frac{f_0}{f_{1m}} = \frac{0.897}{NI/NI_0} \tag{32}$$

$$\frac{z_0}{f_{1m}} = \frac{1.409}{NI/NI_0} - 0.91 \tag{33}$$

5. *Symmetrical Lenses* $D_1 = D_2$; *Gaussian Approximation*

Figure 19 shows reduced curves drawn for a series of symmetrical lenses.

It happens that, as a consequence of the choice of the reduced coordinates, the variation of each of the magnitudes f_1/f_{1m} and f_0/f_{1m} as a function of NI/NI_0 is shown by a unique curve, regardless of D/S; the divergences between the curves drawn for $D/S = 1$ and those corresponding to other values are lower than 1% when NI/NI_0 remains lower than $1.3NI_0$. These reduced curves are suitably shown by the expressions given in the test case $D/S = 0$ (formulas (29) and (30)).

The curves z_0/f_{1m} and z_1/f_{1m} drawn for $(D/S) = 1$ (Fig. 19) are valid for $(D/S) = 2$, but they cease to be valid when (D/S) is lower than 1.

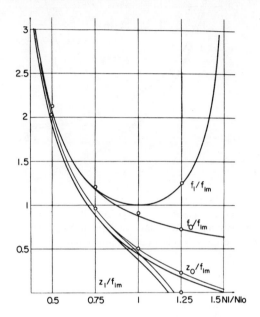

FIG. 19. Symmetrical lenses, reduced curves. The fine-line curves correspond to the extreme case $D/S = 0$.

For $(D/S) = 0.5$, the values begin to correspond to those calculated for $(D/S) = 0$.

The representation of the paraxial properties of symmetrical lenses by the group of reduced curves of Fig. 19 is convenient. This graph, used with that of Fig. 18, or with the expressions (27) and (28), represents the electron-optical parameters of all symmetrical lenses in function of the elements of construction: gap S, diameter D of the holes bored in the polar faces, number of ampere-turns NI and electron energy eV.

6. Dissymmetrical Lenses; Gaussian Approximation

The reduced curves characteristic of different dissymmetrical lenses are the object of Figs. 20–22; each figure corresponds to a value of $(D_1 + D_2)/2S$.

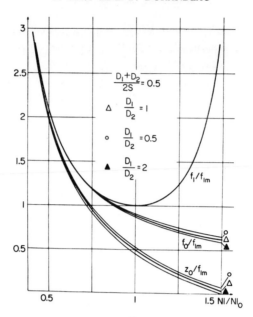

Fig. 20. Dissymmetrical lenses $(D_1 + D_2)/2S = 0.5$; $D_1/D_2 = 0.5$; 1; 2. The curves z_1/f_{1m}, which differ only slighty from those corresponding to symmetrical lenses (Fig. 19), have been omitted in order to avoid overloading the figure.

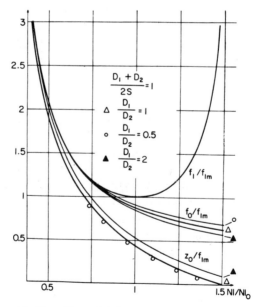

Fig. 21. Dissymmetrical lens $(D_1 + D_2)/2S = 1$; $D_1/D_2 = 0.5$; 1; 2. Same remark as for Fig. 20.

The curve f_1/f_{1m} in function of NI/NI_0 is almost the same for all cases studied ($D_1 + D_2/2S$ between 0.5 and 4; D_1/D_2 between 1/3 and 3). Slight differences appear when NI/NI_0 approaches 1.5, they are visible in the Figs. 20–22, and are of little importance. f_1/f_{1m} may, in actual practice, be taken for a universal curve for symmetrical or dissymmetrical lenses; it is suitably expressed by the formula (29).

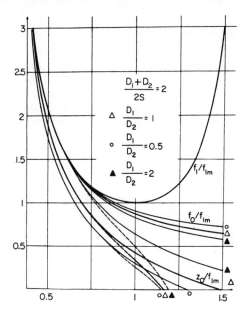

FIG. 22. Dissymmetrical lenses ($D_1 + D_2)/2S = 2$; $D_1/D_2 = 0.5$; 1; 2. The curves in broken lines represent z_1/f_{1m}.

The curve f_0/f_{1m} has been shown in each case[12] for $D_1/D_2 = 2$, 1, $\frac{1}{2}$ which are the limits of the most usual lenses.

The curve z_0/f_{1m} gives the position of the real focal point (Fig. 16). It will be noted that for $(D_1 + D_2)/2S = 1$, it happens that the curve for $D_1/D_2 = 0.5$ almost coincides with that for symmetrical lenses. It is for this reason that it has not been shown here.

The curve z_1/f_{1m} is less important; in practice, it is of little importance whether or not the position is known of a focal point of an asymptotic correspondence, for which f_1 is the important magnitude. It has been shown in

---

[12] D_1 always represents the diameter of the hole on the side of the real osculatory focal point: on the object side of the lens objective of an electron microscope; on the target side for the last lens of an apparatus with electron spot.

the only case in which it diverges considerably from the curve corresponding to that for symmetrical lenses.

All the preceding graphs were carefully drawn with a view to their eventual use.

7. Rotation of Images

Rotation of the images, although less important than the preceding properties, is sometimes useful to know. In the Gaussian approximation the rotation between two frontal planes of abscissas z_0 and z_1 is equal to

$$\theta = \left(\frac{e}{8\,m_0 V^*}\right)^{1/2} \int_{z_0}^{z_1} B\,dz = \left(\frac{e}{8\,m_0}\right)^{1/2} \mu_0 \frac{nI}{(V^*)^{1/2}} \tag{34}$$

or for electrons:

$$\theta = 0.1863\,nI/(V^*)^{1/2}(rd) \tag{35}$$

where nI is the number of ampere-turns between z_0 and z_1.

We use for standard reference rotation θ_0, the rotation corresponding to $NI = NI_0$, namely,

$$\theta_0 = \left(\frac{e}{8\,m_0}\right)^{1/2} \mu_0 \frac{NI_0}{(V^*)^{1/2}} \tag{36}$$

This reference rotation is calculated from Fig. 18 which gives $NI_0/(V^*)^{1/2}$ as a function of $(D_1 + D_2)/2S$. For example, for $(D_1 + D_2)/2S = 1$ and $NI_0/(V^*)^{1/2} = 13.5$,

$$\theta_0 = 2.515 \text{ rad} \quad \text{or} \quad 144°.$$

We use θ to represent the rotation resulting from a lens objective between the plane where the object is located and the image plane, and θ_1 to represent the rotation resulting from a lens projective (asymptotic correspondence), thus obtaining the following expressions:

$$\theta /\theta_0 = nI/NI_0 \tag{37}$$

$$\theta_1/\theta_0 = NI/NI_0 \tag{38}$$

Figure 23 shows, by way of example, how θ/θ_0 and θ_1/θ_0 vary for different symmetrical lenses.

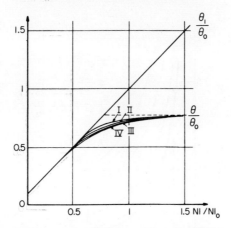

FIG. 23. Rotation of images; symmetrical lenses. Curves I, II, III, and IV, correspond, respectively, for a lens objective to $D/S = 0.5$; 1; 2; and 4. Broken line, extreme case $D/S = 0$.

8. *Aberration Coefficients*

Figure 24 gives the curve C_c/f of the chromatic aberration coefficient on the axis of a lens objective as a function of NI/NI_0. It is, in first approximation, valid for all cases.

Figures 25–27 represent the spherical aberration coefficient on the axis for the lens objective of an electron microscope (or the last lens of a microanalyzer). This coefficient was given for NI/NI_0 contained between 0.5 and 1.5 and an entire series of lenses, in order to permit the calculation of C_s/f_0 by interpolation in all other cases.

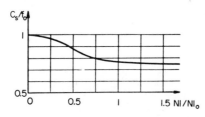

FIG. 24. Chromatic aberration coefficient on the axis for a lens objective: average curve, valid in first approximation for the lenses studied $(D_1 + D_2)/2S$ between 0.5 and 4. D_1/D_2 between $\tfrac{1}{3}$ and 3.

9. *Lenses of Weak Convergence*

The preceding graphs are utilizable almost only for $NI/NI_0 \geq 0.4$. For weaker excitations, it is easy to make the calculation in utilizing simple formulas.

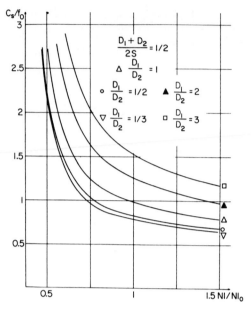

FIG. 25. Aberration coefficient of sphericity on the axis for a lens objective. Lens corresponding to $(D_1 + D_2)/2S = 0.5$.

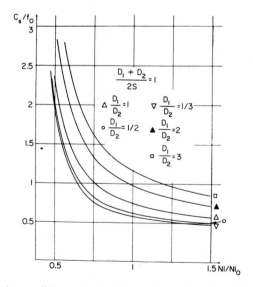

FIG. 26. Aberration coefficient of sphericity on the axis for a lens objective. Lens corresponding to $(D_1 + D_2)/2S = 1$.

It will be noted first of all that, under these conditions, there is no distinction to be made between the different types of correspondence (lens objective and lens projective, for example).

The focal length is correctly given by the formula (29):

$$\frac{f}{f_{1m}} = \frac{0.897}{(NI/NI_0)\sin 2.029(NI/NI_0)} \tag{39}$$

which rapidly reduces, if $NI/NI_0 < 0.2$, to $f/f_{1m} = 0.442/(NI/NI_0)^2$.

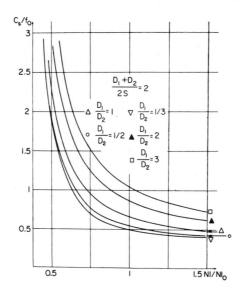

FIG. 27. Aberration coefficient of sphericity on the axis for a lens objective. Lens corresponding to $(D_1 + D_2)/2S = 2$.

The chromatic aberration coefficient on the axis tends very rapidly toward the value f.

Lastly, in regard to the spherical aberration coefficient on the axis, it can be shown (Durandeau et al., 1959a) that, for lenses of low, weak convergence, $C_s D^2/f^3$ [with $D = (D_1 + D_2)/2$] is independent of NI/NI_0. It varies with $(D_1 + D_2)/2S$, as indicated in Fig. 28.

The two graphs shown in the same figure permit ascertaining to what limits of excitation $C_s D^2/f^3$ remains constant, and, eventually, to make eventual corrections.

Lastly, the rotation of the images if given as a matter or course by (38).

10. *Example of Special Designs*

The reader will find examples of the application of magnetic lenses of the electromagnet type in examining the schemes for commercial instruments such as the electron microscope, the scanning microscope, the microanalyzer, the diffracting microscope, and so on. In order to indicate an

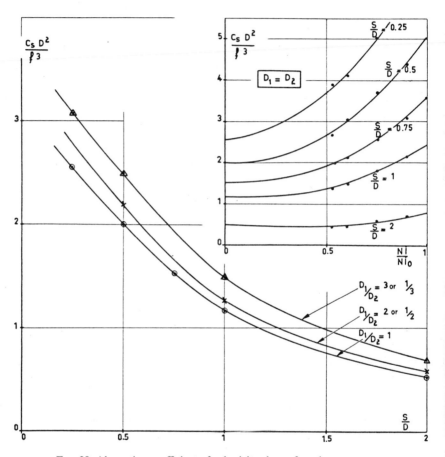

FIG. 28. Aberration coefficient of sphericity: lens of weak convergence.

application for the preceding given conditions, we must determine certain characteristics of the objective and of the projector of an electron microscope.

a. *Objective of an Electron Microscope.* Let λ represent de Broglie's wavelength, for electrons of the beam, and C_s the spherical aberration of the objective of an electron microscope. The best limit of resolution is obtained for the aperture

$$\alpha = A(\lambda/C_s)^{1/4} \tag{40}$$

of the objective, and the theoretical limit of resolution is then equal to

$$\delta = B(\lambda^3 C_s)^{1/4} \tag{41}$$

The values of coefficients A and B depend on different hypotheses (whether the beam is coherent or not, and whether we consider separation of two bright or dark points, and so forth). Moreover, it is of no interest to give a limit of resolution to more than two figures, since the second figure has only an indicative value. Therefore, for A and B, the average simple values are taken:

$$A = 1.3 \qquad B = 0.5$$

It will then be of interest to consider which conditions must be applied in order to obtain the best limit of resolution.

In present-day instruments, the object is in a region of weak induction ($NI/NI_0 \sim 0.4$) or in the vicinity of the first polar face ($NI/NI_0 \sim 0.75$; $z \sim S/2$). The conditions are better in the second case, since they permit obtaining a lower value for C_s. The object is immersed in the field, but the zone of the field preceding the object does not have a significant action.

More recently, in order to reduce C_s, some authors (Ruska, 1965) proposed placing the object in the middle of the gap ($z_0 = 0$; $NI/NI_0 \sim 1.5$). Under these conditions, the zone of the field preceding the object is important and its action on the incident beam is considerable, even troublesome in actual practice. (See, for example, Riecke (1962); an outline of this action may come from a comparison of f_0 and f_1.) But at the price of this difficulty, the resolution may be slightly improved.

In order to obtain some orders of magnitude, the characteristics of several objectives have been determined on the basis of the results noted in the foregoing paragraphs. (See Tables I and II.) A symmetrical lens has been chosen in each case ($D_1/D_2 = 1$) with $D/S = 0.5$, supposing a low saturation of the polar faces, the same in all cases ($B_p = \mu_0 NI/S = 18$ T; $NI/S = 1.430$ At/mm). The correction allowing for incipient saturation does not exceed a few per cent if the polar faces are correctly designed. Similar tables have been set up by different authors (Ruska, 1965). In spite of differences in the bases for calculation, there is no significant difference in the results. In addition, it is easy to show that, for a lens of a given form D_1/D_2 and

TABLE I

COMPARISON OF THE LIMIT OF RESOLUTION OBTAINED FOR TWO DIFFERENT FUNCTIONING CONDITIONS[a]

z_0	NI (At)	S (mm)	f_m (mm)	f_0 (mm)	C_s (mm)	λ (Å)	δ (Å)
0.5 S	0.75 NI_0 3150	2.2	1.14	1.38	1.68	0.037	2.7
0	1.55 NI_0 6496	4.54	2.35	1.41	1.06	0.037	2.4

[a] Lens 1: $z_0 = S/2$; lens 2: $z_0 = 0$. Given conditions common to the two lenses: $D/S = 0.5$; $V = 100$ kV $[NI_0 = 12.6\,(V^*)^{1/2} = 4190$ At]; $B_p = \mu_0 NI/S = 1.8$ Wb/m².

TABLE II

VARIATION OF THE LIMIT OF RESOLUTION WITH ELECTRON ENERGY[a].

V (kV)	NI (At)	S (mm)	f_{1m} (mm)	f_0 (mm)	C_s (mm)	λ (Å)	δ (Å)
100	6496	4.54	2.35	1.41	1.06	0.037	2,4
200	9590	6.70	3.47	2,08	1.56	0.025	2,0
500	16,920	11.83	6.13	3.68	2.76	0.014	1.5
1000	27,600	19.3	10	6	4.5	0.0087	1.16

[a] Given conditions common in all cases: $D/S = 0.5$; $z_0 = 0$; $NI = 1.55\,NI_0$ $[NI_0 = 12.6(V^*)^{1/2}]$; $B_p = \mu NI/S = 1.8$ Wb/m²; $f_{1m} = 0.518\,S$; $f_0 = 0.6 f_{1m}$; $C_s = 0.75 f_0$.

$(D_1 + D_2)/2S$, and fixed functioning conditions $(NI/NI_0 = $ constant$)$

$$\delta \doteq (V^*)^{-1/4} \quad \text{if} \quad B_p = \mu_0 NI/S = \text{const}$$

$$\delta \doteq (B_p)^{-1/4} \quad \text{if} \quad V = \text{const}$$

(It is again assumed that the saturation of the polar faces is relatively low in all cases, namely, that B_p is lower than 1.8 Wb/m².)

It should be noted that the best experimental limit of resolution at present is in the vicinity of 2 Å for an electron microscope with a magnetic objective and $V = 100$ kV.

b. *Projector of an Electron Microscope.* The maximum magnification is obtained for $NI = NI_0$. In Table III are shown, for different values of V,

TABLE III

PROJECTOR OF AN ELECTRON MICROSCOPE[a]

V (kV)	NI_0 (At)	S (mm)	f_{1m} (mm)	D (mm)
100	4190	2.91	1.51	303
200	6190	4.31	2.24	448
500	10,900	7.60	3.95	790
1000	17,800	12.4	6.45	1290

[a] Given conditions common to all cases: $D/S = 0.5$; $NI = NI_0 = 12.6(V^*)^{1/2}$; $B_p = \mu_0 NI/S = 1.8$ Wb/m².

the result of the calculation of the minimum focal distance of a projector and the distance D at which the fluorescent screen must be placed in order to obtain a magnification of $M = 200$.

c. *Orders of Magnitude.* The objective and projective of an electron microscope are lenses that rapidly become very heavy when the number of At is augmented, excessive heating of the coil being avoided. The objective of an electron microscope for 100 kV usually weighs from 40 to 60 kg, whereas the objective of an electron microscope for 1,500,000 V may reach and exceed 700 kg (Dupouy and Perrier, 1962).

2.3.5. Lenses With Permanent Magnets

A. GENERAL CONSIDERATIONS

Figures 29 and 30 present several types of lenses with a permanent magnet. The most obvious advantage of these lenses is that it is not necessary to provide a stabilized source of current to excite them. On the other hand, their power is more difficult to regulate than that of the electromagnet type.

Lenses with permanent magnets are less used than those of the electromagnet type. However, a great number of models has been proposed. Descriptions must be limited to just a few of them.

It should be remembered that for a lens with permanent magnet, the integral $\int_{-\infty}^{+\infty} B \, dz$ lying along the axis of $-\infty$ to $+\infty$ is always zero, and that the rotation of the image is therefore also null.

B. Structures of Some Lenses

1. Simple Lenses

This term designates lenses assimilated to a relatively thin system acting as a single lens. Figures 29–31 show lenses of this type.

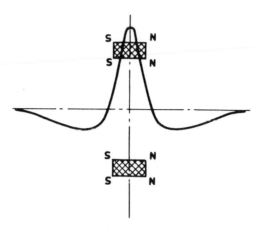

FIG. 29. Magnetized ring.

a. *Magnetized Ring.* The magnetization is, on the average, parallel to the axis of the ring. For the calculation of induction on the axis, this ring can be looked on as two crowns of magnetism with a density of $\sigma = I_n$ The induction and the field inside the ring must be known in order to make this calculation.

Rings of ferrite are often used as low-powered lenses.

b. *Lenses with Magnetic Circuit Having a Single Gap.* Figure 30a,b illustrate two applications. The magnet itself is usually a ring, but it may also be composed of straight bars distributed uniformly around the axis.

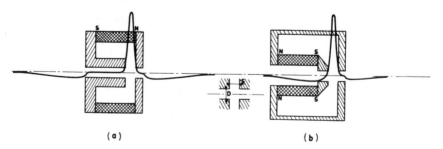

(a)   (b)

FIG. 30. Lenses with a simple gap.

The main field is localized in the gap, but there exists also, and of necessity, a stray field such that $\int_{-\infty}^{+\infty} B \, dz = 0$. A lens of this type is, therefore, always rather thick, whatever the disposition adopted.

c. *Lenses with Two Adjacent Gaps.*   Figure 31a,b show this type of lens and the distribution on the axis. This distribution is dependent upon the geometric parameters S, T, D, given a symmetrical lens and a common diameter for the holes bored in the polar faces (Lenz, 1956; Kimura, 1957).

It will be noted that the distribution of induction on the axis of this lens is similar to that of the electrostatic field on the axis of an electrostatic symmetrical lens of which the electrodes have the same geometrical dimension as the polar faces of the magnetic lens.

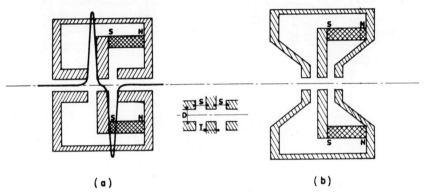

(a)   (b)

FIG. 31. Lenses with two adjacent gaps.

2. *Compound Lenses*

We have seen that for a lens with gap, there is necessarily a stray field. The latter is avoided in the case of the lens with two adjacent gaps.

The same result may be obtained by separating the two gaps. Each gap with the field localized in it is then equivalent to a lens.

Figure 32 shows three examples of such a construction with two or three gaps, respectively.

In the case of Fig. 32a,b, the field becomes localized in the gaps L_1 and L_2 ; L_1 and L_2 can play, respectively, the roles of the objective and the projector of an electron microscope with two lenses. In Fig. 32c, the three lenses L_1, L_2, L_3 can play, respectively, the roles of the objective, the intermediary lense, and the projector. Even more complex systems have been suggested (Müller, 1957; Kimura, 1957). Very compact electron microscopes have been thus realized.

C. CHOICE OF THE PERMANENT MAGNET

1. *Characteristics of Permanent Magnets*

A permanent magnet is characterized by the curves of magnetization
$B(H)$. One of the best steel alloys available today for the realization of
lenses with permanent magnet is Alnico V (Al : 8%; Ni : 13,5%; Co : 24%;

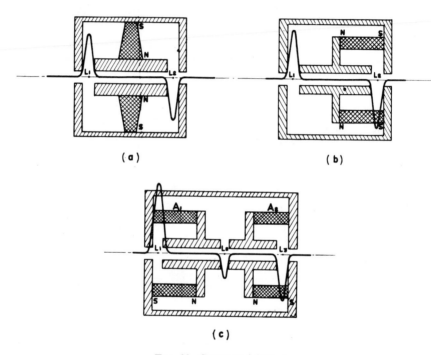

(a)　　　　　　　　　　(b)

(c)

FIG. 32. Compound lenses.

Cu : 3%; Fe : 51%). Its characteristics are shown in Fig. 33:

remanent magnetization　　$B_2 = 1.2$ Wb/m²

coercive force　　$H_c = 45,000$ At/m

The product $(B \cdot H)$ has a maximum value for $B \sim 1$ Wb/m² and
$H \sim 35,000$ At/m.

In order to fix orders of magnitude, it is subsequently assumed that B and
H do not diverge appreciably from these values. The values characteristic
of the alloy utilized will be those used in the design of a lens.

2. *Choice of the Permanent Magnet*

The case considered here is that of lenses having a magnetic circuit. It is supposed that the iron of the magnetic circuit is saturated at no point and that its permeability is high. Under these conditions, the magnetic field H is noticeably null on all points of the soft iron making up the circuit.

The magnet is assumed to be of constant section; L_m designates its length parallel to the magnetization; σ_m its section perpendicular to the magnetization; H_m and B_m the field and the induction, respectively, assumed to be uniform in the magnet (Fig. 33). For the design of a magnetic lens with a permanent magnet, it is necessary to be able to calculate the section and the length of the magnet to be utilized.

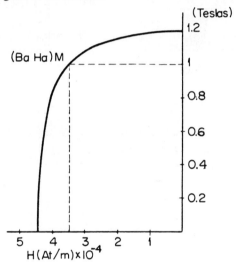

FIG. 33. Curve of magnetization.

The length of the magnet will be obtained in writing Ampere's theorem for a closed circuit through the magnet, the magnetic circuit, and the gap (or the gaps, if there are several).

A number of ampere-turns NI may be made to correspond to the length L_m of the magnet such that

$$NI = H_m L_m \qquad (42)$$

The section of the magnet will be obtained by writing the conservation of the flux of induction along the tube of induction corresponding to the section of the magnet. The flux of induction through the section of the magnet, in the hypothesis of uniform induction, is of course

$$\Phi = B_m \sigma_m \tag{43}$$

L_m and B_m are determined by the properties chosen for the lens (or the lenses) and the profile chosen for the polar faces.

We shall use a simple example corresponding to the lens of Fig. 32b.

In order to simplify the case, the following assumptions are made:

(a) That S/D is small; and

(b) That the polar cores are cylindrical and the polar faces of sections σ_1 and σ_2, respectively, are plane and parallel. Under these conditions, it may be admitted in *a first approximation* that in the gaps L_1 and L_2, induction is uniform

$$B_{p_1} = \mu_0(NI/S_1) \qquad B_{p_2} = \mu_0(NI/S_2) \tag{44}$$

and the flux of induction equal to

$$\Phi_1 = B_{p_1}\sigma_1 = \mu_0 NI(\sigma_1/S_1) \qquad \Phi_2 = B_{p_2}\sigma_2 = \mu_0 NI(\sigma_2/S_2) \tag{45}$$

The number of ampere-turns NI and the values of B_1 and B_2 are determined by the properties desired for lenses L_1 and L_2.

H_m and B_m being the values accepted for the field and the induction in the permanent magnet, L_m and σ_m are respectively determined by relations

$$H_m L_m = NI \qquad B_n \sigma_m = B_p\,\sigma_1 + B_{p_2}\sigma_2 = \mu_0 NI \left(\frac{\sigma_1}{S_1} + \frac{\sigma_2}{S_2}\right) \tag{46}$$

It is preferable to take values of L_m and σ_m slightly higher than those provided for in this elementary calculation in order to account for a possible demagnetization.

D. POSSIBILITY FOR ADJUSTING A LENS

It is useful to be able to adjust a lens with permanent magnet in order to obtain better focusing for example, or to vary the magnification in an electron microscope.

Many arrangements have been suggested. A first solution consists in using mobile magnetic shunts (Fig. 34).

But great adaptability, may also be obtained, in the case of a complex lens with several gaps, by modifying the distribution of the ampere-turns between the two gaps in series.

By way of example, the reader is referred to the construction proposed by different authors (Müller, 1957; Müller and Ruska, 1960; Kimura, 1957; Kimura and Kikuchi, 1959).

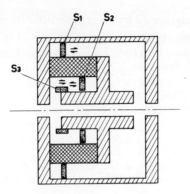

FIG. 34. Magnetic shunts.

GENERAL REFERENCES

Cosslett, V. E. (1946). "Introduction to Electron Optics." Oxford Univ. Press (Clarendon), London and New York.

Dupouy G. (1952). Éléments d'optique électronique, A. Colin (Paris).

Glaser, W. (1952). "Grundlagen der Elektronenoptik." Springer, Berlin.

Glaser, W. (1956). In "Handbuch der Physik" (S. Flügge, ed.), Vol. 33. Springer, Berlin.

Grivet, P. (1965). "Electron Optics," Pergamon Press.

Klemperer, O. (1953). "Electron Optics." Cambridge Univ. Press, London and New York.

Magnan, C. (1961). "Traité de Microscopie Electronique." Hermann, Paris.

Zworikyn, V. K., Morton, G. A., Ramberg, E. G., Hillier, J., and Vance, A.W. (1948). "Electron Optics and the Electron Microscope." Wiley, New York.

REFERENCES

Dosse, J. (1941a). Z. Physik. 117, 316–321.

Dosse, J. (1941b). Z. Physik 117, 437–443.

Dugas, J., Durandeau, P., and Fert, C. (1961). Rev. Opt. 40, 277–305.

Dupouy, G., and Perrier, F. (1962). J. Microscopie 1, 167–192.

Durand, E. (1954). Rev. Opt. 33, 617–629.

Durand, E. (1955). Ann. Phy. (Paris) [12] 10, 883–907.

Durandeau, P. (1956). J. Phys. Radium 17, 18A–25A.

Durandeau, P. and Fert, C. (1957). Rev. Opt. 36, 205–234.

Durandeau, P., Fagot, B., and Fert, C. (1959a). Compt. Rend. 248, 946–949.

Durandeau, P., Fagot, B., Barthere, J., and Laudet, L. (1959b). *J. Phys. Rad.* **20**, 80A 90A.

Gautier, P. (1957). These, Toulouse.

Glaser, W. (1941), *Z. Physik* **117**, 285–314.

Glaser, W. (1952). "Grundlagen der Elektron enoptik." Springer, Berlin.

Grivet, P. (1952). *J. Phys. Radium* **13**, 1A–9A.

Kimura, H. (1957). *Hitachi Rev. Japan* **6**, 52–62.

Kimura, S., and Kikuchi, Y. (1959). *J. Japan. Electron Microscopy* **7**, 45–47.

Laudet, M. (1957). *J. Phys. Radium* **18** 73A–77A.

Lenz, F. (1950a). *Z. Angew. Phys.* **1**, 337–340.

Lenz, F. (1950b). *Z. Angew. Phys.* **2**, 448–453.

Lenz, F. (1956). *Z. Angew. Phys.* **8**, 492–496.

Le Poole, J. B. (1964). *Proc. 3rd Regional Conf. Electron Microscopy Prague. 1964* Vol. 1, pp. 439–440. Publ. Home Czeck. Akad. Sci., Prague.

Liebmann, G. (1951). *Proc. Phys. Soc. (London)* **B64**, 972–977.

Liebmann, G. (1952a). *Proc. Phys. Soc. (London* **B65**, 94–108.

Liebmann, G. (1952b). *Proc. Phys. Soc. (London)* **B65**, 188–192.

Liebmann, G. (1952c). *Proc. Phys. Soc. (London)* **B66**, 448–458.

Liebmann, G. (1955a). *Proc. Phys. Soc. (London)* **B68**, 679–681.

Liebmann, G. (1955b). *Proc. Phys. Soc. (London)* **B68**, 682–685.

Liebmann, G. (1955c). *Proc. Phys. Soc. (London)* **B68**, 737–745.

Liebmann, G., and Grad E. M. (1951). *Proc. Phys. Soc. (London)* **B64**, 956–971.

Ments, M. V., and Le Poole, J. B. (1947). *Appl. Sci. Res.* **B1**.

Morito, N., Tadano, B., and Katagiri, S. (1960). *Proc. 4th Intern. Conf. Electron Microscopy, Berlin, 1958* pp. 51–53. Springer, Berlin.

Müller, K. (1957), *Z. Wiss. Mikroskopie* **63**, 303–327.

Müller, K., and Ruska, E. (1960). *Proc. 4th Intern. Conf. Electron Microscopy, Berlin. 1958* pp. 184–187. Springer, Berlin.

Mulvey, T. (1952). *Proc. Phys. Soc. (London)* **B66**, 441–447.

Reissner, J. H. (1951). *J. Appl. Phys.* **22**, 561–565.

Reissner, J. H., and Dornfeld, E. G. (1950). *J. Appl. Phys.* **21**, 1131–1139.

Riecke, W. D. (1962). *Optik* **19**, 169–207.

Ruska, E. (1954). *Nat. Bur. Std. (U.S.) Circ.* **527**, 389–410.

Ruska, E. (1964). *J. Microscopie* **3**, 357–372.

Ruska, E. (1965). *Optik* **22**, 319–348.

Ruska, E., and Wolff, O. (1956). *Z. Wiss. Mikroskopie* **62**, 465–509.

von Borries, (1952). *Z. Wiss. Mikroskopie* **60**, 329–358.

CHAPTER 2.4

FOCUSING WITH QUADRUPOLES, DOUBLETS, AND TRIPLETS

E. Regenstreif

FACULTÉ DES SCIENCES
RENNES, FRANCE

2.4.1. General Properties of Focusing Systems

A. Matrix of a Drift Space

Consider a field-free region limited by two planes normal to the optical axis (Fig. 1). There are no forces in this region and, therefore, the equation of a ray is

$$x = Az + B \tag{1}$$

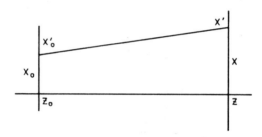

Fig. 1. Particle orbit in a field-free region.

By taking into account the initial conditions, this may be put in the form

$$
\begin{aligned}
x &= x_0 + L x_0' \\
x' &= \quad\quad x_0'
\end{aligned}
\tag{2}
$$

where $L = z - z_0$ represents the length of the drift space. In matrix form

$$
\begin{pmatrix} x \\ x' \end{pmatrix} = \begin{Vmatrix} 1 & L \\ 0 & 1 \end{Vmatrix} \begin{pmatrix} x_0 \\ x_0' \end{pmatrix}
\tag{3}
$$

Consequently, the matrix of a drift space is

$$
m_d = \begin{Vmatrix} 1 & L \\ 0 & 1 \end{Vmatrix}
\tag{4}
$$

B. MATRIX OF A LENS SYSTEM

In linear theory, any lens system may be described by a matrix

$$m_l = \begin{Vmatrix} a & b \\ c & d \end{Vmatrix} \tag{5}$$

where the elements, a, b, c, and d depend on the geometry of the system and its physical parameters (focusing strengths), but not on the incoming or outgoing ray.

If the lens system is preceded by a drift length p and followed by another drift length q, the complete transfer from the "object point" to the "image points" becomes

$$\begin{pmatrix} x \\ x' \end{pmatrix} = \begin{Vmatrix} 1 & q \\ 0 & 1 \end{Vmatrix} \times \begin{Vmatrix} a & b \\ c & d \end{Vmatrix} \times \begin{Vmatrix} 1 & p \\ 0 & 1 \end{Vmatrix} \begin{pmatrix} x_0 \\ x_0' \end{pmatrix} \tag{6}$$

Here, p stands for the object distance, which will be counted positive towards the left starting from the entrance plane, and q stands for the image distance q, which will be counted positive towards the right starting from the exit plane (Fig. 2).

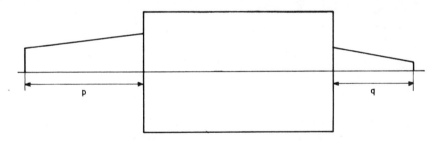

FIG. 2. Definition of object and image distances.

C. OPTICAL CHARACTERISTICS

Multiplying out Eq. (6), we find

$$x = (a + cq)x_0 + [ap + b + q(cp + d)]x_0'$$
$$x' = cx_0 + (cp + d)x_0' \tag{7}$$

These relations reveal all first-order characteristics of the optical system:

1. Object-Image Conjugation

We require x to be independent of x_0'; consequently, the conjugation relation is

$$cpq + ap + dq + b = 0 \tag{8}$$

2. Image Position

Equation (8) immediately gives

$$q = -(ap + b)/(cp + d) \tag{9}$$

3. Magnification

The conjugation relation being satisfied, we have

$$g = x/x_0 = a + cq = a - c(ap + b)/(cp + d)$$

or,

$$g = \Delta/(cp + d)$$

where

$$\Delta = ad - bc \tag{10}$$

is the determinant of the transfer matrix. In what follows, we shall assume $\Delta = 1$, this condition being always satisfied when the space on both sides of the lens combination is at the same potential. The magnification then becomes

$$g = 1/(cp + d) \tag{11}$$

4. Focal Planes

An incoming ray parallel to the axis will, after going through the lens system, intersect the axis at a point whose distance from the exit plane is given by

$$F_i = -a/c \tag{12}$$

Similarly, an outgoing ray which is parallel to the axis originates from a point whose distance from the entry plane is given by

$$F_0 = -d/c \tag{13}$$

5. Principal Planes

These planes are conjugate to each other and defined by unit magnification.

From

$$a + cq = 1$$

we infer

$$H_i = (1 - a)/c \tag{14}$$

whereas

$$cp + d = 1$$

leads to

$$H_0 = (1 - d)/c \tag{15}$$

6. Focal Distances

The focal distance, which is the reciprocal value of the lens strength, is defined as the distance between the focus and the corresponding principal plane. Therefore, we have for the image side

$$f_i = - (a/c) - (1 - a)/c = - 1/c \tag{16}$$

and for the object side

$$f_0 = - (d/c) - (1 - d)/c = - 1/c \tag{17}$$

The equality

$$f_0 = f_i = f$$

is essentially due to the fact that $\varDelta = 1$.

7. Antiprincipal Planes

These planes are again conjugate to each other and defined by negative unit magnification.

From

$$a + cq = - 1$$

we have

$$\bar{H}_i \quad = - (1 + a)/c \tag{18}$$

whereas

$$cp + d = - 1$$

gives

$$\bar{H}_0 \quad = - (1 + d)/c \tag{19}$$

In the case of a symmetrical lens system, $a = d$, and these relations become:

conjugation $$cpq + a(p + q) + b = 0 \tag{20}$$

image distance	q		$= -(ap + b)/(cp + a)$	(21)
magnification	g		$= 1/(cp + a)$	(22)
focal planes	F_i		$= F_0 = -a/c$	(23)
principal planes	H_i		$= H_0 = (1 - a)/c$	(24)
focal distance	f_i		$= f_0 = -(1/c)$	(25)
antiprincipal planes	$\bar{H}_i$		$= \bar{H}_0 = -(1 + a)/c$	(26)

At the risk of repetition, it might be worthwhile to point out again that all image characteristics, i, are counted positive towards the right starting from the exit plane of the lens combination, whereas all object parameters, o, are counted positive towards the left starting from the entrance plane.

D. IDEAL AND PRACTICAL FOCUSING

An efficient focusing system would require, in the electric case, a purely radial field and, in the magnetic case, a purely tangential field (Fig. 3).

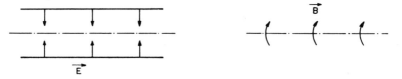

FIG. 3. (a) Focusing electric field; (b) Focusing magnetic field.

In both cases, the restoring force (directed towards the axis) should be proportional to the distance from the axis for optimum focusing conditions.

In the standard axially symmetrical electric lens, **E** has only a small radial component and similarly, in solenoidal focusing devices, **B** has only a small tangential component. In both cases, the focusing forces are weak.

The situation displayed in Fig. 3, with restoring forces proportional to the distance from the axis, has not yet been achieved in practice. However, a way of focusing far superior to the old devices, based on axial symmetry, may be obtained by associating a convergent lens and a divergent lens. The lenses are either electric or magnetic quadrupoles and, in both cases, the field increases almost linearly with the distance from the axis. The price we have to pay for this improvement lies in the appearance of a certain amount of astigmatism or, more generally, the deterioration of the optical quality of the image. However, in many cases and especially in beam transport systems associated with accelerating machines, strong focusing is far more important than image quality.

2.4.2. The Quadrupole Lens

A. FIELD CONFIGURATION

In both the electric and magnetic lenses, the focusing field is obtained by means of four poles. We shall confine ourselves to the magnetic case, breakdown difficulties preventing the use of electrostatic lenses at energies much higher than 10 MeV. Figure 4 shows the field configuration in a four pole geometry and Fig. 5 gives the direction of the B_x and B_y components

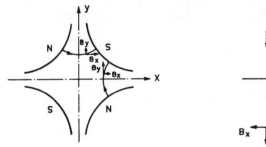
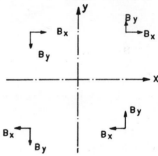

FIG. 4. Quadrupole geometry.   FIG. 5. Field configuration in a quadrupole.

of the field in the four quadrants. The beam is supposed to travel in the z direction, perpendicular to the plane of the paper. It is readily seen that B_x changes sign upon traversal of the xOz plane, whereas B_y changes sign when passing through the yOz plane.

In order to evaluate the field distribution quantitatively, one must solve Laplace's equation $\nabla^2 V = 0$ with the given boundary conditions. This is easily done if the four poles are equilateral hyperbolas. In this case, we find for the potential distribution

$$V = - Gxy \qquad (27)$$

with G constant. The field components are, therefore, proportional to the distance from the axis:

$$B_x = - \partial V/\partial x = Gy \qquad B_y = - \partial V/\partial y = Gx \qquad (28)$$

On the other hand, the magnetic scalar potential is

$$V = \mu_0 nI \tag{29}$$

where nI is the number of exciting ampere-turns per pole; if R_0 denotes the aperture of the quadrupole (Fig. 6), we have

$$G = \mu_0(2nI/R_0^2) \tag{30}$$

Practically, the hyperbolas will never go to infinity and, therefore, higher

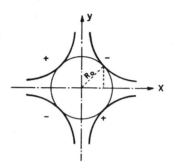

FIG. 6. Available aperture in a quadrupole.

order terms will appear in the expression of the field components; however their contribution will be quite small in general and the field gradient

$$G = \partial B_x/\partial y = \partial B_y/\partial x \tag{31}$$

can be considered as a constant to a fairly high degree of accuracy.

B. FORCES ACTING ON THE BEAM

The magnetic force on a particle of charge q is given by

$$\mathbf{F} = q\mathbf{v} \times \mathbf{B} \tag{32}$$

Since the transverse velocity components are very small compared to the longitudinal velocity v, the components of the force acting on a particle may be written

$$F_x = - qvB_y \qquad F_y = qvB_x \tag{33}$$

Figure 7 shows the direction of the force components in the four quadrants. It is seen that wherever the particle happens to be, the x component will tend to pull it towards the y axis and the y component will tend to push it away from the x axis. Focusing will therefore be achieved in the x plane, whereas defocusing will take place in the y plane.

If one reverses the polarities in Fig. 7, the situation will be reversed, that is, the quadrupole will be focusing in the y plane and defocusing in the x plane.

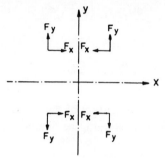

FIG. 7. Force components in a quadrupole.

C. ORBITS IN THE QUADRUPOLE

If one excepts the small energy transfer from longitudinal to transverse motion, the longitudinal velocity v stays constant during the traversal of the quadrupole, for there is no accelerating force to act on the particle. The equations of motion may therefore be written

$$m \frac{d^2 x}{dt^2} = F_x \qquad m \frac{d^2 y}{dt^2} = F_y \tag{34}$$

where m is the relativistic mass of the particle

$$m = \frac{m_0}{[1 - (v^2/c^2)]^{1/2}} \tag{35}$$

(m_0 is the rest mass, and c the velocity of light) which is constant here as v is constant.

Replacing the force components in Eq. (34), we get

$$m \frac{d^2 x}{dt^2} = - qvB_y = - qvGx \qquad m \frac{d^2 y}{dt^2} = qvB_x = qvGy \tag{36}$$

It is usually more interesting to know the equations of the trajectories $x = x(z)$, $y = y(z)$ than the position of the particle as a function of time.

Therefore, using (for v constant) the transformation relations

$$\frac{d^2x}{dt^2} = v^2 \frac{d^2x}{dz^2} \qquad \frac{d^2y}{dt^2} = v^2 \frac{d^2y}{dz^2} \qquad (37)$$

we can write for the differential equations of the orbit

$$\frac{d^2x}{dz^2} + \frac{qG}{mv} x = 0 \qquad \frac{d^2y}{dz^2} - \frac{qG}{mv} y = 0 \qquad (38)$$

Putting

$$k^2 = \frac{qG}{mv} = \frac{qG}{p} = \frac{G}{Br}$$

where p is the (relativistic) momentum of the particle and Br is its magnetic rigidity (momentum per unit charge), we finally have for the equations of motion

$$\frac{d^2x}{dz^2} + k^2x = 0 \qquad \frac{d^2y}{dz^2} - k^2y = 0 \qquad (39)$$

The solutions can be written immediately in the form

$$x = a \cos kz + b \sin kz \qquad y = c \cosh kz + d \sinh kz \qquad (40)$$

D. Transfer Matrices of a Quadrupole

Associating to the equation of the orbit in the focusing plane

$$x = a \cos kz + b \sin kz \qquad (41)$$

the relation one obtains by differentiation

$$x' = dx/dz = - ak \sin kz + bk \cos kz \qquad (42)$$

one can easily express the constants a and b in terms of the initial conditions x_0, x_0' which specify the position and the slope of the ray entering the quadrupole. If we denote the length of the quadrupole by L, we then obtain

$$x = x_0 \cos kL + (x_0'/k) \sin kL \qquad x' = - x_0 k \sin kL + x_0' \cos kL \quad (43)$$

A similar calculation for the defocusing plane gives

$$y = y_0 \cosh kL + (y_0'/k) \sinh kL \qquad y' = y_0 k \sinh kL + y_0' \cosh kL \quad (44)$$

On replacing the dimensionless quantity kL by ϑ

$$kL = \vartheta \tag{45}$$

the transfer matrices can be written in the form

$$T_c = \begin{Vmatrix} \cos \vartheta & (1/k) \sin \vartheta \\ -k \sin \vartheta & \cos \vartheta \end{Vmatrix} \tag{46}$$

for the convergent or focusing plane and

$$T_d = \begin{Vmatrix} \cosh \vartheta & (1/k) \sinh \vartheta \\ k \sinh \vartheta & \cosh \vartheta \end{Vmatrix} \tag{47}$$

for the divergent or defocusing plane.

E OPTICAL PROPERTIES OF THE QUADRUPOLE

All first-order optical properties of the quadrupole can be derived from these matrices. It is sufficient to apply our general relations (20)–(26) noting that, for both matrices, $a = d$.

1. Focal Planes

From Eq. (23) we have for the position of the foci

$$F_{ic} = F_{oc} = F_c = (1/k) \cot \vartheta \qquad F_{id} = F_{od} = F_d = -(1/k) \coth \vartheta \tag{48}$$

2. Focal Distances

From Eq. (25) we get

$$f_c = \frac{1}{k \sin \vartheta} \qquad f_d = -\frac{1}{k \sinh \vartheta} \tag{49}$$

which can be written in nondimensional form

$$f_c/L = \frac{1}{\vartheta \sin \vartheta} \qquad f_d/L = -\frac{1}{\vartheta \sinh \vartheta} \tag{50}$$

The relations (49) and (50) display the strong astigmatic properties of the quadrupole and suggest the use of quadrupole doublets, triplets or multiplets to achieve some amount of stigmatism. Moreover, it is readily seen that a quadrupole can be defocusing in its focusing plane but can never be focusing in its defocusing plane. In the vast majority of cases, however $\vartheta < \pi/2$ and the intersection of the ray with the axis occurs outside the quadrupole.

3. Principal Planes

These are given by Eq. (24)

$$H_{ic} = H_{oc} = H_c = -\frac{1 - \cos \vartheta}{k \sin \vartheta} \qquad H_{id} = H_{od} = H_d = \frac{1 - \cosh \vartheta}{k \sinh \vartheta} \quad (51)$$

4. Image Distances

Equation (21) shows that

$$q_c = \frac{p \cos \vartheta + (1/k) \sin \vartheta}{pk \sin \vartheta - \cos \vartheta} \qquad q_d = -\frac{p \cosh \vartheta + (1/k) \sinh \vartheta}{pk \sinh \vartheta + \cosh \vartheta} \quad (52)$$

It is again possible to rewrite these relations in nondimensional form, which makes them more suitable for graphical representation.
Putting

$$P = p/L \qquad Q = q/L \tag{53}$$

we have

$$Q_c = \frac{1}{\vartheta} \frac{P\vartheta \cot \vartheta + 1}{P\vartheta - \cot \vartheta} \qquad Q_d = -\frac{1}{\vartheta} \frac{P\vartheta \coth \vartheta + 1}{P\vartheta + \coth \vartheta} \tag{54}$$

5. Magnifications

Using the same notations, we find from Eq. (11)

$$g_c = \frac{1}{\cos \vartheta - P\vartheta \sin \vartheta} \qquad g_d = \frac{1}{P\vartheta \sinh \vartheta + \cosh \vartheta} \tag{55}$$

F. Thin-lens Approximation

1. First Definition

A quadrupole is considered to be thin if its longitudinal spacial extension is small compared to its focal length, namely, if

$$L \ll \frac{1}{k \sinh \vartheta} < \frac{1}{k \sin \vartheta} \tag{56}$$

This leads to the inequality

$$kL \ll 1 \tag{57}$$

Under these conditions, the transfer matrices T_c and T_d become

$$T_c = \left\| \begin{matrix} 1 & 0 \\ -\delta & 1 \end{matrix} \right\| \tag{58}$$

$$T_d = \left\| \begin{matrix} 1 & 0 \\ \delta & 1 \end{matrix} \right\| \tag{59}$$

where we have put

$$\delta = k^2 L = (G/Br)L \tag{60}$$

The focal distances are then

$$f_0 = f_c = -f_d = 1/\delta = Br/GL \tag{61}$$

2. Second Definition

A quadrupole is considered to be thin if it does not modify the position of the trajectory but only its slope (Fig. 8); therefore, it acts as a prism (impulse approximation).

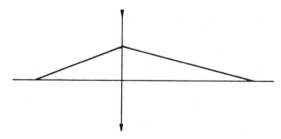

FIG. 8. Particle orbit in a thin lens.

Quantitatively, this means that in the equations of motion

$$\frac{d(dx/dz)}{dz} + \frac{Gx}{Br} = 0 \qquad \frac{d(dy/dz)}{dz} - \frac{Gy}{Br} = 0 \tag{62}$$

x and y can be considered as constants during the traversal of the quadrupole. Upon integration, we find for the change in slope

$$\Delta(dx/dz) = -\frac{GL}{Br} x \qquad \Delta(dy/dz) = +\frac{GL}{Br} y \tag{63}$$

and the focal distances (61) follow immediately.

3. *Third Definition*

A lens can be considered as being thin to the extent that the distance between its principal planes can be considered small compared to its focal length. From the formulas given previously, it follows that this definition is equivalent to the other two.

4. *Second-Order Approximation*

In many cases, the first-order approximation is not sufficient. We can then use a series expansion of $\cos \vartheta$ and $\cosh \vartheta$ and replace them in the formulas giving the optical characteristics.

Carrying out the calculations, we find to the second order

$$f_c = f_0 + \frac{L}{6} \qquad f_d = -f_0 + \frac{L}{6} \tag{64}$$

$$H_{ic} = H_{oc} = H_c = -\frac{L}{2} \left(1 + \frac{L}{12 f_o} \right)$$

$$H_{id} = H_{od} = H_d = -\frac{L}{2} \left(1 - \frac{L}{12 f_o} \right) \tag{65}$$

Figure 9 shows the position of the image principal planes in the two basic directions.

Obviously if, in the formulas given above, L can be neglected with respect to f_0, we again find the simple expressions of the thin-lens theory.

The second-order expressions (64), as well as the thick-lens formulas (49), show that in its focusing plane, a quadrupole is less focusing than the ideal thin lens, whereas in its defocusing plane, the defocusing is stronger than that displayed by the thin-lens approximation.

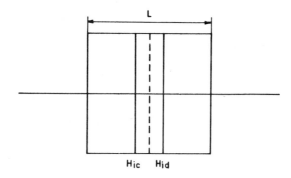

FIG. 9. Position of image principal planes in a quadrupole.

G. EQUIVALENT REPRESENTATIONS

Any quadrupole can be replaced by a thin lens having a drift space on each side (Fig. 10). Writing out the equivalence of the transfer matrices in the focusing plane

$$
\left\|
\begin{array}{cc}
\cos\vartheta & (1/k)\sin\vartheta \\
-k\sin\vartheta & \cos\vartheta
\end{array}
\right\|
=
\left\|
\begin{array}{cc}
1 & s_c \\
0 & 1
\end{array}
\right\|
\times
\left\|
\begin{array}{cc}
1 & 0 \\
-1/f_c & 1
\end{array}
\right\|
\times
\left\|
\begin{array}{cc}
1 & s_c \\
0 & 1
\end{array}
\right\|
\tag{66}
$$

one finds

$$
\frac{1}{f_c} = k\sin\vartheta \qquad s_c = \frac{1-\cos\vartheta}{k\sin\vartheta}
\tag{67}
$$

Whereas the focusing strength of the equivalent lens is the same as that of the original lens, the associated drift length is larger than $L/2$.

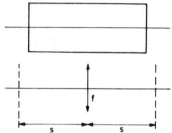

FIG. 10. Equivalent representation of a quadrupole.

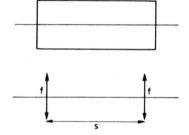

FIG. 11. Equivalent representation of a quadrupole.

By similar considerations, we find in the defocusing plane

$$
\frac{1}{f_d} = -k\sinh\vartheta \qquad s_d = \frac{\cosh\vartheta - 1}{k\sinh\vartheta}
\tag{68}
$$

and, therefore, the virtual drift length is smaller than $L/2$.

In some applications it is useful to replace an actual quadrupole by two thin lenses separated by a drift space of length s (Fig. 11). Writing out again the equality of the transfer matrices, we find

$$
\frac{1}{f_c} = k\,\frac{1-\cos\vartheta}{\sin\vartheta} \qquad s_c = \frac{1}{k}\sin\vartheta
\tag{69}
$$

for the focusing plane, and

$$\frac{1}{f_d} = - k \frac{\cosh \vartheta - 1}{\sinh \vartheta} \qquad s_d = \frac{1}{k} \sinh \vartheta \qquad (70)$$

for the defocusing plane.

2.4.3. The Doublet

A. GENERAL PROPERTIES

Starting from a real object, a simple quadrupole gives a real image point in its converging plane and a virtual image in its diverging plane. However, the practical use of a focusing system requires, in general, a real image in

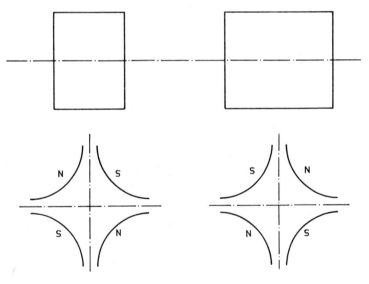

FIG. 12. Basic geometry of a doublet.

both planes. The simplest device achieving this requirement is a doublet (Fig. 12), that is, a set of two quadrupoles of opposite polarity separated by a drift space.

Figure 13 shows the envelope of the beam in the plane where the first quadrupole is diverging and the second is converging, whereas Fig. 14 gives the shape of the same envelope in the plane where the first quadrupole is converging and the second is diverging. The overall effect is that of an AG lens whose focusing strength is generally different in the two basic planes. Although Q_{dc} is different from Q_{cd}, bidirectional focusing is possible. It is indeed seen that in both cases the excursion of the particle is

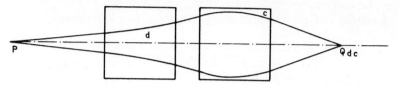

FIG. 13. Beam envelope in the defocusing–focusing plane.

greater in the quadrupole which is focusing; the restoring force being proportional to the excursion of the particle, the focusing effect will win over the defocusing effect.

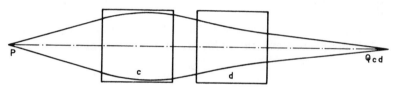

FIG. 14. Beam envelope in the focusing–defocusing plane.

B. THE ANTISYMMETRICAL DOUBLET

We shall call a doublet antisymmetrical if its component quadrupoles are of equal length and gradient. If d denotes the distance between the effective end planes of the quadrupoles (Fig. 15), the transfer matrices in the two basic planes are

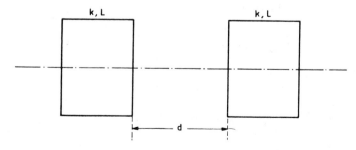

FIG. 15. Parameters of an antisymmetrical doublet.

$$T_{cd} = \begin{Vmatrix} \cosh\vartheta & (1/k)\sinh\vartheta \\ k\sinh\vartheta & \cosh\vartheta \end{Vmatrix} \times \begin{Vmatrix} 1 & d \\ 0 & 1 \end{Vmatrix} \times \begin{Vmatrix} \cos\vartheta & (1/k)\sin\vartheta \\ -k\sin\vartheta & \cos\vartheta \end{Vmatrix} \qquad (71)$$

$$T_{dc} = \begin{Vmatrix} \cos\vartheta & (1/k)\sin\vartheta \\ -k\sin\vartheta & \cos\vartheta \end{Vmatrix} \times \begin{Vmatrix} 1 & d \\ 0 & 1 \end{Vmatrix} \times \begin{Vmatrix} \cosh\vartheta & (1/k)\sinh\vartheta \\ k\sinh\vartheta & \cosh\vartheta \end{Vmatrix} \qquad (72)$$

Carrying out the calculations, we find the two expressions

$$T_{cd} = \begin{Vmatrix} \cos\vartheta\cosh\vartheta - \sin\vartheta\sinh\vartheta - dk\sin\vartheta\cosh\vartheta \\ 1/k\,(\cos\vartheta\sinh\vartheta + \sin\vartheta\cosh\vartheta) + d\cos\vartheta\cosh\vartheta \\ k(\cos\vartheta\sinh\vartheta - \sin\vartheta\cosh\vartheta) - dk^2\sin\vartheta\sinh\vartheta \\ \cos\vartheta\cosh\vartheta + \sin\vartheta\sinh\vartheta + dk\cos\vartheta\sinh\vartheta \end{Vmatrix} \qquad (73)$$

and

$$T_{dc} = \begin{Vmatrix} \cos\vartheta\cosh\vartheta + \sin\vartheta\sinh\vartheta + dk\cos\vartheta\sinh\vartheta \\ 1/k\,(\cos\vartheta\sinh\vartheta + \sin\vartheta\cosh\vartheta) + d\cos\vartheta\cosh\vartheta \\ k(\cos\vartheta\sinh\vartheta - \sin\vartheta\cosh\vartheta) - dk^2\sin\vartheta\sinh\vartheta \\ \cos\vartheta\cosh\vartheta - \sin\vartheta\sinh\vartheta - dk\sin\vartheta\cosh\vartheta \end{Vmatrix} \qquad (74)$$

Comparing the last two transfer matrices and using the general matrix symbolism (5), we infer the following important relations:

$$c_{cd} = c_{dc} = c \qquad b_{cd} = b_{dc} = b \qquad a_{cd} = d_{dc} \qquad d_{cd} = a_{dc}. \qquad (75)$$

One matrix is therefore sufficient to describe the antisymmetrical doublet.

C. OPTICAL PROPERTIES OF THE ANTISYMMETRICAL DOUBLET

Applying the general relations (8)–(19), and using relations (75), we can readily derive the optical characteristics of the antisymmetrical doublet.

1. Focal Distances

From Eq. (16) we have

$$f_{cd} = f_{dc} = \frac{1}{k(\sin\vartheta\cosh\vartheta - \cos\vartheta\sinh\vartheta + dk\sin\vartheta\sinh\vartheta)} \qquad (76)$$

Putting

$$D = d/L \qquad (77)$$

we can write this in the nondimensional form

$$\frac{f_{cd}}{L} = \frac{f_{dc}}{L} = \frac{1}{\vartheta(\sin\vartheta\cosh\vartheta - \cos\vartheta\sinh\vartheta + D\vartheta\sin\vartheta\sinh\vartheta)} \qquad (78)$$

2. Focal Planes

Equation (23) gives for the image foci

$$F_{icd} = \frac{\cos\vartheta\cosh\vartheta - \sin\vartheta\sinh\vartheta - dk\sin\vartheta\cosh\vartheta}{k(\sin\vartheta\cosh\vartheta - \cos\vartheta\sinh\vartheta + dk\sin\vartheta\sinh\vartheta)}$$

$$F_{idc} = \frac{\cos\vartheta\cosh\vartheta + \sin\vartheta\sinh\vartheta + dk\cos\vartheta\sinh\vartheta}{k(\sin\vartheta\cosh\vartheta - \cos\vartheta\sinh\vartheta + dk\sin\vartheta\sinh\vartheta)}$$

(79)

These relations can also be written in the nondimensional form

$$F_{icd}/L = \frac{1}{\vartheta} \cdot \frac{1 - \cot\vartheta\coth\vartheta + D\vartheta\coth\vartheta}{\cot\vartheta - \coth\vartheta - D\vartheta}$$

$$F_{idc}/L = \frac{1}{\vartheta} \cdot \frac{1 + \cot\vartheta\coth\vartheta + D\vartheta\cot\vartheta}{\coth\vartheta - \cot\vartheta + D\vartheta}$$

(80)

or by using Eq. (48)

$$F_{icd} = \frac{F_d(F_c - d) + (1/k^2)}{F_c + F_d - d} \qquad F_{idc} = \frac{F_c(F_d - d) - (1/k^2)}{F_c + F_d - d}$$

(81)

The last two relations show that it is not possible to achieve coincidence of the two foci. $F_{icd} = F_{idc}$ would indeed imply

$$d = 2/[k^2(F_d - F_c)]$$

(82)

which is not possible, since F_d is positive and F_c is generally positive. In exceptional cases, F_c could be negative so that Eq. (82) would be satisfied; however, these cases do not seem to be of interest in beam-transport systems for they would imply very strong focusing and intersection of the ray with the axis occurring inside the lens.\

3. Principal Planes

From Eq. (14) we have

$$H_{icd}/L = \frac{1 - \cos\vartheta\cosh\vartheta + \sin\vartheta\sinh\vartheta + D\vartheta\sin\vartheta\cosh\vartheta}{\vartheta(\cos\vartheta\sinh\vartheta - \sin\vartheta\cosh\vartheta - D\vartheta\sin\vartheta\sinh\vartheta)}$$

$$H_{idc}/L = \frac{1 - \cos\vartheta\cosh\vartheta - \sin\vartheta\sinh\vartheta - D\vartheta\cos\vartheta\sinh\vartheta}{\vartheta(\cos\vartheta\sinh\vartheta - \sin\vartheta\cosh\vartheta - D\vartheta\sin\vartheta\sinh\vartheta)}$$

(83)

Again, it is not possible to bring into coincidence the principal planes in the two basic directions, for $H_{icd} = H_{idc}$ would demand

$$2 \sin \vartheta \sinh \vartheta + D\vartheta(\cos \vartheta \sinh \vartheta + \sin \vartheta \cosh \vartheta) = 0 \qquad (84)$$

which, expressed in other terms, is exactly the same relation as (82).

4. Magnifications

From Eq. (11) we have

$$
\begin{aligned}
\frac{1}{g_{cd}} &= P\vartheta(\cos \vartheta \sinh \vartheta - \sin \vartheta \cosh \vartheta - D\vartheta \sin \vartheta \sinh \vartheta) \\
&\quad + \cos \vartheta \cosh \vartheta + \sin \vartheta \sinh \vartheta + D\vartheta \cos \vartheta \sinh \vartheta \\
\frac{1}{g_{dc}} &= P\vartheta(\cos \vartheta \sinh \vartheta - \sin \vartheta \cosh \vartheta - D\vartheta \sin \vartheta \sinh \vartheta) \\
&\quad + \cos \vartheta \cosh \vartheta - \sin \vartheta \sinh \vartheta - D\vartheta \sin \vartheta \cosh \vartheta
\end{aligned}
\qquad (85)
$$

It is seen that in an antisymmetrical doublet

$$g_{cd} < g_{dc} \qquad (86)$$

The equality of the two magnifications would again lead to Eq. (82) or (84).

5. Image Position

Using Eq. (9) the image positions may be written

$$q_{cd} = F_{icd} + \frac{F_{icd} F_{idc} + t}{p - F_{idc}} \qquad q_{dc} = F_{idc} + \frac{F_{icd} F_{idc} + t}{p - F_{icd}} \qquad (87)$$

with

$$t = -\frac{b}{c} = \frac{F_d - F_c + dk^2 F_c F_d}{k^2(F_c + F_d - d)} \qquad (88)$$

noting that the elements b and c are the same for both transfer matrices (73) and (74).

The image positions can also be written in nondimensional form

$$Q_{cd} = \frac{PF_{icd}/L + T}{P - F_{idc}/L} \qquad Q_{dc} = \frac{PF_{idc}/L + T}{P - F_{icd}/L} \qquad (89)$$

where

$$T = \frac{t}{L^2} = \frac{1}{\vartheta^2} \frac{\cot \vartheta + \coth \vartheta + D\vartheta \cot \vartheta \coth \vartheta}{D\vartheta + \coth \vartheta - \cot \vartheta} \qquad (90)$$

Finally, after replacing in Eq. (89), F_{icd}/L, F_{idc}/L and T by their explicit values, the image distances can be written in a form suitable for graphical plotting of universal curves

$$Q_{cd} = \frac{1}{\vartheta} \frac{P\vartheta(1-\cot\vartheta\,\coth\vartheta+D\vartheta\,\coth\vartheta)-(\cot\vartheta+\coth\vartheta+D\vartheta\cot\vartheta\,\coth\vartheta)}{P\vartheta(\cot\vartheta-\coth\vartheta-D\vartheta)+1+\cot\vartheta\,\coth\vartheta+D\vartheta\cot\vartheta}$$

$$(91)$$

$$Q_{dc} = \frac{1}{\vartheta} \frac{P\vartheta(1+\cot\vartheta\,\coth\vartheta+D\vartheta\cot\vartheta)+(\cot\vartheta+\coth\vartheta+D\vartheta\cot\vartheta\,\coth\vartheta)}{P\vartheta(\coth\vartheta-\cot\vartheta+D\vartheta)+1-\cot\vartheta\,\coth\vartheta+D\vartheta\,\coth\vartheta}$$

6. Characteristics of the Object Space

The object focal planes and the object principal planes can easily be derived from the characteristics of the image space.

One has, indeed, by virtue of Eq. (75)

$$F_{ocd} = -\frac{d_{cd}}{c} = -\frac{a_{dc}}{c} = F_{idc} \qquad F_{odc} = -\frac{d_{dc}}{c} = -\frac{a_{cd}}{c} = F_{icd} \quad (92)$$

On the other hand,

$$H_{ocd} = \frac{1-d_{cd}}{c} = \frac{1-a_{dc}}{c} = H_{idc} \qquad H_{odc} = \frac{1-d_{dc}}{c} = \frac{1-a_{cd}}{c} = H_{icd}$$

7. Stigmatic Operation of an Antisymmetrical Doublet

In general, $q_{cd} \neq q_{dc}$ and the same object point yields two different image points according to whether one considers the cd plane or the dc plane. However, for a given position of the object point, it is possible to choose the parameters of the lens system so as to obtain a one-to-one correspondence.

The condition of stigmatism gives

$$q_{cd} = q_{dc} = q_0 \qquad (94)$$

and writing this out in the two basic planes in terms of the object position

$$q_0 = -\frac{a_{cd}\,p_0 + b}{c\,p_0 + d_{cd}} = -\frac{a_{dc}\,p_0 + b}{c\,p_0 + d_{dc}} \qquad (95)$$

By taking into account Eq. (75), this becomes

$$-q_0 = \frac{a_{cd}\,p_0 + b}{cp_0 + a_{dc}} = \frac{a_{dc}\,p_0 + b}{cp_0 + a_{cd}} = \frac{a_{cd} - a_{dc}}{a_{dc} - a_{cd}}p_0 \qquad (96)$$

and as $a_{cd} \neq a_{dc}$ we have

$$q_0 = p_0 \qquad (97)$$

which is also evident for symmetry reasons. Next, for the equation in p_0, we obtain

$$-p_0 = \frac{a_{cd}\, p_0 + b}{c p_0 + a_{dc}} \qquad (98)$$

so that

$$p_0{}^2 + \frac{a_{cd} + a_{dc}}{c} p_0 + \frac{b}{c} = 0 \qquad (99)$$

or

$$p_0{}^2 - (F_{cd} + F_{dc})p_0 - t = 0 \qquad (100)$$

The solution is therefore

$$p_0 = \frac{F_{cd} + F_{dc}}{2} \pm \left[\frac{(F_{cd} + F_{dc})^2}{4} + t \right]^{1/2} \qquad (101)$$

In nondimensional form, Eq. (101) can be written

$$P_0 = Q_0 = \frac{D\vartheta(\coth \vartheta - \cot \vartheta) - 2 \cot \vartheta \coth \vartheta}{2\vartheta(\cot \vartheta - \coth \vartheta - D\vartheta)} \pm$$

$$\pm \frac{\left\{ [D\vartheta(\cot \vartheta + \coth \vartheta) + 2]^2 + \dfrac{4}{\sin^2 \vartheta \sinh^2 \vartheta} \right\}^{1/2}}{2\vartheta(\cot \vartheta - \coth \vartheta - D\vartheta)} \qquad (102)$$

If p_0 is given, Eq. (100) gives the interlens spacing corresponding to stigmatic operation

$$d = \frac{p_0{}^2(F_c + F_d) - 2p_0 F_c F_d + (F_c - F_0)/k^2}{p_0{}^2 - p_0(F_c + F_d) + F_c F_d} \qquad (103)$$

or, in nondimensional form,

$$D\vartheta = \frac{(P_0\vartheta)^2(\cot \vartheta - \coth \vartheta) + 2P_0\vartheta \cot \vartheta \coth \vartheta + \cot \vartheta + \coth \vartheta}{(P_0\vartheta)^2 + P_0\vartheta(\coth \vartheta - \cot \vartheta) - \cot \vartheta \coth \vartheta} \qquad (104)$$

8. Stigmatism and Magnifications

For stigmatic operation, there is an important relation between the magnifications in the two basic planes.

One has, indeed,

$$g_{cd} \times g_{dc} = \frac{1}{cp_0 + d_{cd}} \cdot \frac{1}{cp_0 + d_{dc}}$$

$$= \frac{1}{cp_0 + a_{dc}} \cdot \frac{1}{cp_0 + a_{cd}}$$

$$= \frac{1}{c^2 p_0^2 + c(a_{cd} + a_{dc})p_0 + a_{cd} a_{dc}}$$

$$= \frac{1}{c \{ cp_0^2 + (a_{cd} + a_{dc})p_0 + [(a_{cd} a_{dc})/c] \}}$$

By using Eqs. (75) and (99), this becomes

$$g_{cd} \times g_{dc} = \frac{1}{c \{ - b + [(a_{cd} a_{dc})/c] \}} = \frac{1}{- bc + a_{cd} d_{cd}}$$

that is

$$g_{cd} \times g_{dc} = 1 \qquad (105)$$

Since the product of the magnifications is equal to unity, there cannot be real stigmatism: a small circle will be deformed into an ellipse because it gets larger in one plane and smaller in the other. This situation of pseudo-stigmatism is not acceptable in electron microscopy, for example, where the requirements on the image quality are high, but it can be quite satisfactory and even interesting in beam-transport systems in which the source may present some inherent astigmatism.

D. THE GENERAL DOUBLET

The main interest of the general doublet, as compared to the antisymmetrical doublet, lies in the possibility of adjusting independently the excitations of the individual quadrupoles and, therefore, to have one more degree of freedom at hand. In most cases, the lengths of the two elements will be equal, but this does not imply any simplification in the mathematical aspects of the problem.

Figure 16 shows the notations used.

1. Transfer matrices

We have, for the two basic planes,

$$T_{cd} = \begin{Vmatrix} \cosh \vartheta_2 & (1/k_2) \sinh \vartheta_2 \\ k_2 \sinh \vartheta_2 & \cosh \vartheta_2 \end{Vmatrix} \times \begin{Vmatrix} 1 & d \\ 0 & 1 \end{Vmatrix} \times \begin{Vmatrix} \cos \vartheta_1 & (1/k_1) \sin \vartheta_1 \\ - k_1 \sin \vartheta_1 & \cos \vartheta_1 \end{Vmatrix}$$

$$(106)$$

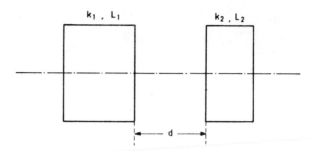

FIG. 16. Parameters of an arbitrary doublet.

$$T_{dc} = \begin{Vmatrix} \cos \vartheta_2 & (1/k_2) \sin \vartheta_2 \\ -k_2 \sin \vartheta_2 & \cos \vartheta_2 \end{Vmatrix} \times \begin{Vmatrix} 1 & d \\ 0 & 1 \end{Vmatrix} \times \begin{Vmatrix} \cosh \vartheta_1 & (1/k_1) \sinh \vartheta_1 \\ k_1 \sinh \vartheta_1 & \cosh \vartheta_1 \end{Vmatrix}$$

(107)

Carrying out the calculations, one finds for the various matrix elements

$$a_{cd} = \cos \vartheta_1 \cosh \vartheta_2 - dk_1 \sin \vartheta_1 \cosh \vartheta_2 - (k_1/k_2) \sin \vartheta_1 \sinh \vartheta_2$$

$$b_{cd} = d \cos \vartheta_1 \cosh \vartheta_2 + (1/k_1) \sin \vartheta_1 \cosh \vartheta_2 + (1/k_2) \cos \vartheta_1 \sinh \vartheta_2$$

$$c_{cd} = k_2 \cos \vartheta_1 \sinh \vartheta_2 - k_1 \sin \vartheta_1 \cosh \vartheta_2 - dk_1 k_2 \sin \vartheta_1 \sinh \vartheta_2$$

$$d_{cd} = \cos \vartheta_1 \cosh \vartheta_2 + dk_2 \cos \vartheta_1 \sinh \vartheta_2 + (k_2/k_1) \sin \vartheta_1 \sinh \vartheta_2$$

(108)

$$a_{dc} = \cosh \vartheta_1 \cos \vartheta_2 + dk_1 \sinh \vartheta_1 \cos \vartheta_2 + (k_1/k_2) \sinh \vartheta_1 \sin \vartheta_2$$

$$b_{dc} = d \cosh \vartheta_1 \cos \vartheta_2 + (1/k_1) \sinh \vartheta_1 \cos \vartheta_2 + (1/k_2) \cosh \vartheta_1 \sin \vartheta_2$$

$$c_{dc} = -k_2 \cosh \vartheta_1 \sin \vartheta_2 + k_1 \sinh \vartheta_1 \cos \vartheta_2 - dk_1 k_2 \sinh \vartheta_1 \sin \vartheta_2$$

$$d_{dc} = \cosh \vartheta_1 \cos \vartheta_2 - dk_2 \cosh \vartheta_1 \sin \vartheta_2 - (k_2/k_1) \sinh \vartheta_1 \sin \vartheta_2$$

(109)

2. Optical Characteristics

All optical characteristics are contained in the matrix elements. We find for the focal distances

$$f_{cd} = \frac{1}{dk_1 k_2 \sin \vartheta_1 \sinh \vartheta_2 + k_1 \sin \vartheta_1 \cosh \vartheta_2 - k_2 \cos \vartheta_1 \sinh \vartheta_2}$$

$$f_{dc} = \frac{1}{dk_1 k_2 \sinh \vartheta_1 \sin \vartheta_2 - k_1 \sinh \vartheta_1 \cos \vartheta_2 + k_2 \cosh \vartheta_1 \sin \vartheta_2}$$

(110)

and, in general, the focal distance will be different.

For the position of the focal planes we find

$$F_{icd} = \frac{1}{k_2} \frac{k_1 - k_2 \cot \vartheta_1 \coth \vartheta_2 + dk_1 k_2 \coth \vartheta_2}{k_2 \cot \vartheta_1 - k_1 \coth \vartheta_2 - dk_1 k_2}$$

$$F_{idc} = \frac{1}{k_2} \frac{k_1 + k_2 \coth \vartheta_1 \cot \vartheta_2 + dk_1 k_2 \cot \vartheta_2}{k_2 \coth \vartheta_1 - k_1 \cot \vartheta_2 + dk_1 k_2}$$

(111)

The abscissas of these foci are different, in general, but it is possible to bring them into coincidence except in the case $k_1 = k_2$, $\vartheta_1 = \vartheta_2$, which corresponds to the antisymmetrical doublet.

By reversing the direction of the incident ray, it is easily seen that the optical characteristics of the object space in the $cd(dc)$ plane can be found simply by using the formulas of the optical characteristics of the image space in the $dc(cd)$ plane and interchanging the subscripts 1 and 2. Therefore, for example,

$$F_{ocd} = \frac{1}{k_1} \frac{k_2 + k_1 \coth \vartheta_2 \cot \vartheta_1 + dk_1 k_2 \cot \vartheta_1}{k_1 \coth \vartheta_2 - k_2 \cot \vartheta_1 + dk_1 k_2}$$

$$F_{odc} = \frac{1}{k_1} \frac{k_2 - k_1 \cot \vartheta_2 \coth \vartheta_1 + dk_1 k_2 \coth \vartheta_1}{k_1 \cot \vartheta_2 - k_2 \coth \vartheta_1 - dk_1 k_2}$$

(112)

3. Stigmatic Operation

The general condition of stigmatism can be written here

$$- q_0 = \frac{a_{cd} p_0 + b_{cd}}{c_{cd} p_0 + d_{cd}} = \frac{a_{dc} p_0 + b_{dc}}{c_{dc} p_0 + d_{dc}}$$

(113)

Putting

$$t_{cd} = - b_{cd}/c_{cd} \qquad t_{dc} = - b_{dc}/c_{dc}$$

(114)

Eq. (113) can be written

$$q_0 = \frac{F_{icd} p_0 + t_{cd}}{p_0 - F_{ocd}} = \frac{F_{idc} p_0 + t_{dc}}{p_0 - F_{odc}}$$

(115)

The last equality gives the value of p_0 for stigmatic operation; the corresponding value of q_0 can then be found from

$$(F_{ocd} - F_{odc})q_0 = (F_{idc} - F_{idc})p_0 + t_{dc} - t_{cd}$$

(116)

which results from Eq. (115).

The product of the two magnifications becomes, under stigmatic conditions,

$$g_{cd} \cdot g_{dc} = \frac{1}{c_{cd}p_0 + d_{cd}} \cdot \frac{1}{c_{dc}p_0 + d_{dc}}$$

$$= \frac{1}{c_{cd}[p_0 + (d_{cd}/c_{cd})]} \cdot \frac{1}{c_{dc}[p_0 + (d_{dc}/c_{dc})]}$$

and this can be written

$$g_{cd} \cdot g_{dc} = \frac{f_{cd}f_{dc}}{(p_0 - F_{ocd})(p_0 - F_{odc})} \tag{117}$$

4. *The Doublet in the Thin-Lens Approximation*

It has been shown that the focusing action of a quadrupole is weaker than the thin-lens approximation would indicate, whereas the defocusing action of a quadrupole is stronger than that indicated by the same approximation. Because the overall focusing effect of a doublet is due to the difference of these partial effects, it follows that the converging power of a doublet might be substantially smaller than the value given by the thin-lens approximation. The thin-lens theory is, however, quite useful in determining general properties of doublets, especially in high-energy work where the focusing power is small, and also in calculating some secondary effects like aberrations. Its correctness can be improved by considering each of the quadrupoles composing the doublet as a thick lens and using the thin-lens approximation only in calculating the properties of the combination.

Let s be the separation of the midplanes of the quadrupoles whose characteristics are L_1, k_1 and L_2, k_2 (Fig. 17). We then have

$$s = d + (L_1 + L_2)/2 \tag{118}$$

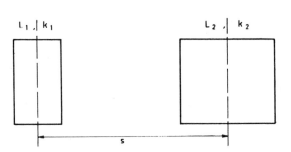

FIG. 17. The doublet as a combination of thin lenses.

On the other hand, it turns out to be convenient to introduce the absolute values of the focusing strengths of the individual quadrupoles and take care of their signs in writing the transfer matrices. Therefore, we shall put

$$\frac{1}{f_{1c}} = k_1 \sin \vartheta_1 \qquad \frac{1}{f_{1d}} = k_1 \sinh \vartheta_1$$

$$\frac{1}{f_{2c}} = k_2 \sin \vartheta_2 \qquad \frac{1}{f_{2d}} = k_2 \sinh \vartheta_2 \tag{119}$$

With these sign conventions, the transfer matrices are

$$T_{cd} = \begin{Vmatrix} 1 & 0 \\ \dfrac{1}{f_{2d}} & 1 \end{Vmatrix} \times \begin{Vmatrix} 1 & s \\ 0 & 1 \end{Vmatrix} \times \begin{Vmatrix} 1 & 0 \\ -\dfrac{1}{f_{1c}} & 1 \end{Vmatrix} \tag{120}$$

$$T_{dc} = \begin{Vmatrix} 1 & 0 \\ -\dfrac{1}{f_{2c}} & 1 \end{Vmatrix} \times \begin{Vmatrix} 1 & s \\ 0 & 1 \end{Vmatrix} \times \begin{Vmatrix} 1 & 0 \\ \dfrac{1}{f_{1d}} & 1 \end{Vmatrix} \tag{121}$$

Multiplying out, we find

$$T_{cd} = \begin{Vmatrix} 1 - \dfrac{s}{f_{1c}} & s \\ -\left(\dfrac{1}{f_{1c}} - \dfrac{1}{f_{2d}} + \dfrac{1}{f_{1c}f_{2d}}\right) & 1 + \dfrac{s}{f_{2d}} \end{Vmatrix} \tag{122}$$

$$T_{dc} = \begin{Vmatrix} 1 + \dfrac{s}{f_{1d}} & s \\ -\left(-\dfrac{1}{f_{1d}} + \dfrac{1}{f_{2c}} + \dfrac{s}{f_{1d}f_{2c}}\right) & 1 - \dfrac{s}{f_{2c}} \end{Vmatrix} \tag{123}$$

5. *Optical Properties Derived from the Thin-Lens Approximation*

From the two matrices we first derive the focal distances of the doublet combination

$$\frac{1}{f_{cd}} = \frac{1}{f_{1c}} - \frac{1}{f_{2d}} + \frac{s}{f_{1c}f_{2d}} \qquad \frac{1}{f_{dc}} = -\frac{1}{f_{1d}} + \frac{1}{f_{2c}} + \frac{s}{d_{1d}f_{2c}} \tag{124}$$

Under the conditions wherein we may apply the thin-lens approximation to the individual quadrupoles, we can drop the subscripts c and d for these elements and write

$$\frac{1}{f_{cd}} = \frac{1}{f_1} - \frac{1}{f_2} + \frac{s}{f_1 f_1} \qquad \frac{1}{f_{dc}} = -\frac{1}{f_1} + \frac{1}{f_2} + \frac{s}{f_1 f_2} \tag{125}$$

with

$$1/f_1 = k_1{}^2 L_1 \qquad 1/f_2 = k_2{}^2 L_2 \tag{126}$$

It is readily seen that the doublet provides bidirectional focusing if

$$|f_1 - f_2| < s \tag{127}$$

For an antisymmetrical doublet

$$f_1 = f_2 = f_0 = 1/k^2 L \tag{128}$$

and

$$f_{cd} = f_{dc} = f_0{}^2/s \tag{129}$$

Considering next the positions of the principal planes, we find from the transfer matrices

$$H_{icd} = -\frac{s}{f_{1c}} f_{cd} \qquad H_{idc} = \frac{s}{f_{1d}} f_{dc} \tag{130}$$

for the image elements and

$$H_{ocd} = \frac{s}{f_{2c}} f_{cd} \qquad H_{odc} = -\frac{s}{f_{2c}} f_{dc} \tag{131}$$

for the object elements.

The distance between corresponding principal planes is therefore

$$\Delta z(H)_{cd} = \frac{s^2}{f_{1c}f_{2d}} f_{cd} \tag{132}$$

in one plane and

$$\Delta z(H)_{dc} = \frac{s^2}{f_{1d}f_{2c}} f_{dc} \tag{133}$$

in the other. The doublet itself can be considered as a thin lens if

$$\frac{\Delta z(H)_{cd}}{f_{cd}} \ll 1 \qquad \frac{\Delta z(H)_{dc}}{f_{dc}} \ll 1 \tag{134}$$

These conditions can be written

$$s^2 \ll f_{1c}f_{2d} \qquad s^2 \ll f_{1d}f_{2c} \tag{135}$$

In the case in which one may drop the subscripts c and d of the individual quadrupoles, the conditions reduce to

$$s^2 \ll f_1 f_2 \tag{136}$$

or

$$s(L_1 L_2)^{1/2} \ll 1/k_1 k_2 \tag{137}$$

For an antisymmetrical doublet, this becomes

$$sL \ll 1/k^2 \tag{138}$$

or, in terms of the beam rigidity,

$$Br \gg sLG \tag{139}$$

E. PRACTICAL USE OF A DOUBLET

A doublet is a major device in a beam-transport system as it can be used for a variety of purposes. Typical examples are focusing of a parallel beam, rendering parallel a divergent beam, focusing of a divergent beam, and focusing of an astigmatic beam. These cases will be considered in more detail in what follows.

1. P-F Problem

This is a common problem encountered in beam transport. A beam, initially parallel, is brought to a focus (Fig. 18) in both the cd and dc planes.

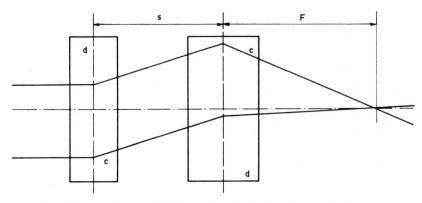

FIG. 18. Focusing an initially parallel beam; thin-lens approximation.

In the thin-lens approximation, the position of the common focus is found from Eqs. (12), (122), and (123) to be

$$F = f_{cd}\left(1 - \frac{s}{f_{1c}}\right) = f_{dc}\left(1 + \frac{s}{f_{1d}}\right) \tag{140}$$

Putting

$$\sigma_1 = \frac{1}{f_{1c}} + \frac{1}{f_{1d}} \qquad \sigma_2 = \frac{1}{f_{2c}} + \frac{1}{f_{2d}}$$

$$\vartheta_1 = \frac{1}{f_{1c}} - \frac{1}{f_{1d}} \qquad \vartheta_2 = \frac{1}{f_{2c}} - \frac{1}{f_{2d}} \tag{141}$$

the solution of Eq. (140) can be written in the form

$$s = 2\,\frac{\sigma_2 - \sigma_1}{\sigma_2\vartheta_1 + [\sigma_1\sigma_2(\vartheta_1{}^2 - \sigma_1{}^2 + \sigma_1\sigma_2)]^{1/2}}$$

$$F = 2\,\frac{\sigma_1}{\sigma_1\vartheta_2 + [\sigma_1\sigma_2(\vartheta_1{}^2 - \sigma_1{}^2 + \sigma_1\sigma_2)]^{1/2}} \tag{142}$$

If one may drop the subscripts c and d of the individual quadrupoles, these relations become

$$s = f_1\left(\frac{f_1 - f_2}{f_1}\right)^{1/2} \qquad F = f_2\left(\frac{f_1}{f_1 - f_2}\right)^{1/2} \tag{143}$$

Therefore, it is necessary that f_1 be larger than f_2. More often, s and F will be given and f_1 and f_2 wanted. From Eq. (143) we find

$$f_1 = (s^2 + sF)^{1/2} \qquad f_2 = \frac{sF}{(s^2 + sF)^{1/2}} \tag{144}$$

The practical design of the two quadrupoles (excitations, geometry, etc.) can proceed from these data.

2. F-P Problem

This is the inverse problem: One wants to render parallel a beam diverging from a point source (Fig. 19). In order to solve this problem, it suffices to interchange the subscripts 1 and 2 in the preceding problem. Thus,

$$s = 2\,\frac{\sigma_1 - \sigma_2}{\sigma_1\vartheta_2 + [\sigma_1\sigma_2(\vartheta_2{}^2 - \sigma_2{}^2 + \sigma_1\sigma_2)]^{1/2}}$$

$$F = 2\,\frac{\sigma_2}{\sigma_2\vartheta_1 + [\sigma_1\sigma_2(\vartheta_2{}^2 - \sigma_2{}^2 + \sigma_1\sigma_2)]^{1/2}} \tag{145}$$

and, if one may drop the subscripts of the individual quadrupoles,

$$s = f_2 \left(\frac{f_2 - f_1}{f_2}\right)^{1/2} \qquad F = f_1 \left(\frac{f_2}{f_2 - f_1}\right)^{1/2} \tag{146}$$

Here, f_2 should be larger than f_1.

If s and F are given (which is the usual case), the focal distances follow from

$$f_1 = \frac{sF}{(s^2 + sF)^{1/2}} \qquad f_2 = (s^2 + sF)^{1/2} \tag{147}$$

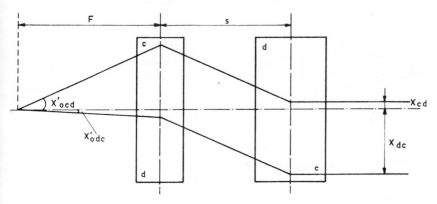

FIG. 19. Rendering parallel an initially divergent beam; thin-lens approximation.

It is sometimes useful to know the width of the emerging parallel beam as a function of the initial divergence. From the transfer matrix, the width is found to be

$$R = x/x_0' = aF + b \tag{148}$$

Particularizing to the two basic planes, we have

$$R_{cd} = (s + F)^{1/2}[(s + F)^{1/2} - s^{1/2}] \qquad R_{dc} = (s + F)^{1/2}[(s + F)^{1/2} + s^{1/2}] \tag{149}$$

The beam is therefore wider in the dc plane.

In many cases, it is necessary to go beyond the first approximation and use the thick-lens approach. With our previous notations we find from the conjugation relation for the case under consideration and considering the two planes (Fig. 20)

$$c_{cd}p + d_{cd} = 0 \qquad c_{dc}p + d_{dc} = 0 \tag{150}$$

$$R_{cd} = a_{cd}p + b_{cd} \qquad R_{dc} = a_{dc}p + b_{dc} \tag{151}$$

Putting[1]

$$a_1 = \cos \vartheta_1 - pk_1 \sin \vartheta_1 \qquad \beta_1 = \cosh \vartheta_1 + pk_1 \sinh \vartheta_1 \qquad (152)$$

$$A_1 = da_1 + p \cos \vartheta_1 + (1/k_1) \sin \vartheta_1 \qquad B_1 = d\beta_1 + p \cosh \vartheta_1 + (1/k)_1 \sinh \vartheta_1$$

using Eqs. (108) and (109), and carrying out the calculations, we find from Eqs. (150)

$$k_2 \tanh \vartheta_2 = - a_1/A_1 \qquad k_2 \tan \vartheta_2 = \beta_1/B_1 \qquad (153)$$

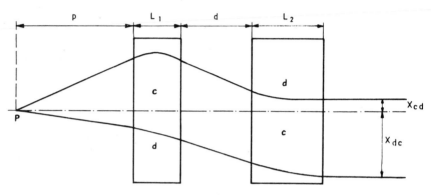

FIG. 20. Rendering parallel an initially divergent beam; thick-lens approach.

The thin-lens approximation proves its usefulness here as it yields approximate values, the knowledge of which considerably facilitates the solution of the system (153). It should be noted that in this set all characteristic parameters of the first quadrupole are localized on the right-hand side whereas the left-hand side contains only the parameters of the second quadrupole.

The problem is simplified if one of the quadrupoles is given both in geometry and focusing strength, the other either in geometry or focusing strength, and the distance d is sought to obtain a parallel beam from an initially divergent one. Combining the last two equations, we have

$$\frac{1}{k_2} (\cot \vartheta_2 + \coth \vartheta_2) = \frac{B_1}{\beta_1} - \frac{A_1}{a_1} \qquad (154)$$

and upon substitution of

$$\frac{B_1}{\beta_1} = d + \frac{p \cosh \vartheta_1 + (1/k_1) \sinh \vartheta_1}{\cosh \vartheta_1 + pk_1 \sinh \vartheta_1} \qquad \frac{A_1}{a_1} = d + \frac{p \cos \vartheta_1 + (1/k_1) \sin \vartheta_1}{\cos \vartheta_1 - pk_1 \sin \vartheta_1} \qquad (155)$$

[1] There is no possible confusion between the matrix element d and the distance between the end faces of the quadrupoles, which is also denoted by d.

we find

$$\frac{1}{k_2}(\cot \vartheta_2 + \coth \vartheta_2)$$

$$= \frac{1}{k_1}\left(\frac{pk_1 \cosh \vartheta_1 + \sinh \vartheta_1}{pk_1 \sinh \vartheta_1 + \cosh \vartheta_1} + \frac{pk_1 \cos \vartheta_1 + \sin \vartheta_1}{pk_1 \sin \vartheta_1 - \cos \vartheta_1}\right) \qquad (156)$$

Here, one has only one transcendental equation to solve and again the parameters of the two quadrupoles are concentrated, respectively, in the two sides of the equation. After solving this equation, d can be found by one or the other of Eqs. (155).

The width of the beam can be found from Eqs. (151). Carrying out the calculations, we find

$$R_{cd} = A_1/\cosh \vartheta_2 \qquad R_{dc} = B_1/\cos \vartheta_2 \qquad (157)$$

Knowledge of these values will permit determination of the lens aperture.

3. F-F Problem

In trying to focus an initially divergent beam, it is again convenient to first use the thin-lens approximation (Fig. 21) to obtain some parameters with which to start in more accurate calculations. In fact, the problem under consideration is one of stigmatism. From the two transfer matrices (122) and (123), we can write the condition of stigmatism in the form

$$F_2 = -\frac{s + F_1\left(1 - \dfrac{s}{f_{1c}}\right)}{1 + \dfrac{s}{f_{2d}} - \dfrac{F_1}{f_{cd}}} = -\frac{s + F_1\left(1 + \dfrac{s}{f_{1d}}\right)}{1 - \dfrac{s}{f_{2c}} - \dfrac{F_1}{f_{dc}}} \qquad (158)$$

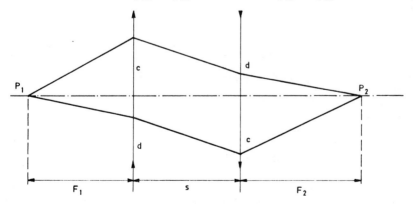

FIG. 21. Focusing an initially divergent beam; thin-lens approximation.

Using the same notations as above and carrying out the calculations, we find

$$F_1 = s / \left\{ \left[\left(\frac{s\sigma_1}{2} \right)^2 + \frac{\sigma_1}{\sigma_2} \right]^{1/2} + \frac{s\vartheta_1}{2} - 1 \right\}$$

$$F_2 = s / \left\{ \left[\left(\frac{s\sigma_2}{2} \right)^2 + \frac{\sigma_2}{\sigma_1} \right]^{1/2} + \frac{s\vartheta_2}{2} - 1 \right\}$$

(159)

In the case in which the subscripts can be dropped, these relations reduce to

$$F_1 = \frac{sf_1}{(s^2 + f_1 f_2)^{1/2} - f_1}$$

$$F_2 = \frac{sf_2}{(s^2 + f_1 f_2)^{1/2} - f_2}$$

(160)

Usually F_1 and F_2 will be known; the required focusing strengths are then obtained from

$$f_1 = F_1 \left[\frac{s(s + F_2)}{(s + F_1)(F_1 + F_2 + s)} \right]^{1/2}$$

$$f_2 = F_2 \left[\frac{s(s + F_1)}{(s + F_2)(F_1 + F_2 + s)} \right]^{1/2}$$

(161)

It may be of some use to know the magnifications under stigmatic operating conditions. From Eqs. (11) and (169) we find

$$g_{cd} = - \frac{(s^2 + f_1 f_2)^{1/2} + s}{f_1} \frac{(s^2 + f_1 f_2)^{1/2} - f_1}{(s^2 + f_1 f_2)^{1/2} - f_2}$$

$$g_{dc} = - \frac{(s^2 + f_1 f_2)^{1/2} - s}{f_1} \frac{(s^2 + f_1 f_2)^{1/2} - f_1}{(s^2 + f_1 f_2)^{1/2} - f_2}$$

(162)

and consequently

$$\frac{g_{cd}}{g_{dc}} = \frac{(s^2 + f_1 f_2)^{1/2} + s}{(s^2 + f_1 f_2)^{1/2} - s}$$

(163)

The magnification is therefore larger in the cd plane.

If $F_1 = F_2 = F_0$, Eqs. (159) become

$$F_0 = s / \left\{ \left[1 + \left(\frac{s\sigma}{2} \right)^2 \right]^{1/2} + \frac{s\vartheta}{2} - 1 \right\}$$

(164)

In the case in which we may drop the individual subscripts, we find for the focusing strengths

$$f_1 = f_2 = f_0 = F_0 \left(\frac{s}{s + 2F_0} \right)^{1/2}$$

(165)

In this case

$$g_{cd} = - \frac{(s^2 + f_0^2)^{1/2} + s}{f_0} \qquad g_{dc} = - \frac{(s^2 + f_0^2)^{1/2} - s}{f_0} \qquad (166)$$

and therefore $g_{cd} \cdot g_{dc} = 1$, as given by the thick-lens theory of the anti-symmetrical doublet [Eq. (105)].

The values we obtain by means of the thin-lens approximation are usually good starting parameters for the more exact procedure of the thick-lens approach.

In the latter case (Fig. 22), using the conditions of stigmatism and our previous notations, we find

$$k_2 \tanh \vartheta_2 = - \frac{a_1 q + A_1}{A_1 q + (a_1/k_2^2)} \qquad k_2 \tan \vartheta_2 = \frac{\beta_1 q + B_1}{B_1 q + (\beta_1/k_2^2)} \qquad (167)$$

The magnifications are then found to be

$$\frac{1}{g_{cd}} = a_1 \cosh \vartheta_2 + A_1 k_2 \sinh \vartheta_2$$

$$\frac{1}{g_{dc}} = \beta_1 \cos \vartheta_2 - B_1 k_2 \sinh \vartheta_2 \qquad (168)$$

If d is unknown, it can be eliminated by means of Eq. (155) as in the P-F case and the condition of stigmatism becomes

$$\frac{1}{k_1} \left(\frac{pk_1 + \tanh \vartheta_1}{pk_1 \tanh \vartheta_1 + 1} + \frac{pk_1 + \tanh \vartheta_1}{pk_1 \tan \vartheta_1 - 1} \right)$$

$$= \frac{1}{k_2} \left(\frac{qk_2 + \tanh \vartheta_2}{qk_2 \tan \vartheta_2 + 1} + \frac{qk_2 + \tanh \vartheta_2}{qk_2 \tan \vartheta_2 - 1} \right)$$

The distance can then be calculated by one or the other of Eqs. (155).

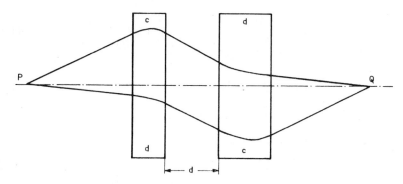

FIG. 22. Focusing an initially divergent beam; thick-lens approach.

4. *Focusing of an Astigmatic Beam*

This problem is encountered in a cyclotron for example where the out-coming beam may have an apparent source which is different in the horizontal and in the vertical plane (Fig. 23). The problem is then to focus the two sources in one point.

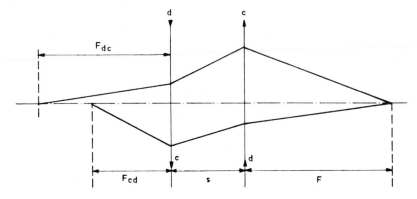

FIG. 23. Focusing of an astigmatic beam; thin-lens approximation.

Considering first the thin-lens approximation, dropping the subscripts of the individual quadrupoles and writing out the conjugation relation (8), one finds for the two planes

$$\left(1 + \frac{s}{f_2} - \frac{1}{f_{cd}} F_{cd}\right) F + s + \left(1 - \frac{s}{f_1}\right) F_{cd} = 0$$

$$\left(1 - \frac{s}{f_2} - \frac{1}{f_{dc}} F_{dc}\right) F + s + \left(1 + \frac{s}{f_1}\right) F_{dc} = 0$$

(169)

Using Eqs. (125) these relations can be written

$$(F + F_{cd} + s)f_1 f_2 + F(F_{cd} + s)f_1 - F_{cd}(F + s)f_2 - sFF_{cd} = 0$$

$$(F + F_{dc} + s)f_1 f_2 - F(F_{dc} + s)f_1 + F_{dc}(F + s)f_2 - sFF_{dc} = 0$$

(170)

Putting

$$F_{cd} + F_{dc} = S \qquad F_{cd} - F_{dc} = D \qquad F_{cd} \cdot F_{dc} = P \qquad (171)$$

the solution of Eqs. (170) can be written

$$f_2 = F\left\{\frac{s(sS + 2P)}{(F + s)[(F + s)S + 2P]}\right\}^{1/2}$$

$$f_1 = \frac{F + s}{(F + s)(S + 2s) + sS + 2P}\left\{sD + \frac{f_2}{F}[(F + s)S + 2P]\right\}$$

(172)

Knowledge of these solutions will greatly facilitate the calculations of the more rigorous thick-lens method.

Putting, in the latter case,

$$\alpha_1 = \cos \vartheta_1 - p_{cd}\, k_1 \sin \vartheta_1 \quad \beta_1 = \cosh \vartheta_1 + p_{dc}\, k_2 \sinh \vartheta_1 \quad (173)$$

$$A_1 = d\alpha_1 + p_{cd} \cos \vartheta_1 + (1/k_1) \sin \vartheta_1 \quad B_1 = d\beta_1 + p_{dc} \cosh \vartheta_1 + (1/k_1) \sinh \vartheta_1$$

one is once again led to the set of transcendental equations

$$k_2 \tanh \vartheta_2 = - \frac{\alpha_1 q + A_1}{A_1 q + (\alpha_1/k_2{}^2)} \qquad k_2 \tan \vartheta_2 = \frac{\beta_1 q + B_1}{B_1 q - (\beta_1/k_2{}^2)} \quad (174)$$

which will have to be solved by numerical procedures.

2.4.4. The Triplet

A. General Properties

Although the doublet achieves a significant improvement with respect to a single quadrupole, and although its construction is inherently simpler than that of a three-lens system, the latter system will be preferred in many cases. The chief disadvantage of the doublet lies in the fact that the variation of a parameter in one of the basic planes may entail an important variation of the parameters in the other plane. If one tries, for example, to adjust the position of the principal plane in the cd direction, the position of the principal plane in the dc direction will equally vary. Supposing that the doublet can be assimilated to a thin lens, the position of this lens may then be quite different in the two perpendicular planes as the variation of the excitation will modify not only the focusing strength of the lens but also its position in space.

In a symmetrical triplet, the principal planes are symmetrical with respect to the median plane of the lens and this applies equally well to the cdc as to the dcd direction and is, to a large extent, independent of the excitation level. If a symmetric triplet can be considered as a thin lens, the position of this equivalent lens is almost fixed in space and independent of excitation.

B. The General Triplet

Figure 24 shows, qualitatively, the behavior of the beam envelope in a general triplet. In most cases, the calculations are carried out in the frame

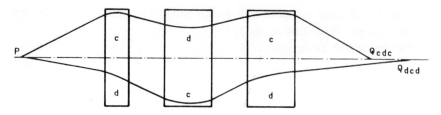

FIG. 24. Beam envelope in a triplet.

of the thin-lens approximation, that is, each quadrupole is considered to be concentrated in its median plane (Fig. 25). If G_1, G_2, and G_3 are the gradients of the magnetic field of the three quadrupoles, one has for the characteristic parameters k and ϑ

$$k_{1,2,3}^2 = \frac{G_{1,2,3}}{Br} \qquad \vartheta_{1,2,3} = k_{1,2,3} L_{1,2,3} \qquad (175)$$

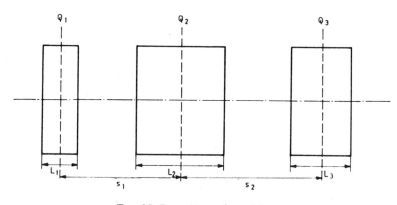

FIG. 25. Parameters of a triplet.

As already mentioned in the case of the doublet, the procedure of the lens association can be improved if we take, for the focal distances, the values we find from the thick lens approach applied to each individual quadrupole. Under these conditions, we have

$$
\begin{aligned}
1/f_{1c} &= k_1 \sin \vartheta_1 & 1/f_{1d} &= k_1 \sinh \vartheta_1 \\
1/f_{2c} &= k_2 \sin \vartheta_2 & 1/f_{2d} &= k_2 \sinh \vartheta_2 \\
1/f_{3c} &= k_3 \sin \vartheta_3 & 1/f_{3d} &= k_3 \sinh \vartheta_3
\end{aligned}
\qquad (176)
$$

All quantities are counted positive in Eq. (176), the proper sign being taken care of in the expressions of the transfer matrices.

C. TRANSFER MATRICES

The triplet being made up of three quadrupoles and two drift spaces, the basic transfer matrices are

$$
T_{cdc} = \begin{Vmatrix} 1 & 0 \\ -\dfrac{1}{f_{3c}} & 1 \end{Vmatrix} \times \begin{Vmatrix} 1 & s_2 \\ 0 & 1 \end{Vmatrix} \times \begin{Vmatrix} 1 & 0 \\ \dfrac{1}{f_{2d}} & 1 \end{Vmatrix} \times \begin{Vmatrix} 1 & s_1 \\ 0 & 1 \end{Vmatrix} \times \begin{Vmatrix} 1 & 0 \\ -\dfrac{1}{f_{1c}} & 1 \end{Vmatrix} \tag{177}
$$

$$
T_{dcd} = \begin{Vmatrix} 1 & 0 \\ \dfrac{1}{f_{3d}} & 1 \end{Vmatrix} \times \begin{Vmatrix} 1 & s_2 \\ 0 & 1 \end{Vmatrix} \times \begin{Vmatrix} 1 & 0 \\ -\dfrac{1}{f_{2c}} & 1 \end{Vmatrix} \times \begin{Vmatrix} 1 & s_1 \\ 0 & 1 \end{Vmatrix} \times \begin{Vmatrix} 1 & 0 \\ \dfrac{1}{f_{1d}} & 1 \end{Vmatrix} \tag{178}
$$

Multiplying out and using the abbreviations

$$
\begin{aligned}
s_i/f_{jc} = x_{ij} \qquad (i = 1, 2, \quad j = 1, 2, 3) \\
s_i/f_{jd} = y_{ij} \qquad (i = 1, 2, \quad j = 1, 2, 3)
\end{aligned} \tag{179}
$$

one finds

$$
T_{cdc} = \begin{Vmatrix} (1 - x_{11})(1 + y_{22}) - x_{21} & \\ s_1 + s_2(1 + y_{12}) & \\ (1 - x_{23})\left[\dfrac{1}{f_{2d}} - \dfrac{1}{f_{1c}}(1 + y_{12})\right] - \dfrac{1}{f_{3c}}(1 - x_{11}) & \\ (1 + y_{12})(1 - x_{23}) - x_{13} & \end{Vmatrix} \tag{180}
$$

$$
T_{dcd} = \begin{Vmatrix} (1 + y_{11})(1 - x_{22}) + y_{21} & \\ s_1 + s_2(1 - x_{12}) & \\ (1 + y_{23})\left[\dfrac{1}{f_{1d}}(1 - x_{12}) - \dfrac{1}{f_{2c}}\right] + \dfrac{1}{f_{3d}}(1 + y_{11}) & \\ (1 - x_{12})(1 + y_{23}) + y_{13} & \end{Vmatrix} \tag{181}
$$

D. OPTICAL PROPERTIES

1. Focal distances

From the transfer matrices, the focal distances can readily be derived

$$
\begin{aligned}
\frac{1}{f_{cdc}} &= (1 - x_{23})\left[\frac{1}{f_{1c}}(1 + y_{12}) - \frac{1}{f_{2d}}\right] + \frac{1}{f_{3c}}(1 - x_{11}) \\
\frac{1}{f_{dcd}} &= (1 + y_{23})\left[\frac{1}{f_{2c}} - \frac{1}{f_{1d}}(1 - x_{12})\right] - \frac{1}{f_{3d}}(1 + y_{11})
\end{aligned} \tag{182}
$$

2. Position of the Focal Planes

$$F_{icdc} = f_{cdc}[(1 - x_{11})(1 + y_{22}) - x_{21}]$$
$$F_{idcd} = f_{dcd}[(1 + y_{11})(1 - x_{22}) + y_{21}]$$

(183)

Similar expressions can be derived for the object focal planes.

3. Principal Planes

Again, from the transfer matrices, we have

$$H_{icdc} = f_{cdc}[(1 - x_{11})(1 + y_{22}) - (1 + x_{21})]$$
$$H_{idcd} = f_{dcd}[(1 + y_{11})(1 - x_{22}) - (1 - y_{21})]$$

(184)

In many cases, it will be simpler to calculate the optical elements by numerical procedures directly from the transfer matrices. Algebraic calculations are more useful in the case of the symmetrical triplet where some specific properties can be derived.

E. The Symmetrical Triplet

In this case we have (Fig. 26)

$$
\begin{aligned}
s_1 &= s_2 = s \\
L_1 &= L_3 = L_e \qquad L_2 = L_i \\
G_1 &= G_3 = G_e \qquad G_2 = G_i \\
k_1 &= k_3 = k_e \qquad k_2 = k_i \\
\vartheta_1 &= \vartheta_3 = \vartheta_e \qquad \vartheta_2 = \vartheta_i
\end{aligned}
$$

(185)

so that the focal lengths of the individual quadrupoles are

$$\frac{1}{f_{ec}} = k_e \sin \vartheta_e \qquad \frac{1}{f_{ed}} = k_e \sinh \vartheta_e$$
$$\frac{1}{f_{ic}} = k_i \sin \vartheta_i \qquad \frac{1}{f_{id}} = k_i \sinh \vartheta_i$$

(186)

It is useful to introduce the dimensionless quantities

$$\frac{s}{f_{ec}} = x_e \qquad \frac{s}{f_{ed}} = y_e$$
$$\frac{s}{f_{ic}} = x_i \qquad \frac{s}{f_{id}} = y_i$$

(187)

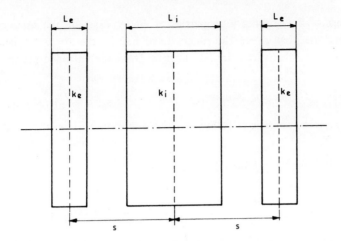

FIG. 26. Parameters of a symmetrical triplet.

1. *Transfer Matrices*

With the above notations, the transfer matrices can be written

$$T_{cdc} = \left\| \begin{matrix} 1 - 2x_e + y_i(1 - x_e) & s(2 + y_i) \\ \dfrac{1 - x_e}{s}[-2x_e + y_i(1 - x_e)] & 1 - 2x_e + y_i(1 - x_e) \end{matrix} \right\| \quad (188)$$

$$T_{dcd} = \left\| \begin{matrix} 1 + 2y_e - x_i(1 + y_e) & s(2 - x_i) \\ \dfrac{1 + y_e}{s}[2y_e - x_i(1 + y_e)] & 1 + 2y_e - x_i(1 + y_e) \end{matrix} \right\| \quad (189)$$

2. *Focal Distances*

From the matrices we have, using nondimensional notation,

$$\frac{s}{f_{cdc}} = (1 - x_e)[2x_e - y_i(1 - x_e)]$$

$$\frac{s}{f_{dcd}} = (1 + y_e)[-2y_e + x_i(1 + y_e)] \quad (190)$$

3. *Position of Focal Planes*

$$\frac{F_{icdc}}{s} = \frac{1 - 2x_e + y_i(1 - x_e)}{(1 - x_e)[2x_e - y_i(1 - x_e)]}$$

$$\frac{F_{idcd}}{s} = \frac{1 + 2y_e - x_i(1 + y_e)}{(1 + y_e)[-2y_e + x_i(1 + y_e)]} \quad (191)$$

Because of the symmetry properties of the lens system, the same expressions give the position of the object focal planes, the abscissas, however, being taken here towards the left starting from the entrance plane.

4. *Position of Principal Planes*

Very simple expressions follow for these quantities from the matrix elements:

$$\frac{H_{icdc}}{s} = \frac{H_{ocdc}}{s} = -\frac{1}{1 - x_e}$$

$$\frac{H_{idcd}}{s} = \frac{H_{odcd}}{s} = -\frac{1}{1 + y_e} \tag{192}$$

Figure 27 shows the position of the two principal planes in the two basic directions for a symmetrical triplet.

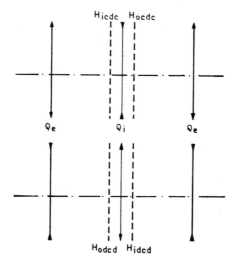

Fɪɢ. 27. Position of principal planes in a symmetrical triplet.

The relations (192) display two properties:

(a) Provided x_e, $y_e \ll 1$, that is, provided the excitation of the outer quadrupole is small, all principal planes coincide with the geometric center plane of the lens system. Under these conditions, the symmetrical triplet can be considered as a thin lens of fixed position.

(b) In order to adjust the optical properties of the triplet, we can vary

x_i, y_i, that is, the excitation of the inner quadrupole, without modifying the position of the principal planes.

These properties have no analogy in doublet behavior.

In writing Eqs. (190) and (191), we have tacitly assumed that

$$2x_e \neq y_i(1 - x_e) \qquad 2y_e \neq x_i(1 + y_e) \qquad (193)$$

the equalities corresponding indeed to infinite focal lengths or afocal systems. Moreover, all higher order terms must be kept in the brackets of expressions (190) giving the focal lengths, otherwise the condition of bidirectional focusing

$$\frac{2y_e}{1 + y_e} < x_i < y_i < \frac{2x_e}{1 - x_e} \qquad (194)$$

cannot be satisfied.

5. Stigmatic Operation of a Symmetrical Triplet

Writing out the condition of stigmatism, we have

$$-q_0 = \frac{a_{cdc}\, p_0 + b_{cdc}}{c_{cdc}\, p_0 + a_{cdc}} = \frac{a_{dcd}\, p_0 + b_{dcd}}{c_{dcd}\, p_0 + a_{dcd}} \qquad (195)$$

Using the same notations as before and putting, moreover,

$$x_i + y_i = 4S_i \qquad x_i - y_i = 4(1 - A_i)$$
$$x_e + y_e = S_e \qquad x_e - y_e = 2 - A_e \qquad (196)$$
$$(2 - x_i)\,(2 + y_i) = 4\lambda \qquad \lambda S_e - 2S_i = S$$

one finds, after carrying out the calculations,

$$p_0 = q_0 = 2s \,\frac{S + S_i}{A_i S_e - A_e(S + S_i) + (S_e^2 S^2 - S_e S)^{1/2}} \qquad (197)$$

For the same reasons as before, all higher order terms must be kept in this expression.

Under operating conditions where the subscripts c and d of the individual quadrupoles can be dropped, we have

$$x_i = y_i \qquad x_e = y_e \qquad (198)$$

The focal lengths can then be written

$$\frac{s}{f_{cdc}} = (1 - x_e)[2x_e - x_i(1 - x_e)]$$

$$\frac{s}{f_{dcd}} = (1 + x_e)[- 2x_e + x_i(1 + x_e)]$$

(199)

Here again, all terms must be kept in the brackets if the condition of bi-directional focusing

$$\frac{2x_e}{1 + x_e} < x_i < \frac{2x_e}{1 - x_e}$$

(200)

is to be satisfied.

Taking into account Eq. (198), Eq. (197) giving the couple of stigmatic points, becomes

$$p_0 = q_0 = s \frac{4x_e - x_i(1 + x_i x_e)}{- 2x_e + x_i(1 + x_i x_e) + } $$
$$+ \{[x_i x_e(2 + x_i x_e) - 4x_e^2]^2 + x_i x_e(2 + x_i x_e) - 4x_e^2\}^{1/2}$$

(201)

As in the general case, all higher order terms must be kept in this expression in order to comply with the bidirectional focusing condition.

2.4.5. The Multiplet

A. INTRODUCTION

A series of AG lenses may be used to guide high-energy particles over large distances. Basically, a single lens suitably placed, for example, a doublet or a triplet, could be used to focus in both directions. However, because of the initial divergence of the beam, a considerable fraction of particles would be lost unless the aperture of the lens is large. It seems more reasonable under these conditions to use a periodic structure of alternatively focusing and defocusing quadrupoles. Another case of interest is a focusing channel where secondary particles, produced by decay in flight of a primary beam, are collected and purified for use in experiments.

Obviously, stability conditions must be satisfied in order to prevent the growth of particle excursions during the transport of the beam inside the channel. If all quadrupoles of the channel are identical in structure, the stability limits are readily established.

B. TRANSFER MATRIX OF A PERIOD

Consider a periodic structure, the basic period of which is made of two AG quadrupoles and the corresponding drift spaces (Fig. 28). With our previous notations the transfer matrices of a period are then

$$P_{cd} = \begin{Vmatrix} 1 & d \\ 0 & 1 \end{Vmatrix} \times \begin{Vmatrix} \cosh \vartheta & (1/k)\sinh \vartheta \\ k\cosh \vartheta & \cosh \vartheta \end{Vmatrix} \times \begin{Vmatrix} 1 & d \\ 0 & 1 \end{Vmatrix} \times \begin{Vmatrix} \cos \vartheta & (1/k)\sin \vartheta \\ -k\sin \vartheta & \cos \vartheta \end{Vmatrix}$$

$$(202)$$

$$P_{dc} = \begin{Vmatrix} 1 & d \\ 0 & 1 \end{Vmatrix} \times \begin{Vmatrix} \cos \vartheta & (1/k)\sin \vartheta \\ -k\sin \vartheta & \cos \vartheta \end{Vmatrix} \times \begin{Vmatrix} 1 & d \\ 0 & 1 \end{Vmatrix} \times \begin{Vmatrix} \cosh \vartheta & (1/k)\sinh \vartheta \\ k\sinh \vartheta & \cosh \vartheta \end{Vmatrix}$$

$$(203)$$

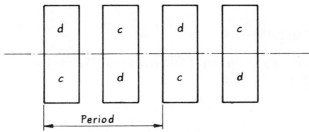

FIG. 28. Periodic focusing–defocusing structure.

Multiplying out, the two matrices may be put in the form

$$P_{cd} = \begin{Vmatrix} \cos \mu + a_{cd}\sin \mu & \beta_{cd}\sin \mu \\ -\dfrac{1 + a_{cd}^2}{\beta_{cd}}\sin \mu & \cos \mu - a_{cd}\sin \mu \end{Vmatrix} \tag{204}$$

and

$$P_{dc} = \begin{Vmatrix} \cos \mu + a_{dc}\sin \mu & \beta_{dc}\sin \mu \\ -\dfrac{1 + a_{dc}^2}{\beta_{cd}}\sin \mu & \cos \mu - a_{cd}\sin \mu \end{Vmatrix} \tag{205}$$

where

$$\cos \mu = \cos \vartheta \cosh \vartheta + dk(\cos \vartheta \sinh \vartheta - \sin \vartheta \cosh \vartheta)$$
$$- \frac{(dk)^2}{2}\sin \vartheta \sinh \vartheta \tag{206}$$

$$a_{cd}\sin \mu = -\sin \vartheta \sinh \vartheta - dk \sin \vartheta \cosh \vartheta - \frac{(dk)^2}{2}\sin \vartheta \sinh \vartheta \tag{207}$$

$$a_{dc}\sin \mu = \sin \vartheta \sinh \vartheta + dk \cos \vartheta \sinh \vartheta + \frac{(dk)^2}{2}\sin \vartheta \sinh \vartheta \tag{208}$$

$$\beta_{cd} \sin \mu = \frac{1}{k} (\sin \vartheta \cosh \vartheta + \cos \vartheta \sinh \vartheta) + d(2 \cos \vartheta \cosh \vartheta + \sin \vartheta \sinh \vartheta)$$

$$+ d^2 k \cos \vartheta \sinh \vartheta \tag{209}$$

$$\beta_{dc} \sin \mu = \frac{1}{k} (\sin \vartheta \cosh \vartheta + \cos \vartheta \sinh \vartheta) + d(2 \cos \vartheta \cosh \vartheta - \sin \vartheta \sinh \vartheta)$$

$$- d^2 k \sin \vartheta \cosh \vartheta \tag{210}$$

The relation

$$\frac{\beta_{cd}}{\beta_{dc}} = \frac{1 + a_{cd}}{1 + a_{dc}} \tag{211}$$

always holds.

C. Stability Conditions

It is readily seen from Eq. (204) and Eq. (205) that the transfer matrix of a period may be written in the form

$$P = I \cos \mu + J \sin \mu \tag{212}$$

where I is the unit matrix

$$I = \begin{Vmatrix} 1 & 0 \\ 0 & 1 \end{Vmatrix} \tag{213}$$

and

$$J = \begin{Vmatrix} \alpha & \beta \\ -\dfrac{1 + \alpha^2}{\beta} & -\alpha \end{Vmatrix} \tag{214}$$

It is also seen from Eq. (212) that if μ is real, that is $|\cos \mu| < 1$, the transfer matrix for N periods (a focusing channel comprising $2N$ quadrupoles and $2N$ drift spaces) becomes

$$P^N = I \cos N\mu + J \sin N\mu \tag{215}$$

On the other hand, if $|\cos \mu| > 1$, Eq. (212) should be written in the form

$$P = I \cosh \mu + J \sinh \mu \tag{216}$$

and therefore

$$P^N = I \cosh N\mu + J \sinh N\mu \tag{217}$$

The stability criterium follows immediately in a form involving the trace of either matrix P_{cd} or P_{dc}

$$|\cos \mu| < 1 \tag{218}$$

that is, from Eq. (206)

$$|\cos \vartheta \cosh \vartheta + dk(\cos \vartheta \sinh \vartheta - \sin \vartheta \cosh \vartheta) - \frac{(dk)^2}{2} \sin \vartheta \sinh \vartheta| < 1 \tag{219}$$

If the focusing strength and the length of the basic quadrupole is given, the inequality (219) determines the allowed range for the interspacing d. Equation (215) may then be used to calculate the excursion and the slope of the beam along the focusing channel.

In the thin-lens approximation (Fig. 29), Eqs. (204) and (205) become

$$P_{cd} = \begin{Vmatrix} 1 - \delta\varDelta - (\delta\varDelta)^2 & 2\varDelta + \delta\varDelta^2 \\ - \delta^2\varDelta & 1 + \delta\varDelta \end{Vmatrix} \tag{220}$$

$$P_{dc} = \begin{Vmatrix} 1 + \delta\varDelta - (\delta\varDelta)^2 & 2\varDelta - \delta\varDelta^2 \\ - \delta^2\varDelta & 1 - \delta\varDelta \end{Vmatrix} \tag{221}$$

Putting

$$\eta = \delta\varDelta \tag{222}$$

we then have

$$\cos \mu = 1 - \frac{\eta^2}{2} \tag{223}$$

$$\alpha_{cd} = -\left(\frac{2+\eta}{2-\eta}\right)^{1/2} \tag{224}$$

$$\alpha_{dc} = \left(\frac{2-\eta}{2+\eta}\right)^{1/2} \tag{225}$$

$$\beta_{cd} = \frac{2\varDelta}{\eta}\left(\frac{2+\eta}{2-\eta}\right)^{1/2} \tag{226}$$

$$\beta_{dc} = \frac{2\varDelta}{\eta}\left(\frac{2-\eta}{2+\eta}\right)^{1/2} \tag{227}$$

and the stability condition simply is

$$\eta < 2 \tag{228}$$

As before, Eq. (215) may be used to calculate the excursion and slope of the beam along the focusing channel.

The conditions (219) and (228), respectively, are necessary; they are, however, insufficient because they give no indication as to what fraction of a given beam (namely, given radius and slope) will be transmitted throughout the channel. This question is best treated by means of phase space considerations (see Fronteau's contribution in this volume).

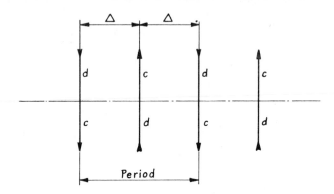

FIG. 29. Periodic structure; thin-lens approximation.

2.4.6. Matching Problems

A. INTRODUCTION

Whereas in standard electron and light optics the concept of focusing is generally used in the restricted sense of precise image formation, individual trajectories are not of predominant importance in beam-transport systems, in which we are chiefly interested in the transmission and conservation of the beam as a whole. Under these circumstances, it is convenient to combine the two main characteristics of the beam, viz. its diameter and its divergence in one parameter, that is, its phase space area which then appropriately describes the overall properties of the beam as it propagates through an optical system.

This procedure is actually borrowed from Hamiltonian mechanics in which one associates to position coordinates describing the configuration of a system of N particles the conjugate momenta coordinates which describe the motion of the system. The f-dimensional configuration space ($f = 3N$) is thus replaced by the $2f$-dimensional "phase space." Liouville's theorem then states that under very general conditions, the density in phase space of an ensemble of points representing a group of particles is invariant as a function of time; in other words, in their phase space the particles behave like an incompressible fluid.

The general theory is treated by Fronteau in this volume (5-4); only the simplest case will be considered very briefly here, the case in which the phase space reduces to the phase plane x, p_x (Fig. 30).

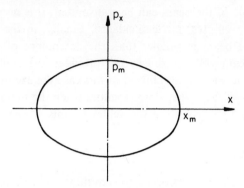

FIG. 30. Phase plane.

B. ACCEPTANCE, EMITTANCE, AND MATCHING

Consider a group of particles going through a focusing system of aperture $2a$. To the lowest order the motion will be sinusoidal, and the excursion of a particle from the axis of the system can be represented by

$$x = x_m \sin \omega t \qquad (229)$$

The conjugate momentum being

$$p_x = m\omega x_m \cos \omega t \qquad (230)$$

the trajectory in phase space will be an ellipse (Fig. 30)

$$\frac{x^2}{x_m{}^2} + \frac{p_x{}^2}{p_m{}^2} = 1 \qquad (231)$$

In fact, this ellipse represents all particles having arbitrarily distributed phases but the same amplitude x_m. It is clear that the system cannot accept particles lying outside the limiting ellipse of semiaxes a and $b = m\omega a$.

Although the coordinate x and the corresponding angular divergence $x' = dx/dz$ are not canonically conjugate, it is customary to draw the phase diagram in x, x' coordinates instead of x, p_x coordinates; in the absence of acceleration this procedure is legitimate. We then call acceptance of the optical system the area of the limiting ellipse drawn in the x, x' plane.

In the same way, we define the emittance of a beam by the limiting area it occupies in the phase plane.

Obviously, when the emittance of the beam exceeds the acceptance of the optical system into which it is injected, a fraction only of the beam will be transmitted. If the emittance of the beam equals the acceptance of the system, the whole of the beam can be transmitted; it is necessary, however to bring the two ellipses into coincidence. Finally, in the case where the emittance of the beam is smaller than the acceptance of the system, the whole of the beam will be transmitted provided the emittance ellipse is brought inside the acceptance ellipse; in this case the excursions performed by the particles will have their lowest values when the two ellipses have the same axis ratio. The procedure for achieving this is known as matching.

C. MATCHING DEVICES

By virtue of Liouville's theorem, the emittance of a beam is an invariant quantity, the area of the corresponding phase-space ellipse is therefore constant. However, its shape and orientation can be modified by rather simple means.

1. *Drift Spaces*

The effect of a drift length L is described by Eq. (3). It is seen that in phase space every point of the limiting ellipse is displaced parallel to the

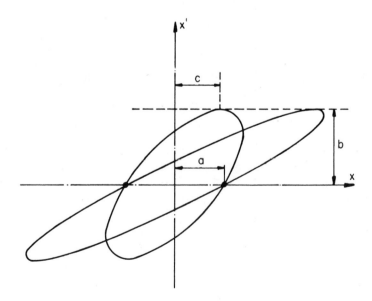

FIG. 31. Effect of a field-free region.

x axis by an amount proportional to the initial divergence of the particle. The ellipse is sheared as a whole, conserving its area and conserving two fixed points (Fig. 31).

2. *Thin Lenses*

The action of a thin lens is given by Eq. (58) or (59). In phase space every point of the ellipse is displaced parallel to the x' axis by an amount proportional to the initial position coordinate of the particle. The ellipse is sheared as a whole, conserving its area and conserving two fixed points (Fig. 32).

By means of an appropriate combination of drift spaces and lenses, any ellipse can be transformed into any other ellipse having the same area but different shape and orientation. Drift spaces and lenses are, therefore, the basic elements of a beam-matching device.

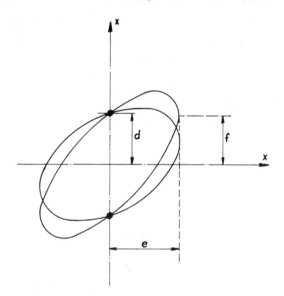

FIG. 32. Effect of a thin lens.

D. COMPLEX REPRESENTATION

The transformation of the phase-space ellipse in going through the matching system can be followed quantitatively, and suitable conditions can be established for achieving matching.

In many cases the use of complex numbers provides an elegant method for writing out the matching conditions.

Any ellipse can be specified by means of two numbers. In going through a drift space (Fig. 31), the convenient quantities are

$$R = a/b \qquad X = c/b \qquad (232)$$

The first of these quantities remains unchanged as the beam goes along the drift length, whereas the second quantity changes by an amount just equal to the length traversed. With obvious notations

$$R_{out} = R_{in} \qquad X_{out} = X_{in} + L \qquad (233)$$

When going through a thin lens, the appropriate parameters to describe the transformation of the ellipse are (Fig. 32)

$$d/e = G \qquad -f/e = B \qquad (234)$$

The first of these parameters does not change as the beam goes through the lens, whereas the second changes by an amount just equal to the strength of the lens:

$$G_{out} = G_{in} \qquad B_{out} = B_{in} + C, \qquad (235)$$

C being the reciprocal value of the focal length.

If R and X are known, the geometry of the ellipse yields

$$G = R/(R^2 + X^2) \qquad B = - X/(R^2 + X^2) \qquad (236)$$

Conversely, if G and B are known, we have

$$R = G/(G^2 + B^2) \qquad X = - B/(G^2 + B^2) \qquad (237)$$

It is appropriate to introduce complex numbers by writing

$$Z = R + jX \qquad Y = G + jB \qquad (238)$$

The passage through a drift length L is then translated by

$$Z_{out} = Z_{in} + jL \qquad (239)$$

whereas the passage through a thin lens of convergence C is represented by

$$Y_{out} = Y_{in} + jC \qquad (240)$$

Equations (236) and (238) can then be written

$$Y = 1/Z \qquad Z = 1/Y \qquad (241)$$

These relations reveal the fact that an optical system composed of drift spaces and lenses has an electrical analogue which is a ladder network composed of a series of inductances and shunt capacitors.

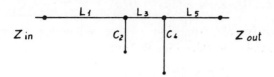

FIG. 33. Elementary example of matching.

Consider, for example, a system (Fig. 33) having three drift spaces and two lenses. By simple inspection we have

$$Z_{\text{out}} = (\{[(Z_{\text{in}} + jL_1)^{-1} + jC_2]^{-1} + jL_3\}^{-1} + jC_4)^{-1} + jL_5 \qquad (242)$$

This relation can be interpreted as a matching condition where Z_{in}, Z_{out} are given, and two of the C's or L's are left to be determined.

If a more realistic two-dimensional matching is desired, Eq. (242) is split in two conditions, one for each plane

$$Z_{\text{out } x} = (\{[(Z_{\text{in } x} + jL_1)^{-1} + jC_2]^{-1} + jL_3\}^{-1} + jC_4)^{-1} + jL_5$$

$$Z_{\text{out } y} = (\{[(Z_{\text{in } y} + jL_1)^{-1} - jC_2]^{-1} + jL_3\}^{-1} - jC_4)^{-1} + jL_5 \qquad (243)$$

with four of the L's or C's as unknowns. A fast digital computer proves most helpful in setting up and solving the problem.

APPENDIX

THE GENERAL MULTIPLET

Although this paper is mainly concerned with quadrupoles, doublets, and triplets, we shall briefly treat the case of the general quadruplet, and quintuplet.

Consider a system made of n quadrupole lenses and $n-1$ drift spaces. Let

$$Q_i = \begin{Vmatrix} a_i & b_i \\ c_i & a_i \end{Vmatrix} \qquad (A1)$$

be the transfer matrix of the ith quadrupole and

$$D_{i,i+1} = \begin{Vmatrix} 1 & L_{i,i+1} \\ 0 & 1 \end{Vmatrix} \qquad (A2)$$

the transfer matrix of the drift space located between the quadrupoles i and $i + 1$.

The direct calculation of the product

$$M = Q_n \times \text{-----} \; Q_{i+1} \times D_{i,i+1} \times Q_i \times \text{-----} \; Q_1 \tag{A3}$$

leads to unmanageable expressions. There is however a procedure that allows considerable simplification. We put

$$L_{i,i+1} + \frac{a_i}{c_i} + \frac{a_{i+1}}{c_{i+1}} = X_{i,i+1} \qquad (i = 2, 3, 4, ..., n - 2) \tag{A4}$$

for the body of the multiplet and

$$X_{1,2} \quad - \frac{1}{a_1 c_1} = \bar{X}_{1,2} \tag{A5}$$

$$X_{n-1,n} - \frac{1}{a_n c_n} = \bar{X}_{n-1,n} \tag{A6}$$

to describe the two ends. If one does not deal with a great number of lenses, it is also appropriate to simplify the writing by putting

$$X_{1,2} = X, \qquad X_{2,3} = Y, \qquad X_{3,4} = Z, \qquad X_{4,5} = T \tag{A7}$$

With these notations one finds

a) *for the triplet*

$$M_3 = \begin{Vmatrix} A_3 & B_3 \\ C_3 & D_3 \end{Vmatrix} \tag{A8}$$

with

$$A_3 = c_1 c_2 a_3 \left(X\bar{Y} - \frac{1}{c_2{}^2} \right)$$

$$D_3 = a_1 c_2 c_3 \left(\bar{X}Y - \frac{1}{c_2{}^2} \right)$$

$$C_3 = a_1 c_2 a_3 \left(XY - \frac{1}{c_2{}^2} \right) \tag{A9}$$

$$B_3 = a_1 c_2 a_3 \left(\bar{X}\bar{Y} - \frac{1}{c_2{}^2} \right)$$

b) *for the quadruplet*

$$M_4 = \begin{Vmatrix} A_4 & B_4 \\ C_4 & D_4 \end{Vmatrix} \tag{A10}$$

with

$$A_4 = c_1 c_2 c_3 a_4 \left(XY\bar{Z} - \frac{\bar{Z}}{c_2^2} - \frac{X}{c_3^2} \right)$$

$$D_4 = a_1 c_2 c_3 c_4 \left(\bar{X}YZ - \frac{Z}{c_2^2} - \frac{\bar{X}}{c_3^2} \right)$$

$$C_4 = c_1 c_2 c_3 c_4 \left(XYZ - \frac{Z}{c_2^2} - \frac{X}{c_3^2} \right) \qquad \text{(A11)}$$

$$D_4 = a_1 c_2 c_3 a_4 \left(\bar{X}Y\bar{Z} - \frac{\bar{Z}}{c_2^2} - \frac{\bar{X}}{c_3^2} \right)$$

c) *for the quintuplet*

$$M_5 = \begin{Vmatrix} A_5 & B_5 \\ C_5 & D_5 \end{Vmatrix} \qquad \text{(A12)}$$

with

$$A_5 = c_1 c_2 c_3 c_4 a_5 \left[\left(XY - \frac{1}{c_2^2} \right) \left(Z\bar{T} - \frac{1}{c_4^2} \right) - \frac{X\bar{T}}{c_3^2} \right]$$

$$D_5 = a_1 c_2 c_3 c_4 c_5 \left[\left(\bar{X}Y - \frac{1}{c_2^2} \right) \left(ZT - \frac{1}{c_4^2} \right) - \frac{\bar{X}T}{c_3^2} \right]$$

$$C_5 = c_1 c_2 c_3 c_4 c_5 \left[\left(XY - \frac{1}{c_2^2} \right) \left(ZT - \frac{1}{c_4^2} \right) - \frac{XT}{c_3^2} \right] \qquad \text{(A13)}$$

$$B_5 = a_1 c_2 c_3 c_4 a_5 \left[\left(\bar{X}Y - \frac{1}{c_2^2} \right) \left(Z\bar{T} - \frac{1}{c_4^2} \right) - \frac{\bar{X}\bar{T}}{c_3^2} \right]$$

Relatively simple expressions may also be obtained for the sextuplet and the septuplet. Obviously, considerable simplification occurs in the case of symmetry or antisymmetry.

These formulas may be used to achieve arbitrary transformations in phase space. A particular and simple case in the unity matrix

$$M = \begin{Vmatrix} \pm 1 & b \\ 0 & \pm 1 \end{Vmatrix}, \qquad \text{(A14)}$$

the upper sign corresponding to reproduction and the lower to turn-over. The conditions for achieving this are readily found from the expressions given above

a) *for the triplet*

$$L_{1,2} + \frac{a_1}{c_1} + \frac{a_2}{c_2} = \mp \frac{c_3}{c_1 c_2}$$

$$L_{2,3} + \frac{a_2}{c_2} + \frac{a_3}{c_3} = \pm \frac{c_1}{c_2 c_3} \qquad \text{(A15)}$$

b) *for the quadruplet*

$$c_1 c_2 \left(L_{1,2} + \frac{a_1}{c_1} + \frac{a_2}{c_2} \right) = \mp\, c_3 c_4 \left(L_{3,4} + \frac{a_3}{c_3} + \frac{a_4}{c_4} \right)$$

$$L_{2,3} + \frac{a_2}{c_2} + \frac{a_3}{c_3} = \frac{1}{c_2{}^2 \left(L_{1,2} + \dfrac{a_1}{c_1} + \dfrac{a_2}{c_2} \right)} + \frac{1}{c_3{}^2 \left(L_{3,4} + \dfrac{a_3}{c_3} + \dfrac{a_4}{c_4} \right)}$$

(A16)

c) *for the quintuplet*

$$\left(L_{1,2} + \frac{a_1}{c_1} + \frac{a_2}{c_2} \right) \left(L_{2,3} + \frac{a_2}{c_2} + \frac{a_3}{c_3} \right) \left(L_{3,4} + \frac{a_3}{c_3} + \frac{a_4}{c_4} \right)$$

$$= \frac{1}{c_2{}^2} \left(L_{3,4} + \frac{a_3}{c_3} + \frac{a_4}{c_4} \right) + \frac{1}{c_3{}^2} \left(L_{1,2} + \frac{a_1}{c_1} + \frac{a_2}{c_2} \right) \mp \frac{c_5}{c_1 c_2 c_3 c_4}$$

$$\left(L_{2,3} + \frac{a_2}{c_2} + \frac{a_3}{c_3} \right) \left(L_{3,4} + \frac{a_3}{c_3} + \frac{a_4}{c_4} \right) \left(L_{4,5} + \frac{a_4}{c_4} + \frac{a_5}{c_5} \right)$$

$$= \frac{1}{c_3{}^2} \left(L_{4,5} + \frac{a_4}{c_4} + \frac{a_5}{c_5} \right) + \frac{1}{c_4{}^2} \left(L_{2,3} + \frac{a_2}{c_2} + \frac{a_3}{c_3} \right) \pm \frac{c_1}{c_2 c_3 c_4 c_5}$$

(A17)

Here again considerable simplification occurs in the case of symmetry or antisymmetry. The conditions (A15), (A16), or (A17) will generally have to be applied to both planes of an AG system. In addition, in most cases one would like to have stigmatism and real object and image points, and this puts other constraints on the matrix element *b*.

BIBLIOGRAPHY

This survey covers the period 1958 to January 1965.

Asner, A., Bossard, P., and Rohner, F. (1964). A new "splitpole" quadrupole lens. *CERN* **64-5**.

Auberson, G. (1961). Transport de faisceaux: aperçu des méthodes de calcul. CERN MPS/Int. DL 61-36.

Blewett, J. P. (1959). The focal properties of certain quadrupole lenses. BNL/Int. JPB13.

Braunersreuther, E., Chabaud, V., Delorme, C., and Morpurgo, M. (1961). Lentilles quadrupolaires magnétiques constituant le dispositif de focalisation du faisceau de mésons μ du CERN. *CERN* **61-12**.

Chamberlain, O. (1960). Optics of high-energy beams. *Ann. Rev. Nucl. Sci.* **10**, 161.

Chih-Shu, L., and Carr, H. E. (1962). Electrostatic quadrupole lens pair for mass spectrometers. *Rev. Sci. Instr.* **33**, 8 and 823.

Citron, A., Morpurgo, M., and Overas, H. (1963). The high intensity muon beam with low pion contamination at the CERN synchro-cyclotron. *CERN* **63-35**.

Courant, E. D., and Cool, R. (1959). Transport and separation of beams from an

A. G. synchrotron. *Proc. Intern. Conf. High-Energy Accelerators, CERN, Geneva, 1959*, p. 403. CERN, Geneva, Switzerland.

Courant, E. D., and Marshall, L. (1960). Mass separation of high-energy particles in quadrupole lens focusing systems. *Rev. Sci. Instr.* **31**, 193.

Crewe, A. V. (1960). Quadrupole magnets and their uses. ANL/PAD/Int. AVC-3.

De Raad, B. (1962). Methods to calculate beam transport systems. CERN AR/Int. GS 62–5.

Gardner, J. W., King, N. M., and Whiteside, D. (1962). Design studies for the Nimrod external proton beam. NIRL/R/12.

Geibel, J. A., and Auberson, G. (1962). Note sur l'acceptance angulaire d'un doublet. CERN MPS/DL 62–10.

Geibel, J. A., and King, N. M. (1963). Given transformation matrices related to a quadrupole and drifts lengths. CERN MPS/EP 63–2.

Gouiran, R. (1962). Les faisceaux secondaires issus des grands accélérateurs. *Ind. Atomiques* **1**/2.

Grivet, P., and Septier, A. (1958). Les lentilles quadrupolaires magnétiques. *CERN* **58-25**.

Grivet, P., and Septier, A. (1960). Les lentilles quadrupolaires magnétiques I et II. *Nucl. Instr. Methods* **6**, 126 and 243.

Hereward, H. G. (1959). Properties of particle beams in optical matching systems in terms of phase plane ellipses. CERN PS/Int. TH 59-5.

Hereward, H. G. (1963). Effect of quadrupoles in the CPS; methods of calculation. CERN MPS/DL 63-9.

Kern, W., and Steffen, K. G. (1961). Some new aspects in beam transport development-Part I: A fast analogue computer for beam envelopes and particle trajectories. *Proc. Intern. Conf. High-Energy Accelerators, Brookhaven, 1961*. Brookhaven Natl. Lab., New York.

King, N. M. (1961). Theoretical beam-handling studies at the Rutherford Laboratory. *Proc. Intern. Conf. High-Energy Accelerators, Brookhaven, 1961*. Brookhaven Natl. Lab., New York.

King, N. M. (1962). Some focusing properties of quadrupole doublets. CERN/MPS/EP/22.

King, N. M. (1963). Doublet configuration representing given transformation matrices. CERN MPS/EP 63–4.

King, N. M. (1963). Focusing with quadrupole triplets; thin lens approximation. CERN MPS/EP 63-6.

King, N. M. (1964). Theoretical techniques of high-energy beam design. *Progr. Nucl. Phys.* **9**, 73.

Langeseth, B., Pluym, G., and de Raad, B. (1960). Magnetic measurements on the beam transport quadrupoles for the CERN proton synchrotron. CERN PS/Int. EA 60–5.

Livingood, J. J. (1961). "Principles of Cyclic Particle Accelerators." Van Nostrand, Princeton, New Jersey.

Luckey, D. (1961). Beam optics. *In* "Techniques of High-Energy Physics" (D. M. Ritson, ed.), Wiley (Interscience), New York.

Madsen, J. H.B. (1964). Focusing and magnification ratio adjustments with quadrupole triplets and doublets. CERN MPS/EP 64-5.

Montague, B. W. (1960). Phase space analogue computer for beam matching problems. *CERN* **60-24**.

Morpurgo, M. (1964). Design and construction of quadrupole lenses with large elliptical aperture. *CERN* **64-34**.

O'Connell, J. S., Morrison, R. C., and Stewart, J. R. (1964). The quadrupole triplet as a momentum spectrometer. *Nucl. Instr. Methods* **30**, 229.

Ramm, C. A. (1960). Some features of beam handling equipment for the CERN proton synchrotron. *Proc. Intern. Conf. Instr. High-Energy Phys.*, *Lawrence Radiation Lab.*, 1960. Wiley (Interscience), New York.

Sedman, E. G. (1962). Beam transport: practical design and setting up. CERN MPS/DL 62–23.

Septier A. (1960). Les lentilles magnétiques quadrupolaires sans fer; réalisations de répartitions d'induction à gradient constant. *CERN* **60-6**.

Septier, A. (1961). Strong focusing lenses. *Advan. Electron. Electron Phys.* **14**, 85.

Steffen, K. G. (1961). Some new aspects in beam transport development - Part II: A quadrupole magnet with non-circular aperture and linearized end fringing fields. *Proc. Intern. Conf. High-Energy Accelerators, Brookhaven, 1961.* Brookhaven Natl. Lab., New York.

Steffen, K. G. (1965). "High-Energy Beam Optics." Wiley, New York (in press).

Steffen, K. G., Hultschig H., and Kern, W. (1960). Use of generalized amplitude and phase functions in designing beam transport systems. *Proc. Intern. Conf. Instr. High-Energy Phys.*, *Lawrence Radiation Lab.*, *1960.* Wiley (Interscience), New York.

Tollefsrud, B., and Baconnier, Y. (1965). Étude et réalisation de quadrupoles d'injection. CERN MPS/POW 65–1.

Weiss, M. (1963). Considerations on the beam transport system of the preinjector for the 50 MeV linac. CERN MPS/Lin 63–3.

Wilson, E.J.N. (1962). A quadrupole magnet of large acceptance, CERN MPS/EP 62–29.

Wilson, E.J.N. (1962). Quadrupole design for high acceptance. CERN MPS/EP 62–30.

Yagi, K. (1964). A new broad range spectrometer with a uniform-field sectorial magnet and a preceding quadrupole-magnet doublet. *Nucl. Instr. Methods* **31**, 173.

CHAPTER 2.5

LENS ABERRATIONS

P. W. Hawkes

THE CAVENDISH LABORATORY
PETERHOUSE, CAMBRIDGE, ENGLAND

2.5.1. Introduction

The quality of any electron optical system depends not only upon the wavelength of the electrons, but also upon the aberrations from which it may suffer. These aberrations can arise for a number of different reasons. If the accelerating potential and the lens excitations fluctuate about their mean values, *chromatic aberration* will mar the image. If the properties of the system are investigated, using a more exact approximation to the refractive index than is employed in the Gaussian approximation, we find that *geometrical aberrations* affect both the quality and the fidelity of the Gaussian image. When the properties of the system are analyzed using the non-relativistic approximation, the disparities between the relativistic and non-relativistic results can, for accelerating potentials up to about 100 kV, be conveniently regarded as a *relativistic aberration*. If, finally, the properties of the system are calculated on the assumption that mechanically, the latter is perfect—that the machining and alignment of all its parts are faultless—the properties of any real system will disagree, to a greater or lesser extent, with the calculated values; this we call the *mechanical aberration*. These are

411

the most important types of aberration in electron optical systems, unless there are regions where the electron current density is very high; in such systems, the *space-charge aberration* produced by the interaction between the electron charges may have to be considered.

In this chapter we shall be concerned primarily with the geometrical aberrations; nevertheless, we shall occasionally indicate the origins of certain types of mechanical aberration, for the latter are only the geometrical aberrations of systems which would, but for certain imperfections, display the symmetry properties of some more regular class of systems. Of the geometrical aberrations, we shall analyze the aperture aberration in the greatest detail. This type of aberration is, by definition, unaffected by the position of the object relative to the axis; unlike all the other geometrical aberrations, therefore, it affects the image of an object placed on the axis. Furthermore, we shall see that the spherical aberration of a system consisting solely of conventional round lenses can never be eliminated.

We shall consider two types of system: *round systems* (which consist only of axially symmetric electrostatic and magnetic lenses) and *rectilinear orthogonal systems* (which, for brevity, we shall call "*quadrupole systems*"). The latter are composed of magnetic and electrostatic round lenses, and magnetic and electrostatic quadrupole lenses; by definition, the optic axis is straight, and the quadrupoles and round lenses must be situated and oriented in such a way that the *orthogonality condition* is satisfied. In the simplest such arrangement, magnetic round lenses are absent, and there exists a pair of planes, intersecting along the z axis, which pass through all the electrodes and bisect the angles between all the pole-pieces. In theory, more complex orthogonal systems can be constructed, but in practice, almost all quadrupole systems possess these symmetry planes, unless magnetic round lenses are included. In this case, the round magnetic lens fields must not overlap the quadrupole fields, and the quadrupolar section of the system does again possess symmetry planes.

Two mathematical techniques are available for analyzing the aberrations: the *characteristic function method*, introduced into electron optics by Glaser and Sturrock, which we shall employ here, and the *trajectory method*. Both, naturally, lead to the same results, although considerable ingenuity is sometimes required to demonstrate that they are the same, and the two methods entail a comparable amount of labor. The use of characteristic functions does, however, have the advantage that any interrelations between the aberration coefficients are immediately obvious, whereas an additional effort is required to establish them when the trajectory method is employed.

2.5.2. Fermat's Principle; Characteristic Functions; Perturbation Theory

Fermat's principle states that the optical path that a ray of light will follow between two arbitrary points is an extremum. If $n(x, y, z)$ denotes the refractive index at any point, then

$$\delta \int n \, ds = 0$$

In electron optics, a similar principle is valid, and the refractive index n is related to the electrostatic potential $\varphi(x, y, z)$ and the vector potential $\mathbf{A}(x, y, z)$ by the formula

$$n = [\varphi(1 + \varepsilon\varphi)]^{1/2} - \eta \, \mathbf{A} \cdot \mathbf{s} \tag{2.1}$$

in which $\mathbf{s}$ is a unit vector, tangent to the path of the electron at the point (x, y, z); ε and η are constants, defined by $\varepsilon = e/2m_0c^2 = 0.978 \times 10^{-6}$ C sec^2 kg^{-1} m^{-2} and $\eta = (e/2m_0)^{1/2} = 2.966 \times 10^5$ C$^{1/2}$ kg$^{-1/2}$; $-e$ and m_0 are the charge and rest mass, respectively, of an electron, and c is the velocity of light in free space.

It is convenient [though not essential: see, for example, Picht's work (1963)] to transform the equation $\delta \int n \, ds = 0$ into the form

$$\delta \int m \, dz = 0 \tag{2.2}$$

where

$$m = [\varphi(1 + \varepsilon\varphi)(1 + x'^2 + y'^2)]^{1/2} - \eta(A_x x' + A_y y' + A_z) \tag{2.3}$$

The functions $\varphi(x, y, z)$, $A_x(x, y, z)$, $A_y(x, y, z)$ and $A_z(x, y, z)$ can be expanded as power series in the off-axial coordinates x, y, the coordinate z figuring only in the coefficients, as we shall see later. Inserting these expansions into m and expanding the square roots as power series also, we find that m is composed of a series of terms which we group according to their degree in x, y and their derivatives. Thus

$$m = m^{(0)} + m^{(2)} + m^{(4)} + \cdots + m^{(2n)} + \cdots$$

in which $m^{(2n)}$ consists of all the terms of degree $2n$. The explicit dependence of $m^{(0)}$, $m^{(2)}$, and $m^{(4)}$ upon the potentials is set out in Section 2.5.4. Let us now consider how to work in terms of these quantities.

We introduce the notation

$$p = \partial m/\partial x' \qquad q = \partial m/\partial y' \tag{2.4}$$

Suppose now that the optical path along an arbitrary curve joining two points $(x_\alpha, y_\alpha, z_\alpha)$ and $(x_\beta, y_\beta, z_\beta)$ is denoted by S. Thus

$$S = \int_{z_\alpha}^{z_\beta} m \, dz$$

The optical path, S^*, along a neighboring curve joining two points lying in the same planes, $z = z_\alpha$ and $z = z_\beta$, will be given by

$$S^* = S + \delta S$$

where

$$\delta S = \int_{z_\alpha}^{z_\beta} \left(\delta x \, \frac{\partial m}{\delta x} + \delta y \, \frac{\partial m}{\partial y} + \delta x' \, \frac{\partial m}{\partial x'} + \delta y' \, \frac{\partial m}{\partial y'} \right) dz$$

which can be integrated to give

$$\delta S = (p_\beta \, \delta x_\beta + q_\beta \, \delta y_\beta) - (p_\alpha \, \delta x_\alpha + q_\alpha \, \delta y_\alpha)$$
$$- \int_{z_\alpha}^{z_\beta} \left\{ \delta x \left(\frac{dp}{dz} - \frac{\partial m}{\partial x} \right) + \delta y \left(\frac{dq}{dz} - \frac{\partial m}{\partial y} \right) \right\} dz$$

If the path of integration is a ray, δS must vanish provided the end points remain unaltered, and hence

$$dp/dz = \partial m/\partial x \qquad dq/dz = \partial m/\partial y \qquad (2.5)$$

These are the Euler equations of the variational equation (2.2). The value of S when the path of integration is a ray will be denoted by $V_{\alpha\beta}$. Thus

$$V_{\alpha\beta} = \int_{z_\alpha}^{z_\beta} m \, dz \qquad (2.6)$$

so that

$$\delta V_{\alpha\beta} = p_\beta \, \delta x_\beta + q_\beta \, \delta y_\beta - p_\alpha \, \delta x_\alpha - q_\alpha \, \delta y_\alpha \qquad (2.7)$$

which shows that $V_{\alpha\beta}$ is a function of the x and y coordinates of the end points; in fact, $V_{\alpha\beta}$ is essentially Hamilton's *point characteristic function*, since we understand the suffix $\alpha\beta$ to indicate that z_α and z_β are the limits of integration.

Suppose now that $m(x, y, z)$ is replaced by a slightly different function, $m(x, y, z) + m^{\mathrm{I}}(x, y, z)$; it can be shown[1] that if we write

$$V_{\alpha\beta}^{\mathrm{I}} = \int_{z_\alpha}^{z_\beta} m^{\mathrm{I}}(x, y, z) \, dz \qquad (2.8)$$

[1] A detailed proof is to be found in Sturrock (1951).

where the path of integration is the same as the original path, then

$$\delta V_{\alpha\beta}^{\text{I}} = (p_\beta^{\text{I}}\,\delta x_\beta + q_\beta^{\text{I}}\,\delta y_\beta - x_\beta^{\text{I}}\,\delta p_\beta - y_\beta^{\text{I}}\,\delta q_\beta)$$
$$- (p_\alpha^{\text{I}}\,\delta x_\alpha + q_\alpha^{\text{I}}\,\delta y_\alpha - x_\alpha^{\text{I}}\,\delta p_\alpha - y_\alpha^{\text{I}}\,\delta q_\alpha) \tag{2.9}$$

Provided the system does not produce a stigmatic image of the point $(x_\alpha, y_\alpha, z_\alpha)$ at $(x_\beta, y_\beta, z_\beta)$, we can elect to keep these points stationary, so that $x_\alpha^{\text{I}} = y_\alpha^{\text{I}} = x_\beta^{\text{I}} = y_\beta^{\text{I}} = 0$. The arguments of the function $V_{\alpha\beta}^{\text{I}}$ are thus x_α, y_α, x_β and y_β, and we can deduce that

$$\delta V_{\alpha\beta}^{\text{I}} = p_\beta^{\text{I}}\,\delta x_\beta + q_\beta^{\text{I}}\,\delta y_\beta - p_\alpha^{\text{I}}\,\delta x_\alpha - q_\alpha^{\text{I}}\,\delta y_\alpha \tag{2.10}$$

whence

$$p_\alpha^{\text{I}} = -\frac{\partial V_{\alpha\beta}^{\text{I}}}{\partial x_\alpha} \qquad q_\alpha^{\text{I}} = -\frac{\partial V_{\alpha\beta}^{\text{I}}}{\partial y_\alpha}$$
$$p_\beta^{\text{I}} = +\frac{\partial V_{\alpha\beta}^{\text{I}}}{\partial x_\beta} \qquad q_\beta^{\text{I}} = +\frac{\partial V_{\alpha\beta}^{\text{I}}}{\partial y_\beta} \tag{2.11a}$$

These differential relations are the perturbed counterparts of those we can deduce from Eq. (2.7), namely,

$$p_\alpha = -\frac{\partial V_{\alpha\beta}}{\partial x_\alpha} \qquad q_\alpha = -\frac{\partial V_{\alpha\beta}}{\partial y_\alpha}$$
$$p_\beta = +\frac{\partial V_{\alpha\beta}}{\partial x_\beta} \qquad q_\beta = +\frac{\partial V_{\alpha\beta}}{\partial y_\beta} \tag{2.11b}$$

The function $V_{\alpha\beta}^{\text{I}}$ that we have obtained by setting $x_\alpha^{\text{I}} = y_\alpha^{\text{I}} = x_\beta^{\text{I}} = y_\beta^{\text{I}} = 0$ is known as the *point* perturbation characteristic function, since its arguments are point coordinates; alternatively, we could have written $p_\alpha^{\text{I}} = q_\alpha^{\text{I}} = p_\beta^{\text{I}} = q_\beta^{\text{I}} = 0$ (*angle* perturbation characteristic function), or $p_\alpha^{\text{I}} = q_\alpha^{\text{I}} = x_\beta^{\text{I}} = y_\beta^{\text{I}} = 0$ or $p_\beta^{\text{I}} = q_\beta^{\text{I}} = x_\alpha^{\text{I}} = y_\alpha^{\text{I}} = 0$ (*mixed* perturbation characteristic functions). Except in special cases where one or other of these functions cannot be used, and we are forced to employ one of the remainder, we are at liberty to use whichever type of characteristic function provides the required information most readily. Throughout the present chapter, the *point* function will be employed.

To what use are these equations to be put? To answer this question, we must consider how the function m and its perturbation, m^{I}, can be related to the practical notions of Gaussian imagery and geometrical aberrations.

To obtain the Gaussian optical properties of an optical or electron optical system—the positions of the principal planes, foci and nodal planes—we reject all but the lowest terms in x and y in the refractive index function m;

these are the terms comprising $m^{(2)}$, and if in Eqs. (2.5) we write $m = m^{(2)}$, we obtain the Gaussian equations of motion.[2] The "small change" in m (from m to $m + m^{\mathrm{I}}$) will thus in fact be the change from the Gaussian approximation to the third-order approximation: $m^{\mathrm{I}} = m^{(4)}$. The small change therefore represents an advance from a crude approximation to a more refined one. The same is true when the relativistic "aberration" is being evaluated, but not when we calculate the chromatic aberration arising from variations in the accelerating potential; here, a genuine change is involved, for while we still write $m = m^{(2)}$, the variation is $m^{\mathrm{I}} = \partial m^{(2)}/\partial\Phi$.

As they stand, Eqs. (2.11a) do not provide us with the kind of information we should like to possess; we should prefer to know where a ray will intersect some arbitrary image plane, given that the ray emerges from a particular image point and is limited by the size of the opening in the aperture plane.

Let $z = z_0$ be the object plane and $z = z_a$ the aperture plane; $z = z_c$ is an arbitrary current plane, which in a round system usually lies in the immediate vicinity of the Gaussian image plane. After substituting first $z_\alpha = z_0$, $z_\beta = z_c$ and then $z_\alpha = z_a$, $z_\beta = z_c$ in Eq. (2.10) and proceeding to the limit, we find

$$p_c^{\mathrm{I}}\frac{\partial x_c}{\partial x_a} + q_c^{\mathrm{I}}\frac{\partial y_c}{\partial x_a} - x_c^{\mathrm{I}}\frac{\partial p_c}{\partial x_a} - y_c^{\mathrm{I}}\frac{\partial q_c}{\partial x_a} = \frac{\partial V_{oc}^{\mathrm{I}}}{\partial x_a}$$

$$p_c^{\mathrm{I}}\frac{\partial x_c}{\partial y_a} + q_c^{\mathrm{I}}\frac{\partial y_c}{\partial y_a} - x_c^{\mathrm{I}}\frac{\partial p_c}{\partial y_a} - y_c^{\mathrm{I}}\frac{\partial q_c}{\partial y_a} = \frac{\partial V_{oc}^{\mathrm{I}}}{\partial y_a}$$

$$p_c^{\mathrm{I}}\frac{\partial x_c}{\partial x_o} + q_c^{\mathrm{I}}\frac{\partial y_c}{\partial x_o} - x_c^{\mathrm{I}}\frac{\partial p_c}{\partial x_o} - y_c^{\mathrm{I}}\frac{\partial q_c}{\partial x_o} = \frac{\partial V_{ac}^{\mathrm{I}}}{\partial x_o}$$

$$p_c^{\mathrm{I}}\frac{\partial x_c}{\partial y_o} + q_c^{\mathrm{I}}\frac{\partial y_c}{\partial y_o} - x_c^{\mathrm{I}}\frac{\partial p_c}{\partial y_o} - y_c^{\mathrm{I}}\frac{\partial q_c}{\partial y_o} = \frac{\partial V_{ac}^{\mathrm{I}}}{\partial y_o}$$

(2.12)

Solving these simultaneous equations for p_c^{I}, q_c^{I}, x_c^{I}, and y_c^{I}, we obtain the effect of m^{I} upon the positions and slopes of rays in the current plane. Henceforward, we shall write $V^{(4)}$ instead of V^{I} and $m^{(4)}$ instead of m^{I} to emphasize that the third-order (or primary) geometrical aberrations are in

---

[2] In the trajectory method, the Gaussian equations of motion are obtained by writing $m^{(2)}$ for m in Eqs. (2.5), and the aberrations, by writing $m^{(2)} + m^{(4)}$ and subtracting the Gaussian solution. The differential equation for the aberrations is solved with the aid of the "variation of parameters" method. Considerable care is necessary if this method is employed to calculate the fifth-order aberrations; if $m^{(2)} + m^{(4)} + m^{(6)}$ is substituted for m, an incomplete result will be obtained unless it is remembered that the third-order approximation is now the best available.

question. $V^{(4)}$ is obtained by integrating $m^{(4)}$ with respect to z; x, y and their derivatives appear in $m^{(4)}$, and for these quantities, we substitute the expressions obtained in the Gaussian approximation: the solutions of the Gaussian equations of motion. The nature of these solutions depends upon the form of $m^{(2)}$. If the system is round, x and y can appear only in the combinations $(x^2 + y^2)$, $(x'^2 + y'^2)$, and $(xy' - x'y)$, and by introducing a rotating system of coordinates, the last term can be eliminated; there exists a coordinate system, therefore, in which

$$p = Cx' \qquad q = Cy' \tag{2.13}$$

and the equations of motion are

$$\frac{d}{dz}(Cx') = \bar{C}x \qquad \frac{d}{dz}(Cy') = \bar{C}y \tag{2.14}$$

in which C and $\bar{C}$ are the functions of z which appear in $m^{(2)}$:

$$m^{(2)} = \tfrac{1}{2}C(x'^2 + y'^2) + \tfrac{1}{2}\bar{C}(x^2 + y^2) \tag{2.15}$$

The general solution of each of the differential equations (2.14) consists of a linear combination of any two linearly independent solutions; for the latter, we select the functions $g(z)$, $h(z)$ which satisfy the boundary conditions

$$g(z_o) = h(z_a) = 1 \qquad g(z_a) = h(z_o) = 0 \tag{2.16}$$

(see Fig. 1a). The general solutions are thus

$$x(z) = x_o g(z) + x_a h(z) \qquad y(z) = y_o g(z) + y_a h(z) \tag{2.17}$$

and from (2.12)

$$x_c^{\mathrm{I}} = \frac{h_c(\partial V_{ac}^{\mathrm{I}}/\partial x_o) - g_c(\partial V_{oc}^{\mathrm{I}}/\partial x_a)}{C(gh' - g'h)}$$

$$y_c^{\mathrm{I}} = \frac{h_c(\partial V_{ac}^{\mathrm{I}}/\partial y_o) - g_c(\partial V_{oc}^{\mathrm{I}}/\partial y_a)}{C(gh' - g'h)} \tag{2.18}$$

$$p_c^{\mathrm{I}} = \frac{h_c'(\partial V_{ac}^{\mathrm{I}}/\partial x_o) - g_c'(\partial V_{oc}^{\mathrm{I}}/\partial x_a)}{gh' - g'h}$$

$$q_c^{\mathrm{I}} = \frac{h_c'(\partial V_{ac}^{\mathrm{I}}/\partial y_o) - g_c'(\partial V_{oc}^{\mathrm{I}}/\partial y_a)}{gh' - g'h} \tag{2.19}$$

In orthogonal systems, the potential distribution is such that the equations of motion separate. The general expression for $m^{(2)}$ in this case, after any necessary coordinate transformations have been made, will be

$$m^{(2)} = \tfrac{1}{2} C(x'^2 + y'^2) + \tfrac{1}{2} (\bar{C}_x x^2 + \bar{C}_y y^2) \qquad (2.20)$$

so that once again, $p = Cx'$, $q = Cy'$, but now the equations of motion are

$$\frac{d}{dz}(Cx') = \bar{C}_x x \qquad \frac{d}{dz}(Cy') = \bar{C}_y y \qquad (2.21)$$

The general solutions of these equations can be written in the form

$$x(z) = x_0 g_x(z) + x_a h_x(z) \qquad y(z) = y_0 g_y(z) + y_a h_y(z) \qquad (2.22)$$

where $g_x(z)$ and $g_y(z)$ satisfy the same boundary conditions as $g(z)$ in Eqs. (2.16), and $h_x(z)$, $h_y(z)$ as $h(z)$ (see Fig. 1b).

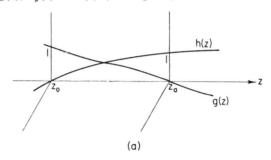

(a)

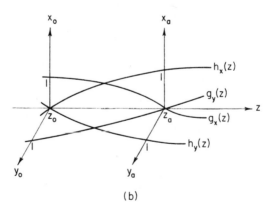

(b)

FIG. 1. (a) The functions $g(z)$, $h(z)$. (b) the functions $g_x(z)$, $g_y(z)$, $h_x(z)$, $h_y(z)$.

From Eqs. (2.21), it is clear that $C(g_x h_x' - g_x' h_x)$, which we denote by k_x, is constant, and likewise $C(g_y h_y' - g_y' h_y)$ or k_y is constant; for round systems, we write $C(gh' - g'h) = k$. From Eqs. (2.12), we find that for orthogonal systems

$$k_x x_c^{\mathrm{I}} = h_{xc} \frac{\partial V_{ac}^{\mathrm{I}}}{\partial x_o} - g_{xc} \frac{\partial V_{oc}^{\mathrm{I}}}{\partial x_a}$$

$$k_y y_c^{\mathrm{I}} = h_{yc} \frac{\partial V_{ac}^{\mathrm{I}}}{\partial y_o} - g_{yc} \frac{\partial V_{oc}^{\mathrm{I}}}{\partial y_a} \tag{2.23}$$

$$k_x p_c^{\mathrm{I}} = C\left(h_{xc}' \frac{\partial V_{ac}^{\mathrm{I}}}{\partial x_o} - g_{xc}' \frac{\partial V_{oc}^{\mathrm{I}}}{\partial x_a} \right)$$

$$k_y q_c^{\mathrm{I}} = C\left(h_{yc}' \frac{\partial V_{ac}^{\mathrm{I}}}{\partial y_o} - g_{yc}' \frac{\partial V_{oc}^{\mathrm{I}}}{\partial y_a} \right) \tag{2.24}$$

There are circumstances in which complex coordinates prove to be more convenient than Cartesians; if we write $u = x + iy$, $s = p + iq$ and denote the complex conjugate of u by $\bar{u}$ and of s by $\bar{s}$, Eqs. (2.18) and (2.19) become

$$ku_c^{\mathrm{I}} = 2h_c \frac{\partial V_{ac}^{\mathrm{I}}}{\partial \bar{u}_o} - 2g_c \frac{\partial V_{oc}^{\mathrm{I}}}{\partial \bar{u}_a} \tag{2.25a}$$

$$ks_c^{\mathrm{I}} = 2Ch_c' \frac{\partial V_{ac}^{\mathrm{I}}}{\partial \bar{u}_o} - 2Cg_c' \frac{\partial V_{oc}^{\mathrm{I}}}{\partial \bar{u}_a} \tag{2.25b}$$

There is more than one way of writing Eqs. (2.23) and (2.24) in complex coordinates. A form that is often convenient is the following, in which $\frac{1}{2}(k_x + k_y)$ is denoted by $\bar{k}$, $\frac{1}{2}(k_x - k_y)$ by k^*, $(h_x + h_y)$ by $\bar{h}$, $(h_x - h_y)$ by h^*, $(g_x + g_y)$ by $\bar{g}$ and $(g_x - g_y)$ by g^*.

$$(\bar{k}^2 - k^{*2})u_c^{\mathrm{I}} = (\overline{hk} - h^*k^*)\frac{\partial V_{ac}^{\mathrm{I}}}{\partial \bar{u}_o} + (h^*\bar{k} - \bar{h}k^*)\frac{\partial V_{ac}^{\mathrm{I}}}{\partial u_o}$$

$$+ (\overline{gk} - g^*k^*)\frac{\partial V_{oc}^{\mathrm{I}}}{\partial \bar{u}_a} + (g^*\bar{k} - \bar{g}k^*)\frac{\partial V_{oc}^{\mathrm{I}}}{\partial u_a} \tag{2.26a}$$

$$(\bar{k}^2 - k^{*2})s_c^{\mathrm{I}} = C\left\{ (\bar{h}'\bar{k} - h'^*k^*)\frac{\partial V_{ac}^{\mathrm{I}}}{\partial \bar{u}_o} + (h'^*\bar{k} - \bar{h}'k^*)\frac{\partial V_{ac}^{\mathrm{I}}}{\partial u_o} \right.$$

$$\left. + (\bar{g}'\bar{k} - g'^*k^*)\frac{\partial V_{oc}^{\mathrm{I}}}{\partial \bar{u}_a} + (g'^*\bar{k} - \bar{g}'k^*)\frac{\partial V_{oc}^{\mathrm{I}}}{\partial u_a} \right\} \tag{2.26b}$$

If the aberrations are to be expressed in terms of x_0', y_0' instead of x_a, y_a, a slightly different approach is necessary; cf. Glaser (1952) and Hawkes (1965/6).

2.5.3. Symmetry

A knowledge of its characteristic function is adequate to provide us with full information about any given electron optical system. This function possesses a concrete physical meaning: it is a measure of the length of the optical path between two points along a ray. If, therefore, the system displays symmetry properties, the characteristic function must reflect these faithfully. It is thus natural to wonder whether we cannot deduce the nature of the imagery and aberrations from a knowledge of its symmetry alone.

This proves not to be difficult. An elaborate exposition of the technique is to be found in Chako (1957); Amboss (1959) has employed it to analyze the mechanical aberrations of round electron optical systems. Using complex coordinates ($u = x + iy$, $\bar{u}$ is the complex conjugate of u), we can express the result that the point of intersection of a ray with an arbitrary current plane, u_c, is a function of its points of intersection with the object and aperture planes, u_0 and u_a, thus:

$$u_c = u_c(u_o, \bar{u}_o; u_a, \bar{u}_a)$$

We expand u_c as a power series, and obtain

$$u_c = \sum_{\alpha,\beta,\gamma,\delta \geq 0} (\alpha\beta\gamma\delta)u_o^\alpha \bar{u}_o^\beta u_a^\gamma \bar{u}_a^\delta \tag{3.1}$$

in which $(\alpha\beta\gamma\delta)$ is a real or complex function of z. In order to understand how the various terms are to be interpreted, consider the Gaussian imagery, for which $\alpha + \beta + \gamma + \delta = 1$, and hence

$$u_c = (1000)u_o + (0100)\bar{u}_o + (0010)u_a + (0001)\bar{u}_a$$

or

$$x_c = \text{Re}(1000 + 0100)x_o - \text{Im}(1000 - 0100)y_o$$
$$+ \text{Re}(0010 + 0001)x_a - \text{Im}(0010 - 0001)y_a$$
$$y_c = \text{Im}(1000 + 0100)x_o + \text{Re}(1000 - 0100)y_o$$
$$+ \text{Im}(0010 + 0001)x_a + \text{Re}(0010 - 0001)y_a$$

In an orthogonal system, x_c depends only on x_o and x_a, and y_c only upon y_o and y_a, so that (1000), (0100), (0010) and (0001) are all real. In a round system, the coefficients of x_o and x_a in x_c must be the same as those of y_o and y_a, respectively, in y_c; (0100) and (0001) therefore vanish altogether. Thus (1000) corresponds to the magnification in a round system, and (0010) measures the distance between the current plane and the Gaussian image plane where (0010) vanishes. In an orthogonal system, there will in general be no stigmatic[3] image plane; a beam emerging from a point object and bounded by a circular aperture will collapse into two line foci, where $(0010) = \pm (0001)$. Furthermore, the magnification may be different in the x direction and the y direction: only if (0100) vanishes will the magnification be orthomorphic, and it is therefore natural to call (0100) the (first-order, or primordial) *distortion*.

Round Systems. If the ray passing through the points u_o and u_a intersects the current plane at u_c, then the ray that intersects the object and aperture planes at $u_o e^{i\theta}$ and $u_a e^{i\theta}$ must intersect the current plane at $u_c e^{i\theta}$:

$$u_c e^{i\theta} = \Sigma \, (\alpha\beta\gamma\delta)u_o{}^\alpha \, \bar{u}_o{}^\beta \, u_a{}^\gamma \, \bar{u}_a{}^\delta \, \exp i(\alpha - \beta + \gamma - \delta)\theta$$

This will be true for all values of θ only if $\alpha - \beta + \gamma - \delta = 1$, so that

$$
\begin{aligned}
u_c = \; & (1000)u_o + (0010)u_a \\
& + (2100)u_o{}^2\bar{u}_o + (2001)u_o{}^2\bar{u}_a + (1011)u_o u_a \bar{u}_a \\
& + (0021)u_a{}^2\bar{u}_a + (1110)u_o\bar{u}_o u_a + (0120)\bar{u}_o u_a{}^2
\end{aligned}
\tag{3.2}
$$

As anticipated, only (1000) and (0010) figure in the primordial terms. Of the six possibly complex primary aberrations, (2100) measures the *distortion*; (0021) the *spherical aberration* or *aperture aberration*; (2001) corresponds to *astigmatism*, and (1110) to *field curvature*; (1011) and (0120) will prove to be related, and together produce the aberration known as *coma*. To derive any relationships between the aberration coefficients, we must consider the form of the perturbation characteristic function, $V_{\alpha\beta}^{(4)}$. This function must be expressible in terms of the rotational invariants of the system, $u_o\bar{u}_o$, $u_o\bar{u}_a$, $\bar{u}_o u_a$ and $u_a\bar{u}_a$. As before, we use a four-symbol notation $(pqrs)$ to designate the coefficient of $u_o{}^p\bar{u}_o{}^q u_a{}^r\bar{u}_a{}^s$. Clearly

$$
V_{\alpha\beta}^{(4)} = \begin{pmatrix} u_o{}^2 \\ u_o u_a \\ u_a{}^2 \end{pmatrix} \begin{pmatrix} 2200 & 2101 & 2002 \\ 1210 & 1111 & 1012 \\ 0220 & 0121 & 0022 \end{pmatrix} \begin{pmatrix} \bar{u}_o{}^2 \\ \bar{u}_o \bar{u}_a \\ \bar{u}_a{}^2 \end{pmatrix}
\tag{3.3a}
$$

Since this expression must be real, $(pqrs)u_o{}^p\bar{u}_o{}^q u_a{}^r\bar{u}_a{}^s = (\overline{pqrs})u_o{}^q\bar{u}_o{}^p u_a{}^s\bar{u}_a{}^r$ and hence

$$(pqrs) = (\overline{qpsr}) \tag{3.4}$$

---

[3] In orthogonal systems, we must distinguish between first-order, or primordial astigmatism, and the third-order aberration of the same name; it is convenient to say that a system with no first-order astigmatism is *stigmatic*, and that a system corrected for third-order astigmatism is *anastigmatic*.

where $\overline{(pqrs)}$ denotes the conjugate complex of $(pqrs)$. Substituting this into $V_{\alpha\beta}^{(4)}$, and labeling real coefficients (r), we obtain

$$V_{\alpha\beta}^{(4)} = \begin{pmatrix} u_0{}^2 \\ u_0 u_a \\ u_a{}^2 \end{pmatrix} \begin{pmatrix} 2200(r) & 2101 & 2002 \\ \overline{2101} & 1111(r) & 1012 \\ \overline{2002} & \overline{1012} & 0022(r) \end{pmatrix} \begin{pmatrix} \bar{u}_0{}^2 \\ \bar{u}_0 \bar{u}_a \\ \bar{u}_a{}^2 \end{pmatrix} \qquad (3.3b)$$

Invoking Eq. (2.25a), we obtain the relations between the coefficients $(\alpha\beta\gamma\delta)$, which figure in u_c, and the coefficients $(pqrs)$, which appear in $V_{\alpha\beta}^{(4)}$ $(\alpha + \beta + \gamma + \delta = 3$ and $p + q + r + s = 4)$. In order to distinguish between the coefficients in $V_{oc}^{(4)}$ and $V_{ac}^{(4)}$, we enclose the latter within square brackets:

$$V_{oc}^{(4)} = \Sigma \, (pqrs)u_0{}^p \bar{u}_0{}^q u_a{}^r \bar{u}_a{}^s; \quad V_{ac}^{(4)} = \Sigma \, [pqrs]u_0{}^p \bar{u}_0{}^q u_a{}^r \bar{u}_a{}^s$$

We find

$$k(2100) = 4h_c[2200] - 2g_c(2101)$$
$$k(2001) = 2h_c[2101] - 4g_c(2002)$$
$$k(1110) = 4h_c[\overline{2101}] - 2g_c(1111)$$
$$k(1011) = 2h_c[1111] - 4g_c(1012) \qquad (3.5)$$
$$k(0120) = 4h_c[\overline{2002}] - 2g_c(\overline{1012})$$
$$k(0021) = 2h_c[\overline{1012}] - 4g_c(0022)$$

In the Gaussian image plane, $z = z_i$, only the second column of terms on the right-hand side of expressions (3.5) remains, since $h_i = 0$. If, furthermore, magnetic lenses are not present, the characteristic function must be invariant under a change from right-handed to left-handed axes, so that $(pqrs) = (qpsr)$; since $(pqrs) = \overline{(qpsr)}$, all the coefficients must, for round electrostatic systems, be real.

We now consider the effect of each of the aberrations in turn on the Gaussian image of a fixed-object point, formed by a beam limited by a circular stop.

Spherical Aberration

$$u_i{}^I = (0021)u_a{}^2 \bar{u}_a$$

We write $u = re^{i\theta}$ and $(\alpha\beta\gamma\delta) = |\,\alpha\beta\gamma\delta\,| \, \exp i \, \langle\alpha\beta\gamma\delta\rangle$. For a circular stop of radius r_A, we have $0 \le r_a \le r_A$, and we consider the pencil of rays that intersect the aperture plane around the circle of radius r_a. We find

$$u_i^{\mathrm{I}} = r_a^3 \mid 0021 \mid \exp i\theta_a = -\, 4\, r_a^3 g_i \mid 0022 \mid \exp i\theta_a / k$$

which represents a circle, radius $r_a^3 \mid 0021 \mid$, centered on the Gaussian image point.

In the neighborhood of the Gaussian image plane, we must consider the influence of the dominant term containing $h(z)$, namely hu_a. At a distance δz from z_i, therefore, we have

$$u_c^{\mathrm{I}} = r_a^3 \mid 0021 \mid \exp i\theta_a - h_i' \,\delta z\, r_a \exp i\theta_a$$

$$= r_a \exp i\theta_a (\mid 0021 \mid r_a^2 - h_i' \,\delta z)$$

in which we have made use of the fact, to be proved later, that marginal rays intersect the axis nearer to the system than axial rays (see Fig. 2). Rays that

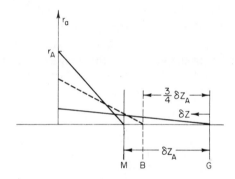

FIG. 2. The formation of the spherical aberration disc. M is the marginal image plane, G, the Gaussian image plane, and B, the plane of best focus.

intersect the aperture plane around the periphery of the stop, $r_a = r_A$, cross the axis at a distance

$$\delta z_A = \mid 0021 \mid r_A^2 / h_i'$$

from the Gaussian image plane, and further analysis (for a proof of this, see Klemperer, 1953; Sturrock, 1955, p. 101) reveals that the cross section of the beam is smallest in the plane distant $\tfrac{3}{4}\,\delta z_A$ from the Gaussian image plane, where the radius of the aberration disk is $\tfrac{1}{4}\, r_A^3 \mid 0021 \mid$. The aberration disk in this "plane of best focus" is called the "circle of least confusion." The distance δz_A is known as the longitudinal spherical aberration.

Astigmatism and Field Curvature:

$$u_i^{\mathrm{I}} = (2001)u_o^2\bar{u}_a + (1110)u_o\bar{u}_o u_a$$

We transform u_i^{I}, by writing the complex numbers in terms of modulus and argument, into the form

$$u_i^{\mathrm{I}} = r_o^2 r_a \{ \mid 2001 \mid \exp i(2\theta_o + \langle 2001 \rangle - \theta_a)$$
$$+ \mid 1110 \mid \exp i\theta_a \}$$

(in the Gaussian image plane), or

$$u_i^{\mathrm{I}} \exp(- i\varphi) = \{ \mid 2001 \mid \exp(- i\,\Theta_a) + \mid 1110 \mid \exp i\,\Theta_a \} r_o^2 r_a$$

in which

$$\Theta_a = \theta_a - \theta_o - \tfrac{1}{2}\langle 2001 \rangle \qquad \varphi = \theta_o + \tfrac{1}{2}\langle 2001 \rangle$$

which represents a tilted ellipse, semiaxes $r_o^2 r_a(\mid 1110 \mid \pm \mid 2001 \mid)$. Outside the Gaussian image plane, the aberration figure remains elliptical, except in the two planes in which the ellipse collapses into a straight line.

In a general plane (in the vicinity of the image plane), the lengths of the semiaxes are

$$(r_o^2 \mid 1110 \mid + h_i'\,\delta z + r_o^2 \mid 2001 \mid)r_a$$
$$(r_o^2 \mid 1110 \mid + h_i'\,\delta z - r_o^2 \mid 2001 \mid)r_a$$

which vanish for

$$\delta z_1 = - r_o^2 \frac{\mid 2001 \mid + \mid 1110 \mid}{h_i'}$$

$$\delta z_2 = r_o^2 \frac{\mid 2001 \mid - \mid 1110 \mid}{h_i'}$$

The semiaxes are equal in length for

$$\delta z_3 = - \frac{r_o^2 \mid 1110 \mid}{h_i'}$$

and in this, the *surface of least confusion*, the ellipse degenerates into a circle of radius $r_o^2 r_a \mid 2001 \mid$. The distance between the two *line foci*, $\delta z_2 - \delta z_1$, is called the *astigmatic difference*, and is equal to

$$2r_o^2 \mid 2001 \mid /h_i'$$

(see Fig. 3).

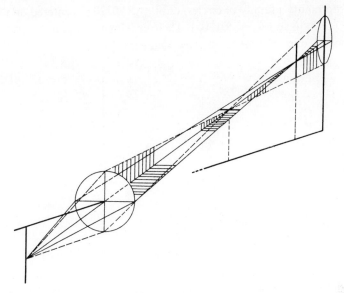

FIG. 3. Astigmatism in a beam emerging from an object point off the optic axis.

This aberration can be canceled by introducing a controlled astigmatism which reduces the axes of the ellipse to zero; this is the principle behind the *stigmator*, in which the first-order astigmatism of a weak quadrupole lens, or in practice, of a more complex element capable of producing a quadrupole potential distribution with any desired orientation, is introduced to counter the astigmatism of the round system. This astigmatism may be simply the geometrical aberration we have been discussing, or it may be a mechanical aberration of the round system.

Coma

$$u_i^{\mathrm{I}} = (1011)u_o u_a \bar{u}_a + (0120)\bar{u}_o u_a^2$$

In the Gaussian image plane, $(1011) = 2(\overline{0120})$, so that

$$u_i^{\mathrm{I}} = \{2(\overline{0120})u_o \bar{u}_a + (0120)\bar{u}_o u_a\}u_a$$
$$= r_o r_a^2 \mid 0120 \mid \{2 \exp i(\theta_o - \langle 0120 \rangle) + \exp i(2\theta_a - \theta_o + \langle 0120 \rangle)\}$$

so that

$$u_i^{\mathrm{I}} \exp(- i\varphi) = r_o r_a^2 \mid 0120 \mid (2 + \exp 2i\,\Theta_a)$$

in which

$$\Theta_a = \theta_a - \theta_o + \langle 0120 \rangle \qquad \varphi = \theta_o - \langle 0120 \rangle$$

This represents a family of circles, radii $r_o r_a{}^2 \,|\, 0120 \,|$, centered on the points $2r_o r_a{}^2 \,|\, 0120 \,|\, \exp i(\theta_o - \langle 0120\rangle)$. Their envelope is a pair of straight lines, inclined to one another at $60°$ (see Fig. 4).

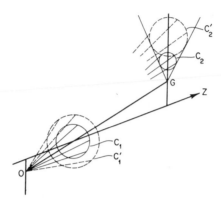

FIG. 4. The coma pattern produced by a beam emerging from an off-axial object point and passing through a circular stop. The pencil of rays that intersects the aperture plane around the circle C_1 produces the circle C_2 and likewise the pencil through C_1' intersects the image plane in C_2'; the complete circle C_2 (or C_2') is described if only a semi-circle of rays reaches C_1 (or C_1').

Distortion

$$u_i{}^{\mathrm{I}} = (2100)u_o{}^2 \bar{u}_o$$

This aberration is in a certain sense the counterpart of spherical aberration if the roles of the object and aperture planes are interchanged. [See Hawkes (1965a) for a fuller discussion of these relationships.] Its effect is not to blur the image at all—a point object produces a point image—but to render it unfaithful: the image point does not coincide with the Gaussian image point.

If we consider an annular object, $r_o = $ const, we find

$$u_i = (1000)u_o + (2100)u_o{}^2 \bar{u}_o$$
$$= u_o(\,|\, 1000 \,|\, + \,|\, 2100 \,|\, r_o{}^2 \exp i \,\langle 2100\rangle)$$

so that the circular Gaussian image is expanded or contracted into a circle of radius

$$r_o \,|\, (\,|\, 1000 \,|\, + \,|\, 2100 \,|\, r_o{}^2 \exp i \,\langle 2100\rangle) \,|$$

The effect of this aberration is more clearly demonstrated if we consider a rectangular object, formed from the straight lines $x_o = \pm X_o$, and $y_o = \pm Y_o$. Denoting the magnification, (1000), by M and the distortion (2100) by $D \exp i\tilde{D}$, we obtain

$$u_i = (M + Dr_o^2 \exp i\tilde{D})u_o$$

If the system is electrostatic, $\tilde{D}$ vanishes and $r_i \exp i\theta_i = (M + Dr_o^2)r_o\exp i\theta_o$ or

$$r_i = (M + Dr_o^2)r_o \qquad \theta_i = \theta_o$$

The distortion scales the image anisotropically, but homocentrically: the image of each object point is shifted radially from the Gaussian image point either outwards ($D > 0$) giving pin-cushion distortion (Fig. 5a) or inwards

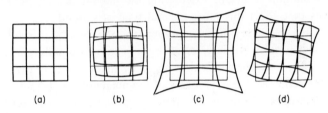

FIG. 5. (a) The object: a square grid; (b) barrel distortion; (c) pin-cushion distortion; (d) pocket-handkerchief distortion.

($D < 0$) giving barrel distortion (Fig. 5b). For magnetic systems, $\tilde{D}$ does not vanish, and

$$r_i = r_o[(M^2 + D^2r_o^4 + 2MDr_o^2 \cos \tilde{D})]^{1/2}$$

$$\theta_i = \theta_o + \tan^{-1}\left(\frac{Dr_o^2 \sin \tilde{D}}{M + Dr_o^2 \cos \tilde{D}}\right)$$

so that the image point is not merely moved in the radial direction, but is also rotated; the resultant distortion has been aptly styled "pocket-handkerchief distortion" by Sturrock (Fig. 5c).

Quadrupole Systems. Although we shall deal primarily with orthogonal systems, so designed that two perpendicular planes exist in which electrons travel without experiencing any expulsive force, we first briefly consider systems in which there is merely two-fold symmetry. By this, we mean that if the ray passing through the points u_o and u_a intersects the current plane at u_c, then the ray that intersects the object and aperture planes at $-u_o$ and

$- u_a$ (that is, $u_o \exp i\pi$ and $u_a \exp i\pi$) must intersect the current plane at $- u_c$:

$$- u_c = \Sigma \ (\alpha\beta\gamma\delta) u_o{}^\alpha \bar{u}_o{}^\beta u_a{}^\gamma \bar{u}_a{}^\delta (- 1)^{\alpha - \beta + \gamma - \delta}$$

None of the possible third-order aberrations is excluded, for $\alpha - \beta + \gamma - \delta$ has only to be odd. Physically the system will consist of quadrupoles, set in arbitrary orientations about a straight axis. Even the first-order properties are extremely complex [for a methodical analysis of these complex types of systems, see, for example, Carathéodory (1937) or Luneburg (1964)] and would, in general, render the system totally useless. However, any slight errors in the azimuthal alignment of a system designed to possess planes of symmetry would reduce it to a system of this kind, so that in calculating the mechanical aberrations of a system intended to have symmetry planes, we should in fact be calculating the magnitudes of the additional aberrations that affect this more general type of system.

If the system possesses planes of symmetry, the electrodes of all the electrostatic quadrupoles are intersected by a pair of mutually perpendicular planes, midway between which lie a second pair of mutually perpendicular planes intersecting all the pole-pieces of the magnetic quadrupoles (see Fig. 6). The electrodes of any electrostatic quadrupole may be excited asymmetrically and need not be equidistant from the axis, always provided that the symmetry is maintained. The presence of these symmetry planes considerably simplifies the aberrations, since all the coefficients $(pqrs)$ in

$$V_{\alpha\beta}^{(4)} = \sum_{p+q+r+s=4} (pqrs) u_o{}^p \bar{u}_o{}^q u_a{}^r \bar{u}_a{}^s$$

are real.

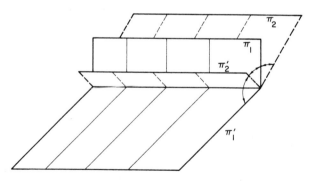

FIG. 6. The mutual orientations of the planes bisecting the electrodes and pole-pieces of the quadrupoles in a twist-free orthogonal system; if Π_1 and $\Pi_1{}'$ bisect the electrodes of all the electrostatic elements, then Π_2 and $\Pi_2{}'$ must bisect the pole-pieces of all the magnetic elements. (Each of the angles is equal to 45°.)

In orthogonal systems, it is simpler to derive the inter-relations between the aberration coefficients in terms of Cartesian coordinates than the complex numbers u_o and u_a. We have

$$
V_{a\beta}^{(4)} = \begin{bmatrix} x_o^2 \\ y_o^2 \\ x_a^2 \\ y_a^2 \\ x_o x_a \end{bmatrix}' \begin{bmatrix} 4000 & 2200 & 2020 & 2002 & 3010 & 2101 \\ 0 & 0400 & 0220 & 0202 & 1210 & 0301 \\ 0 & 0 & 0040 & 0022 & 1030 & 0121 \\ 0 & 0 & 0 & 0004 & 1012 & 0103 \\ 0 & 0 & 0 & 0 & 0 & 1111 \end{bmatrix} \begin{bmatrix} x_o^2 \\ y_o^2 \\ x_a^2 \\ y_a^2 \\ x_o x_a \\ y_o y_a \end{bmatrix}
\tag{3.6}
$$

so that the aberrations in a current plane, $z = z_c$, are given by

$$
\begin{aligned}
k_x x_c^{\mathrm{I}} = h_{xc}\{ & 4[4000]x_o^3 + 2[2200]x_o y_o^2 \\
& + [1030]x_a^3 + [1012]x_a y_a^2 \\
& + x_a(3[3010]x_o^2 + [1210]y_o^2) + 2[2101]x_o y_o y_a \\
& + 2x_o([2020]x_a^2 + [2002]y_a^2) + [1111]y_o x_a y_a\} \\
- g_{xc}\{ & (3010)x_o^3 + (1210)x_o y_o^2 \\
& + 4(0040)x_a^3 + 2(0022)x_a y_a^2 \\
& + 2x_a((2020)x_o^2 + (0220)y_o^2) + (1111)x_o y_o y_a \\
& + x_o(3(1030)x_a^2 + (1012)y_a^2) + 2(0121)y_o x_a y_a\}
\end{aligned}
\tag{3.7a}
$$

$$
\begin{aligned}
k_y y_c^{\mathrm{I}} = h_{yc}\{ & 2[2200]x_o^2 y_o + 4[0400]y_o^3 \\
& + [0121]x_a^2 y_a + [0103]y_a^3 \\
& + y_a([2101]x_o^2 + 3[0301]y_o^2) + 2[1210]x_o y_o x_a \\
& + 2y_o([0220]x_a^2 + [0202]y_a^2) + [1111]x_o x_a y_a\} \\
- g_{yc}\{ & (2101)x_o^2 y_o + (0301)y_o^3 \\
& + 2(0022)x_a^2 y_a + 4(0004)y_a^3 \\
& + 2y_a((2002)x_o^2 + (0202)y_o^2) + (1111)x_o y_o x_a \\
& + y_o((0121)x_a^2 + 3(0103)y_a^2) + 2(1012)x_o x_a y_a\}
\end{aligned}
\tag{3.7b}
$$

and writing

$$
\begin{aligned}
x_c^{\mathrm{I}} &= \sum_{\alpha+\beta+\gamma+\delta=3} (\alpha\beta\gamma\delta)x_o^\alpha y_o^\beta x_a^\gamma y_a^\delta \\
y_c^{\mathrm{I}} &= \sum_{\alpha+\beta+\gamma+\delta=3} (\alpha\beta\gamma\delta)x_o^\alpha y_o^\beta x_a^\gamma y_a^\delta
\end{aligned}
\tag{3.8}
$$

we find that the various types of aberration are given by the following expressions.

Distortions

$$x_c^I = (3000)x_o^3 + (1200)x_o y_o^2$$
$$y_c^I = (0300)y_o^3 + (2100)x_o^2 y_o$$

(3.9)

in which

$$(3000) = 4\frac{h_{xc}}{k_x}[4000] - \frac{g_{xc}}{k_x}(3010)$$

$$(1200) = 2\frac{h_{xc}}{k_x}[2200] - \frac{g_{xc}}{k_x}(1210)$$

$$(0300) = 4\frac{h_{yc}}{k_y}[0400] - \frac{g_{yc}}{k_y}(0301)$$

$$(2100) = 2\frac{h_{yc}}{k_y}[2200] - \frac{g_{yc}}{k_y}(2101)$$

(3.10)

Astigmatisms (terms of first degree in the aperture coordinates)

$$x_c^I = (2010)x_o^2 x_a + (0210)y_o^2 x_a + (1101)x_o y_o y_a$$
$$y_c^I = (2001)x_o^2 y_a + (0201)y_o^2 y_a + (1110)x_o y_o x_a$$

(3.11)

in which

$$(2010) = 3\frac{h_{xc}}{k_x}[3010] - 2\frac{g_{xc}}{k_x}(2020)$$

$$(0210) = \frac{h_{xc}}{k_x}[1210] - 2\frac{g_{xc}}{k_x}(0220)$$

$$(1101) = 2\frac{h_{xc}}{k_x}[2101] - \frac{g_{xc}}{k_x}(1111)$$

(3.12a)

and

$$(2001) = \frac{h_{yc}}{k_y}[2101] - 2\frac{g_{yc}}{k_y}(2002)$$

$$(0201) = 3\frac{h_{yc}}{k_y}[0301] - 2\frac{g_{yc}}{k_y}(0202)$$

$$(1110) = 2\frac{h_{yc}}{k_y}[1210] - \frac{g_{yc}}{k_y}(1111)$$

(3.12b)

Coma and Anticoma (terms of second degree in the aperture coordinates)

$$x_c^I = (1020)x_o x_a^2 + (1002)x_o y_a^2 + (0111)y_o x_a y_a$$
$$y_c^I = (0120)y_o x_a^2 + (0102)y_o y_a^2 + (1011)x_o x_a y_a$$

(3.13)

in which

$$(1020) = 2 \frac{h_{xc}}{k_x} [2020] - 3 \frac{g_{xc}}{k_x} (1030)$$

$$(1002) = 2 \frac{h_{xc}}{k_x} [2002] - \frac{g_{xc}}{k_x} (1012) \qquad (3.14a)$$

$$(0111) = \frac{h_{xc}}{k_x} [1111] - 2 \frac{g_{xc}}{k_x} (0121)$$

and

$$(0120) = 2 \frac{h_{yc}}{k_y} [0220] - \frac{g_{yc}}{k_y} (0121)$$

$$(0102) = 2 \frac{h_{yc}}{k_y} [0202] - 3 \frac{g_{yc}}{k_y} (0103) \qquad (3.14b)$$

$$(1011) = \frac{h_{yc}}{k_y} [1111] - 2 \frac{g_{yc}}{k_y} (1012)$$

Aperture Aberrations

$$x_c{}^I = (0030)x_a{}^3 + (0012)x_a y_a{}^2$$
$$y_c{}^I = (0003)y_a{}^3 + (0021)x_a{}^2 y_a \qquad (3.15)$$

in which

$$(0030) = \frac{h_{xc}}{k_x} [1030] - 4 \frac{g_{xc}}{k_x} (0040)$$

$$(0012) = \frac{h_{xc}}{k_x} [1012] - 2 \frac{g_{xc}}{k_x} (0022)$$

$$(3.16)$$

$$(0003) = \frac{h_{yc}}{k_y} [0103] - 4 \frac{g_{yc}}{k_y} (0004)$$

$$(0021) = \frac{h_{yc}}{k_y} [0121] - 2 \frac{g_{yc}}{k_y} (0022)$$

Aberration Figures. We now consider the effect of each of these aberrations in turn upon a beam emerging from a fixed object point (x_o, y_o), and bounded by a circular stop in the aperture plane, of radius r_a; the distortions, of course, need a separate treatment, being independent of the aperture coordinates.

Astigmatisms

$$x_c = g_{xc}x_o + h_{xc}x_a + \varkappa x_a + \lambda y_a$$
$$y_c = g_{yc}y_o + h_{yc}y_a + \mu y_a + \nu x_a$$

in which

$$\varkappa = (2010)x_o{}^2 + (0210)y_o{}^2; \qquad \lambda = (1101)x_oy_o$$
$$\mu = (2001)x_o{}^2 + (0201)y_o{}^2; \qquad \nu = (1110)x_oy_o$$

Thus

$$x_c - g_{xc}x_o = r_a\{(h_{xc} + \varkappa)\cos\theta_a + \lambda\sin\theta_a\}$$
$$\overline{y_c - g_{yc}y_o} = r_a\{(h_{yc} + \mu)\sin\theta_a + \nu\cos\theta_a\}$$

which represents an ellipse, inclined to the coordinate axes Ox_c, Oy_c in the current plane, with semiaxes

$$\tfrac{1}{2}\, r_a\{[(h_{xc} + h_{yc} + \varkappa + \mu)^2 + (\nu - \lambda)^2]^{1/2}$$
$$\pm\,[(h_{xc} - h_{yc} + \varkappa - \mu)^2 + (\nu + \lambda)^2]^{1/2}\}$$

centered on the point $(g_{xc}x_o, g_{yc}y_o)$. If the system produces a stigmatic, orthomorphic[4] image, in the plane $z = z_i$, the astigmatism will be given in this plane by

$$x_i - g_{xi}x_o = r_a(\varkappa\cos\theta_a + \lambda\sin\theta_a)$$
$$y_i - g_{yi}y_o = r_a(\mu\sin\theta_a + \lambda\cos\theta_a)$$

(since $\lambda = \nu$) and the semiaxes of the ellipse simplify to

$$\tfrac{1}{2}\, r_a(\varkappa + \mu \pm [(\varkappa - \mu)^2 + 4\lambda^2]^{1/2}$$

Coma and Anticoma

$$x_c = g_{xc}x_o + h_{xc}x_a + px_a{}^2 + ry_a{}^2 + sx_ay_a$$
$$y_c = g_{yc}y_o + h_{yc}y_a + qy_a{}^2 + Rx_ay_a + Sx_a{}^2$$

with $p = (1020)x_o$, $q = (0102)y_o$, $r = (1002)x_o$, $R = (1011)x_o$, $s = (0111)y_o$ and $S = (0120)y_o$.

In the image plane of a stigmatic orthomorphic system, $R = 2r$ and $s = 2S$.
In the general case,

$$x_c - g_{xc}x_o = h_{xc}x_a + r_a{}^2(p\cos^2\theta_a + r\sin^2\theta_a + s\cos\theta_a\sin\theta_a)$$
$$y_c - g_{yc}y_o = h_{yc}y_a + r_a{}^2(q\sin^2\theta_a + R\sin\theta_a\cos\theta_a + S\cos^2\theta_a)$$

and the possible shapes of the aberration figure vary very widely. Writing

---

[4] Free of first-order distortion: $h'_{xi} = h'_{yi}$.

$U_c = (x_c - g_{xc}x_o) + i(y_c - g_{yc}y_o)$, and considering only the image plane of systems free of first-order astigmatism, we find

$$U_c = \tfrac{1}{2} r_a{}^2 \left(\frac{p + r}{2} + i\, \frac{S + q}{2} \right)$$
$$+ r_a{}^2 \{ a \exp[i(2\theta_a + \tilde{a})] + \beta \exp[- i(2\theta_a - \tilde{\beta})] \}$$

in which

$$a \exp(i\tilde{a}) = \frac{p - r + R}{4} - i\, \frac{s - S + q}{4}$$

$$\beta \exp(i\tilde{\beta}) = \frac{p - r - R}{4} + i\, \frac{s + S - q}{4}$$

Writing $\varphi = - \tfrac{1}{2}(\tilde{a} + \tilde{\beta})$ and $\Theta_a = \theta_a + \tfrac{1}{4}(\tilde{a} - \tilde{\beta})$, we find

$$U_c \exp(- i\varphi) = \gamma r_a{}^2 + r_a{}^2 [a \exp(2i\Theta_a) + \beta \exp(- 2i\Theta_a]$$

where γ denotes

$$\frac{1}{2} \exp(- i\varphi) \left(\frac{p + r}{2} + i\, \frac{S + q}{2} \right)$$

which for varying values of r_a represents a family of ellipses, semiaxes $\tfrac{1}{2}(a \pm \beta)$, enveloped by a pair of straight lines. In Fig. 7, several examples of these aberration patterns, obtained by Burfoot (1954, 1956), are illustrated; Burfoot defines four types of coma, and the connection between his aberration coefficients and p, q, $R(= 2r)$ and $s(= 2S)$ is as follows:

Coma

$$\frac{p + 3r}{12x_o} + \frac{3S + q}{12y_o}$$

Conjugate Coma

$$\frac{p + 3r}{12x_o} - \frac{3S + q}{12y_o}$$

Elliptical Coma

$$\frac{p - 3r}{12x_o} - \frac{3S - q}{12y_o}$$

Conjugate Elliptical Coma

$$\frac{p - 3r}{12x_o} + \frac{3S - q}{12y_o}$$

Aperture Aberrations

$$x_c = x_o g_{xc} + x_a\{h_{xc} + (30)x_a{}^2 + (12)y_a{}^2\}$$

$$y_c = y_o g_{yc} + y_a\{h_{yc} + (03)y_a{}^2 + (21)x_a{}^2\}$$

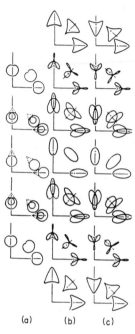

(a) (b) (c)

FIG. 7. Third-order comas. (a) Conjugate coma; (b) elliptical coma; (c) conjugate elliptical coma. The fourth row down shows the aberration figures in the Gaussian image plane, and the other rows, the same figures as they appear in other current planes. (From Burfoot, 1956.)

in which we have denoted $(00\gamma\delta)$ by $(\gamma\delta)$; in the image plane of a stigmatic orthomorphic system, $(12) = (21)$. In general, we have

$$x_c - x_o g_{xc} = x_a \left\{ h_{xc} + \frac{(30) + (12)}{2} r_a{}^2 + \frac{(30) - (12)}{2} r_a{}^2 \cos 2\theta_a \right\}$$

$$y_c - y_o g_{yc} = y_a \left\{ h_{yc} + \frac{(21) + (03)}{2} r_a{}^2 + \frac{(21) - (03)}{2} r_a{}^2 \cos 2\theta_a \right\}$$

or

$$x_c - x_o g_{xc} = \varkappa x_a + \lambda x_a \cos 2\theta_a$$

$$y_c - y_o g_{yc} = \mu y_a + \nu y_a \cos 2\theta_a$$

with

$$\varkappa = h_{xc} + \frac{(30) + (12)}{2} r_a{}^2 \qquad \mu = h_{yc} + \frac{(03) + (21)}{2} r_a{}^2$$

$$\lambda = \frac{(30) - (12)}{2} r_a{}^2 \qquad \nu = \frac{(21) - (03)}{2} r_a{}^2$$

Considering $\varkappa$ and μ alone, the aberration figure is an ellipse, semiaxes

$$r_a\varkappa, \qquad r_a\mu$$

For λ and ν alone, the image point describes a four-lobed rosette (Fig. 8). It has, however, become usual to divide the contributions to x_c, y_c into

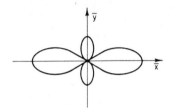

FIG. 8. The rosette aberration in a general plane corresponding to the aperture aberration coefficients λ and ν.

$\bar{x}$ denotes $(x_c - x_0 g_{xc})$ and $\bar{y}$ denotes $(y_c - y_0 g_{yc})$.

three different groups (when the system is stigmatic and orthomorphic). If we write

$$4\alpha = (30) + (03) + 2(12) \qquad 2\beta = (30) - (03) \qquad 4\gamma = (30) + (03) - 2(12)$$

we obtain

$$x_c - x_0 g_{xc} = x_a(\alpha r_a{}^2 + \beta x_a{}^2 + \gamma r_a{}^2 \cos 2\theta_a)$$
$$y_c - y_0 g_{yc} = y_a(\alpha r_a{}^2 - \beta y_a{}^2 - \gamma r_a{}^2 \cos 2\theta_a)$$

The coefficient α now characterizes an aberration identical with the familiar spherical aberration of round lenses; β is known as the "star" aberration coefficient and γ, as "rosette." [The reason for this choice of names (cf. Burfoot, 1954) can be seen from Fig. 9.]

Distortions

$$x_c = g_{xc}x_0 + h_{xc}x_a + x_0\{(30)x_0{}^2 + (12)y_0{}^2\}$$
$$y_c = g_{yc}y_0 + h_{yc}y_a + y_0\{(03)y_0{}^2 + (21)x_0{}^2\}$$

436 P. W. HAWKES

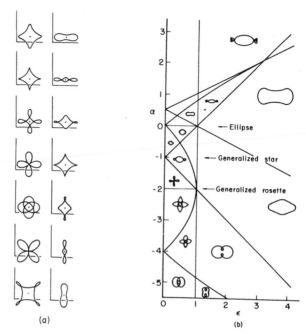

(a) (b)

FIG. 9. (a) Rosette aberrations (*left*) and star aberration (*right*). (Cf. Fig. 7.) (From Burfoot, 1956.) (b) Classification of the aperture aberration patterns. The aperture aberration patterns produced when the beam is restricted by an elliptical aperture have been considered by Meads. In scaled coordinates, $X = (1 + \varepsilon) \cos \theta + \alpha \sin^2\theta \cos \theta$, and $Y = (1 - \varepsilon) \sin \theta + \alpha \cos^2\theta \sin \theta$. This figure shows how the shape of the aberration pattern depends upon the values of the parameters α and ε. (After Fig. 7 of Meads, 1963.)

Since the distortion does not blur the image at all, we disregard the terms in x_a and y_a and consider only the effect of the contributions containing x_o and y_o; $(\alpha\beta)$ now denotes $(\alpha\beta00)$. We can analyze the effects of these distortions in various ways; Burfoot (1954) writes

$$x_c - g_{xc}x_o = x_o\{k_1(x_o{}^2+y_o{}^2)+k_2(x_o{}^2-y_o{}^2)+k_3(x_o{}^2-y_o{}^2)+k_4(x_o{}^2+y_o{}^2)\}$$
$$y_c - g_{yc}y_o = y_o\{k_1(x_o{}^2+y_o{}^2)-k_2(x_o{}^2-y_o{}^2)+k_3(x_o{}^2-y_o{}^2)-k_4(x_o{}^2+y_o{}^2)\}$$

and the coefficients k_i and $(\alpha\beta)$ are connected by the relations

$$k_1 = \frac{(30) + (03) + (12) + (21)}{4} \qquad k_2 = \frac{(30) + (03) - (12) - (21)}{4}$$

$$k_3 = \frac{(30) - (03) - (12) + (21)}{4} \qquad k_4 = \frac{(30) - (03) + (12) - (21)}{4}$$

The coefficient k_1 produces the familiar barrel or pin-cushion distortion; k_2 produces an aberration pattern similar in shape but undistorted along the lines $x = \pm y$; k_3 and k_4 produce the effects that Burfoot designates "hammock" distortions (Fig. 10). Alternatively, we can consider not a rectangular grid in the object plane, but an annulus or family of annuli.

$$x_c - g_{xc}x_0 = r_0{}^3 \cos\theta_0 \left\{ \frac{(30) + (12)}{2} + \frac{(30) - (12)}{2} \cos 2\theta_0 \right\}$$

$$y_c - g_{yc}x_0 = r_0{}^3 \sin\theta_0 \left\{ \frac{(21) + (03)}{2} + \frac{(21) - (03)}{2} \cos 2\theta_0 \right\}$$

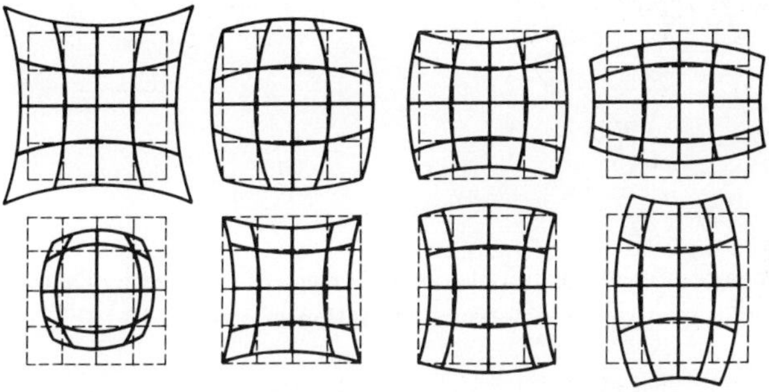

FIG. 10. Distortions in quadrupole systems. (From Burfoot, 1956.)

If we write $2\varkappa = (30) + (12)$, $2\lambda = (21) + (03)$, $2\mu = (30) - (12)$ and $2\nu = (21) - (03)$, and consider first the effects of $\varkappa$ and λ alone, we find

$$x_c = x_0(g_{xc} + \varkappa r_0{}^2) \qquad y_c = y_0(g_{yc} + \lambda r_0{}^2)$$

so that the distortion converts the object annulus into an ellipse. For μ and ν, we have

$$x_c = x_0(g_{xc} + \mu r_0{}^2 \cos 2\theta_0) \qquad y_c = y_0(g_{yc} + \nu r_0{}^2 \cos 2\theta_0)$$

and the image of a circle will be identical to the trace produced in a current plane by an astigmatic beam affected by aperture aberration.[5]

Octopole Systems. A symmetrically excited octopole has no effect upon the Gaussian imagery of a system containing round lenses and octopoles (or quadrupoles and octopoles). Since octopoles are essential compo-

---

[5] For further discussion of aberrations patterns, see Hawkes (1965b) where many references to earlier work are listed; a convenient method of displaying these patterns is described in Amboss (1959) and Hawkes and Cosslett (1962).

nents of a system designed to be corrected for spherical aberration by abandoning rotational symmetry (Fig. 11), it is convenient to consider their aberrations separately, and we therefore consider the aberrations of a system consisting of round lenses and octopoles. If a ray which passes through u_o

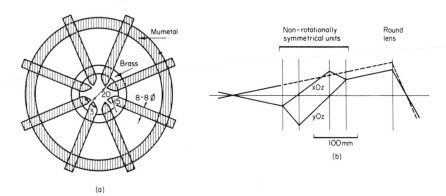

FIG. 11. The combined quadrupole-octopole element (a) used by Deltrap (1964a,b) in a recent attempt to correct the aperture aberration of a round lens (b).

and u_a intersects a current plane at u_c, then clearly the ray which intersects the object plane at $u_o \exp(i\pi/2)$ and the aperture plane at $u_a \exp(i\pi/2)$ must intersect the current plane at $u_c \exp(i\pi/2)$. With $u_c = \Sigma \ (\alpha\beta\gamma\delta)u_o{}^\alpha \bar{u}_o{}^\beta u_a{}^\gamma \bar{u}_a{}^\delta$, we find $\alpha - \beta + \gamma - \delta = 1$ or -3; thus

$$u_c = (1000)u_o + (0010)u_a$$
$$+ \text{ round lens aberrations} \tag{3.17}$$
$$+ (0300)\bar{u}_o{}^3 + (0201)\bar{u}_o{}^2\bar{u}_a + (0102)\bar{u}_o\bar{u}_a{}^2 + (0003)\bar{u}_a{}^3$$

We consider only the aperture aberration in detail; the others can be analyzed very straightforwardly by the usual methods. If the octopole is electrostatic, (0003) will be real if the axes intersect the electrodes and if the remainder of the system is electrostatic; we have

$$x_c{}^{\text{I}} = (0003)x_a{}^3 - 3(0003)x_a y_a{}^2 = x_a(0003) \ (x_a{}^2 - 3y_a{}^2)$$
$$y_c{}^{\text{I}} = (0003)y_a{}^3 - 3(0003)x_a{}^2 y_a = y_a(0003) \ (y_a{}^2 - 3x_a{}^2)$$
$$\tag{3.18}$$

If a system does not possess planes of symmetry, either because magnetic lenses are present or because the different octopoles are not aligned in the same meridian plane, (0003) will not be real, and additional aberrations will appear, of the form

$$x_c^{\mathrm{I}} = - y_a \, \mathrm{Im}(0003) \, (y_a^2 - 3x_a^2)$$
$$y_c^{\mathrm{I}} = x_a \, \mathrm{Im}(0003) \, (x_a^2 - 3y_a^2)$$

$$(3.19)$$

If octopoles and quadrupoles are present together, we cannot deduce any information about the aberrations of the former by these techniques, since the system as a whole possesses quadrupole symmetry.

The aperture aberrations of a system consisting of round electrostatic lenses and octopoles (magnetic or electrostatic) are thus of the form

$$x_i^{\mathrm{I}} = \alpha x_a r_a^2 + \beta x_a r_a^2 - 4\beta x_a y_a^2$$
$$y_i^{\mathrm{I}} = \alpha y_a r_a^2 + \beta y_a r_a^2 - 4\beta x_a^2 y_a$$

in which α denotes the spherical aberration coefficient of the round lens system, and $\beta = (0003)$. From this, we see immediately that octopoles alone cannot be used to annul the spherical aberration of a round lens, for if $\beta = - \alpha$, there will be a residual aberration of the form

$$x_i^{\mathrm{I}} = - 2\beta r_a^3 \sin 2\theta_a \cos \theta_a$$
$$y_i^{\mathrm{I}} = - 2\beta r_a^3 \sin 2\theta_a \sin \theta_a$$

which represents a rosette (Fig. 12). Since the maximum length of each lobe

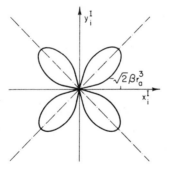

FIG. 12. The rosette aperture aberration which remains when octopoles are incorporated into a round system.

is $2\alpha r_a^3$, we might say that the octopole cancels the spherical aberration in the directions Ox_i, Oy_i, and doubles it in the directions at $45°$ to these axes.

2.5.4. The Aberration Coefficients

The notation used by Glaser (1956, p. 162 ff.) in his essay on *Elektronen-und Ionenoptik* in the *Handbuch der Physik* is a most convenient one, and the symbols employed below for the most part correspond to his usage.[6] Let z be the optic axis of a round or quadrupole system; (X, Y) are the off-axial Cartesian coordinates of the point which has cylindrical polar coordinates (r, ψ). Expanding the electrostatic potential $\varphi(X, Y, z)$ as a power series in X and Y, and applying Laplace's equation, $\nabla^2\varphi = 0$, we find that

$$\varphi(X, Y, z) = \Phi - \tfrac{1}{4}\Phi''(X^2 + Y^2) + \tfrac{1}{64}\Phi^{(iv)}(X^2 + Y^2)^2 - \cdots$$
$$+ \tfrac{1}{4}D(X^2 - Y^2) - \tfrac{1}{48}D''(X^2 + Y^2)(X^2 - Y^2) + \cdots$$
$$+ PXY - \tfrac{1}{12}P''(X^2 + Y^2)XY + \cdots \qquad (4.1)$$
$$+ D_1(X^4 - 6X^2Y^2 + Y^4) - \cdots$$
$$+ 4P_1XY(X^2 - Y^2) - \cdots$$

or in cylindrical polars

$$\varphi(r, \psi, z) = \Phi - \tfrac{1}{4}\Phi''r^2 + \tfrac{1}{64}\Phi^{(iv)}r^4$$
$$+ (\tfrac{1}{4}Dr^2 - \tfrac{1}{48}D''r^4)\cos 2\psi$$
$$+ (\tfrac{1}{2}Pr^2 - \tfrac{1}{24}P''r^4)\sin 2\psi \qquad (4.2)$$
$$+ D_1r^4\cos 4\psi + P_1r^4\sin 4\psi$$

In these expressions, Φ, D, P, D_1, and P_1 are functions of z only; dashes thus signify differentiation with respect to z.

An analogous expression gives the magnetic scalar potential $\varphi_m(X, Y, z)$. Thus

$$\varphi_m(X, Y, z) = \Phi_m - \tfrac{1}{4}\Phi_m''(X^2 + Y^2) + \tfrac{1}{64}\Phi_m^{(iv)}(X^2 + Y^2)^2$$
$$+ \tfrac{1}{4}\Delta(X^2 - Y^2) - \tfrac{1}{48}\Delta''(X^2 + Y^2)(X^2 - Y^2)$$
$$+ QXY - \tfrac{1}{12}Q''(X^2 + Y^2)XY \qquad (4.3)$$
$$+ \Delta_1(X^4 - 6X^2Y^2 + Y^4)$$
$$+ 4Q_1XY(X^2 - Y^2)$$

and we shall need the components of the magnetic vector potential

---

[6] The aberrations of quadrupole systems were first calculated by A. Melkich, a pupil of J. Picht, who employs an alternative notation; the relations between these different notations are set out in the appendix to Hawkes (1965b).

$A(X, Y, z)$. Writing $\Omega = - d\Phi_m/dz$, we find

$$A_X = -\Omega Y + \tfrac{1}{4}\Omega'' Y(X^2 + \tfrac{1}{3}Y^2)$$
$$+ \tfrac{1}{4}\Delta' Y(X^2 - \tfrac{1}{3}Y^2) + \tfrac{1}{2}Q'XY^2$$

$$A_Y = 0$$

$$A_z = -\tfrac{1}{2}\Omega'XY - \tfrac{1}{2}\Delta XY + \tfrac{1}{2}Q(X^2 - Y^2)$$
$$+ \tfrac{1}{16}\Omega'''XY(X^2 + \tfrac{1}{3}Y^2) + \tfrac{1}{12}\Delta''X^3 Y$$
$$- \tfrac{1}{48}Q''(X^4 - 6X^2Y^2 - Y^4)$$
$$- 4\Delta_1 XY(X^2 - Y^2)$$
$$+ Q_1(X^4 - 6X^2Y^2 + Y^4)$$

$$(4.4)$$

The functions Ω, Δ, Q, Δ_1, and Q_1 are functions of z alone. (Glaser, 1956, denotes Ω by B_z.)

Each of the functions Φ, D, P, D_1 and P_1 which appear in $\Phi(X, Y, z)$ and Ω, Δ, Q, Δ_1 and Q_1 which figure in $\Phi_m(X, Y, z)$ corresponds to the potential distribution of a particular type of electron optical lens or corrector element.[7] In a rotationally symmetrical electrostatic system, the potential must be independent of the angle, ψ, so that the function $\Phi(z)$ is sufficient to characterize the potential distribution throughout the system; setting $X = Y = 0$, we see that

$$\varphi(0,0,z) = \Phi(z)$$

so that $\Phi(z)$ represents the potential distribution along the z axis. Likewise, in a magnetic system, $\Omega(z)$ fully characterizes the potential distribution, and since

$$\Omega(z) = -\frac{d\Phi_m}{dz} = -\left(\frac{\partial \varphi_m}{\partial z}\right)_{X=Y=0}$$

we see immediately that $\Omega(z)$ represents the distribution of magnetic induction along the z axis.

---

[7] Many authors, particularly those of the French school, employ normalized functions instead of $\Phi(z)$, $D(z)\ldots$; writing $D(z) = D_{max}k(z)$, for example, all the analysis can be performed in terms of the function $k(z)$, and the quadrupole strengths appear only as multipliers. This is very convenient if a system consists of several lens or corrector elements, all producing fields of different strengths but substantially the same shape.

In another nomenclature that is common, all the terms in 2ψ are collectively called the φ_2 (or Φ_2) terms, all those in 4ψ the φ_4 (or Φ_4) terms and all the terms independent of ψ the φ_0 (or Φ_0) terms; we thus speak of φ_0-elements, φ_2-elements, and φ_4-elements and this is convenient when lenses producing round, quadrupole and octopole fields simultaneously are being employed (for example, an asymmetrically excited quadrupole).

The functions $D(z)$ and $P(z)$ describe the potential distribution within symmetrical electrostatic quadrupoles, and $\Delta(z)$ and $Q(z)$, the potential within symmetrical magnetic quadrupoles. Which of the functions will be necessary in any given situation depends upon the orientations of the quadrupoles with respect to the coordinate axes. Consider first the electrostatic case. In a symmetrical quadrupole, two of the electrodes are held at a po-

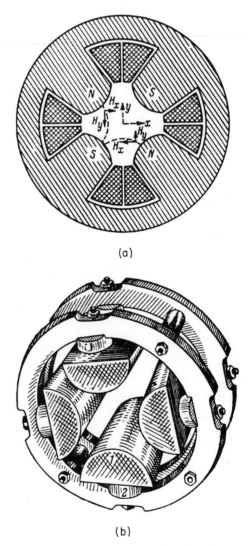

(a)

(b)

FIG. 13. Typical electrostatic and magnetic quadrupoles. (From V. M. Kel'man and S. Ya. Yavor, "Электронная Оптика" Moscow and Leningrad, 1963.)

tential $+ V$, say, and two at a potential $- V$; there is thus electrical symmetry about the two planes Π_1, Π_2 through the electrodes. Furthermore, the quadrupole must be geometrically symmetrical, so that in transverse section, it possesses four planes of mirror symmetry, and in longitudinal section, it is symmetrical about its midplane (Fig. 13). It is obvious that only odd multiples of 2ψ can appear in the potential function φ, and hence P and D are sufficient fully to describe the potential (if the expansion is halted at terms in r^4). If the planes Π_1 and Π_2 contain the coordinate axes, OX and OY, then $D(z)$ will be adequate to describe the potential throughout the quadrupole; if the coordinate axes lie in planes at $45°$ to Π_1 and Π_2, $P(z)$ will give a complete description. In any other orientation, both P and D will be necessary. Thus, if a system contains several electrostatic quadrupoles arranged so that all their electrodes lie in the same pair of planes, which we also select to define the directions of the X and Y axes, the function $D(z)$ will define the potential throughout the whole system. Magnetic quadrupoles can be analyzed in the same way, and we find that if the planes through the pole-pieces also contain the axes OX and OY, the potential is characterized by $\Delta(z)$, and if the axes lie midway between these planes, the potential is fully characterized by $Q(z)$.

In φ and φ_m, the functions P_1, Δ_1, Q_1, and D_1 appear only in combination with terms of fourth degree in X and Y; in polar coordinates, they are modulated by $\cos 4\psi$ or $\sin 4\psi$, and they describe the potential distribution within symmetrical octopoles (Fig. 14). For electrostatic octopoles, the distinction between $D_1(z)$ and $P_1(z)$ is similar to that between $D(z)$ and $P(z)$: if the coordinate axes pass through the electrodes, $D_1(z)$ fully describes the potential, if they pass between them, $P_1(z)$ alone is adequate. A corresponding result is true of $\Delta_1(z)$ and $Q_1(z)$ for magnetic quadrupoles.

The functions $\Phi(z)$, $D(z)$, $\ldots$, $\Delta_1(z)$, $Q_1(z)$ can be obtained either by measurement or by computation. Furthermore, the results of either procedure may indicate that some analytical model is a satisfactorily close approximation to reality; this is particularly convenient when the aberrations are being calculated, as the derivatives of the functions are required. We must now consider what combinations of these functions will produce orthogonal systems, and how the "components" of m, namely $m^{(0)}$, $m^{(2)}$ and $m^{(4)}$, are related to them.

We have

$$m = [\varphi(1 + \varepsilon\varphi)(1 + X'^2 + Y'^2)]^{1/2} - \eta(A_X X' + A_Y Y' + A_z)$$

in which η denotes $(e/2m_0)^{1/2}$ and $\varepsilon = e/2m_0 c^2$. Substituting the expressions

444   P. W. HAWKES

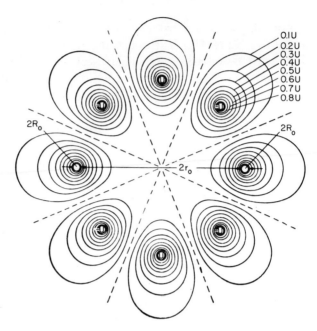

FIG. 14. The two-dimensional potential distribution within an octopole consisting of eight (long) circular cylinders. (From A. M. Strashkevich, "Электронная Оптика Электростатических Полей не обладающих Осевой Симметрией" Moscow, 1959.)

for $\varphi(X, Y, z)$ and $\mathbf{A}(X, Y, z)$, and using Sturrock's (1955, p. 64) partial integration rule[8], we find

$$m^{(0)} = [\Phi(1 + \varepsilon\Phi)]^{1/2} \tag{4.5}$$

$$m^{(2)} = -\frac{1 + 2\varepsilon\Phi}{8[\Phi(1 + \varepsilon\Phi)]^{1/2}}\, \Phi''(X^2 + Y^2)$$

$$+ \left[\frac{1 + 2\varepsilon\Phi}{8[\Phi(1 + \varepsilon\Phi)]^{1/2}}\, D - \frac{1}{2}\eta Q\right](X^2 - Y^2)$$

$$+ \left[\frac{1 + 2\varepsilon\Phi}{2[\Phi(1 + \varepsilon\Phi)]^{1/2}}\, P + \frac{1}{2}\eta\Delta\right]XY \tag{4.6}$$

$$+ \tfrac{1}{2}[\Phi(1 + \varepsilon\Phi)]^{1/2}\, (X'^2 + Y'^2)$$

$$- \tfrac{1}{2}\eta\Omega\, (XY' - X'Y)$$

[8] "If the variational function [m] may be expressed as a sum of functions, one of which is of the form $df/dz \cdot g$, where f and g are functions of x_i [namely, X and Y] and z only, then this term may be replaced by $- f dg/dz$."

and

$$
\begin{aligned}
m^{(4)} = \; & F(X^2 + Y^2)^2 + G(X^2 - Y^2)^2 \\
& + HXY(X^2 + Y^2) + IXY(X^2 - Y^2) \\
& + J(X^4 - Y^4) + K(X'^2 + Y'^2)(X^2 + Y^2) \\
& + L(X'^2 + Y'^2)(X^2 - Y^2) + M(X'^2 + Y'^2)XY \\
& + N(X'^2 + Y'^2)^2 + RXY(XY' - X'Y) \\
& + SXY \frac{d}{dz}(X^2 - Y^2) + TXY \frac{d}{dz}(X^2 + Y^2)
\end{aligned}
\tag{4.7}
$$

in which

$$
\begin{aligned}
F = \; & \frac{1}{128} \frac{1 + 2\varepsilon\Phi}{[\Phi(1 + \varepsilon\Phi)]^{1/2}} \Phi^{(iv)} - \frac{1}{128} \frac{(\Phi'')^2}{[\Phi(1 + \varepsilon\Phi)]^{3/2}} - \frac{1 + 2\varepsilon\Phi}{2[\Phi(1+\varepsilon\Phi)]^{1/2}} D_1 \\
& - \frac{1}{32} \frac{P^2}{[\Phi(1 + \varepsilon\Phi)]^{3/2}} + \eta Q_1
\end{aligned}
$$

$$
G = \frac{1 + 2\varepsilon\Phi}{[\Phi(1 + \varepsilon\Phi)]^{1/2}} D_1 - \frac{1}{128} \frac{D^2}{[\Phi(1 + \varepsilon\Phi)]^{3/2}} + \frac{1}{32} \frac{P^2}{[\Phi(1 + \varepsilon\Phi)]^{3/2}} - 2\eta Q_1
$$

$$
H = -\frac{1}{24} \frac{1 + 2\varepsilon\Phi}{[\Phi(1 + \varepsilon\Phi)]^{1/2}} P'' + \frac{1}{16} \frac{P\Phi''}{[\Phi(1 + \varepsilon\Phi)]^{3/2}} - \frac{1}{12} \eta\Delta''
$$

$$
I = \frac{2(1 + 2\varepsilon\Phi)}{[\Phi(1 + \varepsilon\Phi)]^{1/2}} P_1 - \frac{1}{16} \frac{DP}{[\Phi(1 + \varepsilon\Phi)]^{3/2}} - \frac{1}{16} \eta\Omega''' + 4\eta\Delta_1
$$

$$
J = -\frac{1}{96} \frac{1 + 2\varepsilon\Phi}{[\Phi(1 + \varepsilon\Phi)]^{1/2}} D'' + \frac{1}{64} \frac{D\Phi''}{[\Phi(1 + \varepsilon\Phi)]^{3/2}} + \frac{1}{48} \eta Q''
$$

$$
K = -\frac{1}{16} \frac{1 + 2\varepsilon\Phi}{[\Phi(1 + \varepsilon\Phi)]^{1/2}} \Phi''
\tag{4.8}
$$

$$
L = \frac{1}{16} \frac{1 + 2\varepsilon\Phi}{[\Phi(1 + \varepsilon\Phi)]^{1/2}} D
$$

$$
M = \frac{1}{4} \frac{1 + 2\varepsilon\Phi}{[\Phi(1 + \varepsilon\Phi)]^{1/2}} P
$$

$$
N = -\frac{1}{8} [\Phi(1 + \varepsilon\Phi)]^{1/2}
$$

$$
R = \frac{1}{4} \eta Q'
$$

$$
S = -\frac{1}{8} \eta\Omega''
$$

$$
T = -\frac{1}{8} \eta\Delta'
$$

From Eq. (4.6), we see that $\bar{p} = \partial m^{(2)}/\partial X'$ and $\bar{q} = \partial m^{(2)}/\partial Y'$ are of the form

$$\bar{p} = [\Phi(1 + \varepsilon\Phi)]^{1/2}X' + \tfrac{1}{2}\eta\Omega Y$$
$$\bar{q} = [\Phi(1 + \varepsilon\Phi)]^{1/2}Y' - \tfrac{1}{2}\eta\Omega X$$

but if we introduce a rotating coordinate system (Fig. 15), we can eliminate the terms in X and Y. We have

$$X = r \cos \psi \qquad Y = r \sin \psi$$

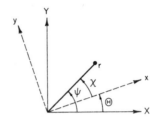

FIG. 15. The connection between the stationary coordinate system (X, Y, z) and the rotating system (x, y, z); Θ is a function of z.

and writing

$$x = r \cos \chi \qquad y = r \sin \chi \qquad \psi = \chi + \Theta$$

we find (Glaser, 1956, p. 176) that if

$$\frac{d\Theta}{dz} = \frac{\eta}{2} \frac{\Omega}{[\Phi(1 + \varepsilon\Phi)]^{1/2}} \qquad (4.9)$$

the expression for $m^{(2)}$ becomes

$$
\begin{aligned}
m^{(2)} = {} & \left\{ -\frac{1}{8} \frac{1 + 2\varepsilon\Phi}{[\Phi(1 + \varepsilon\Phi)]^{1/2}} \Phi'' - \frac{\eta^2}{8} \frac{\Omega^2}{[\Phi(1 + \varepsilon\Phi)]^{1/2}} \right\} (x^2 + y^2) \\
& + \left\{ \left(\frac{1}{8} \frac{1 + 2\varepsilon\Phi}{[\Phi(1 + \varepsilon\Phi)]^{1/2}} D - \frac{1}{2}\eta Q \right) \cos 2\Theta \right. \\
& \left. + \left(\frac{1}{4} \frac{1 + 2\varepsilon\Phi}{[\Phi(1 + \varepsilon\Phi)]^{1/2}} P + \frac{1}{4}\eta\varDelta \right) \sin 2\Theta \right\} (x^2 - y^2) \\
& + \left\{ -\left(\frac{1}{4} \frac{1 + 2\varepsilon\Phi}{[\Phi(1 + \varepsilon\Phi)]^{1/2}} D - \eta Q \right) \sin 2\Theta \right. \\
& \left. + \left(\frac{1}{2} \frac{1 + 2\varepsilon\Phi}{[\Phi(1 + \varepsilon\Phi)]^{1/2}} P + \frac{1}{2}\eta\varDelta \right) \cos 2\Theta \right\} xy \\
& + \frac{1}{2} [\Phi(1 + \varepsilon\Phi)]^{1/2} (x'^2 + y'^2)
\end{aligned}
\qquad (4.10)
$$

The quantities $p = \partial m^{(2)}/\partial x'$ and $q = \partial m^{(2)}/\partial y'$ are now in the simple form

$$p = [\Phi(1 + \varepsilon\Phi)]^{1/2}x' \qquad q = [\Phi(1 + \varepsilon\Phi)]^{1/2}y' \qquad (4.11)$$

The differential equations of motion, $p' = \partial m^{(2)}/\partial x$, $q' = \partial m^{(2)}/\partial y$ will only be separated into a differential equation for x (from which y is absent) and a differential equation for y (from which x is absent) if the term in xy in $m^{(2)}$ vanishes. A system that possesses this property is called an *orthogonal system*[9]; from Eq. (4.10), we can see that the coefficient of xy vanishes if

$$\tan 2\Theta = \frac{\left(\dfrac{1 + 2\varepsilon\Phi}{[\Phi(1 + \varepsilon\Phi)]^{1/2}} P + \eta\Delta \right)}{\left(\dfrac{1 + 2\varepsilon\Phi}{2[\Phi(1 + \varepsilon\Phi)]^{1/2}} D - 2\eta Q \right)} \qquad (4.12)$$

Equations (4.12) and (4.9) together comprise the general *orthogonality condition*.

In practice, the orthogonality condition can be satisfied in one way only. Unless $\Theta(z)$ is effectively constant over the region occupied by the quadrupole fields, a complicated balance of electric and magnetic quadrupole field distributions has to be created and maintained; problems of alignment and exact field-shaping make this virtually impossible. Practical systems therefore either do not contain magnetic round lenses or if they do, their fields do not overlap the quadrupole fields. If $\Theta(z)$ is constant throughout the quadrupole region, but not equal to 0 or $\pi/4$, a balance of electric and magnetic fields is again necessary, and in practical systems, the quadrupoles are always disposed about the axis in such a way that either $P(z) = \Delta(z) \equiv 0$ while $D(z)$ or $Q(z)$ or both do not vanish ($\Theta = 0$) or, alternatively, $D(z) = Q(z) \equiv 0$ while $P(z)$, $\Delta(z)$ or both are finite ($\Theta = \pi/4$). Physically, all this has a very simple meaning.[10] Since the equations of motion separate, it will be possible for electrons to follow trajectories $x = x(z)$, $y = 0$ and $y = y(z)$, $x = 0$: there exists a pair of mutually perpendicular surfaces from which electrons will not be expelled once they are travelling over them. If $\Theta(z)$ is constant, these surfaces degenerate into planes, and clearly all the electrostatic quadrupoles must be aligned in such a way that these planes

[9] The necessary conditions for electron optical systems to be orthogonal were first deduced by Cotte (1938).

[10] The different types of orthogonal system are listed by Dušek (1959; extracted from his 1958 dissertation). Dušek also discusses the first-order imagery of rectilinear orthogonal systems extremely thoroughly in general terms. Systems that do not satisfy the orthogonality condition are analyzed by Rose (1966/7).

intersect their electrodes, while all the magnetic quadrupoles must be arranged so that their pole-pieces lie midway between the planes (Fig. 6). If the coordinate axes lie in these planes, $P(z)$ and $\Delta(z)$ vanish everywhere; if they are inclined to them at 45°, $D(z)$ and $Q(z)$ vanish. If a round magnetic field is present, either in front of or behind all the quadrupoles, these remarks remain valid; the orthogonal planes become curved surfaces within the magnetic lens field, but elsewhere remain planar. If the magnetic lens is placed between two quadrupoles or groups of quadrupoles, however, the quadrupole orientations will be different on either side, and the difference in azimuth will be given by

$$\Delta\Theta = \frac{\eta}{2} \int \frac{\Omega}{[\Phi(1 + \varepsilon\Phi)]^{1/2}} \, dz$$

Henceforward, we shall discuss principally "twist-free" systems (Θ = constant), and for convenience, we select the axes to give $\Theta = 0$. The function $m^{(4)}$ is of the form

$$m^{(4)}(x, y, z) =$$

$$\left\{ \frac{1 + 2\varepsilon\Phi}{128[\Phi(1 + \varepsilon\Phi)]^{1/2}} \Phi^{(iv)} - \frac{(\Phi'')^2}{128[\Phi(1 + \varepsilon\Phi)]^{3/2}} \right.$$

$$\left. - \frac{1 + 2\varepsilon\Phi}{2[\Phi(1 + \varepsilon\Phi)]^{1/2}} D_1 + \eta Q_1 \right\}(x^2 + y^2)^2$$

$$+ \left\{ - \frac{D^2}{128[\Phi(1 + \varepsilon\Phi)]^{3/2}} + \frac{1 + 2\varepsilon\Phi}{[\Phi(1 + \varepsilon\Phi)]^{1/2}} D_1 - 2\eta Q_1 \right\}(x^2 - y^2)^2$$

$$+ \left\{ \frac{1 + 2\varepsilon\Phi}{[\Phi(1 + \varepsilon\Phi)]^{1/2}} P_1 + 2\eta\Delta_1 \right\} 2xy(x^2 + y^2)$$

$$+ \left\{ - \frac{1 + 2\varepsilon\Phi}{96[\Phi(1 + \varepsilon\Phi)]^{1/2}} D'' + \frac{D\Phi''}{64[\Phi(1 + \varepsilon\Phi)]^{3/2}} + \frac{1}{48}\eta Q'' \right\}(x^4 - y^4)$$

$$- \frac{1 + 2c\Phi}{16[\Phi(1 + \varepsilon\Phi)]^{1/2}} \Phi''(x^2 + y^2)(x'^2 + y'^2)$$

$$+ \frac{1 + 2\varepsilon\Phi}{16[\Phi(1 + \varepsilon\Phi)]^{1/2}} D(x^2 - y^2)(x'^2 + y'^2)$$

$$- \frac{1}{8}[\Phi(1 + \varepsilon\Phi)]^{1/2}(x'^2 + y'^2)^2$$

$$+ \frac{1}{4}\eta Q xy(xy' - x'y) \tag{4.13}$$

Although the quadrupole potentials are now fully characterized by $D(z)$ and $Q(z)$, all four octopole functions, $D_1(z)$, $Q_1(z)$, $\Delta_1(z)$, and $P_1(z)$, are still present: the orthogonality condition restricts only the Gaussian imagery of the lenses, whereas octopoles affect only their aberrations. They can therefore be set in any orientation, but as we have seen, the number of aberrations increases if $\Delta_1(z)$ and $P_1(z)$ do not vanish.

It is only rarely that we are interested in the aberrations of round systems other than in the vicinity of the Gaussian image plane. In quadrupole systems, however, we may wish to exploit the first-order astigmatism and use one of the line foci for some purpose (Le Poole, 1964; Bok et al., 1964); the nature of the aberrations of stigmatic and astigmatic systems must hence be carefully distinguished. First, therefore, we shall analyze the general case, and subsequently discuss the various special cases of practical importance.

The equations of motion (2.21) are now

$$[\Phi(1 + \varepsilon\Phi)]^{1/2}\, x'' + \frac{1}{2}\frac{(1 + 2\varepsilon\Phi)}{[\Phi(1 + \varepsilon\Phi)]^{1/2}}\,\Phi'x'$$
$$+ \frac{1}{4}\left\{\frac{(1 + 2\varepsilon\Phi)}{[\Phi(1 + \varepsilon\Phi)]^{1/2}}\,(\Phi'' - D) + 4\eta Q\right\} x = 0 \tag{4.14a}$$

$$[\Phi(1 + \varepsilon\Phi)]^{1/2}\, y'' + \frac{1}{2}\frac{(1 + 2\varepsilon\Phi)}{[\Phi(1 + \varepsilon\Phi)]^{1/2}}\,\Phi'y'$$
$$+ \frac{1}{4}\left\{\frac{(1 + 2\varepsilon\Phi)}{[\Phi(1 + \varepsilon\Phi)]^{1/2}}\,(\Phi'' + D) - 4\eta Q\right\} y = 0 \tag{4.14b}$$

Only in the absence of any rotationally symmetrical electrostatic fields ($\Phi = $ constant) do these equations simplify into the form

$$x'' - \beta(z)x = 0 \tag{4.15a}$$

$$y'' + \beta(z)y = 0 \tag{4.15b}$$

with

$$\beta(z) = \frac{(1 + 2\varepsilon\Phi)D - 4\eta Q[\Phi(1 + \varepsilon\Phi)]^{1/2}}{4\Phi(1 + \varepsilon\Phi)} \tag{4.16}$$

The general paraxial solutions

$$x(z) = x_o g_x(z) + x_a h_x(z)$$
$$y(z) = y_o g_y(z) + y_a h_y(z)$$

now satisfy Eqs. (4.14a) and (4.14b), respectively. Substituting these solutions into the expression for $m^{(4)}$ (Eq. 4.13), we obtain

$$
\begin{aligned}
m^{(4)} =\ & \bar{a}x_o^4 + \bar{b}y_o^4 + \bar{c}x_a^4 + \bar{d}y_a^4 \\
& + \bar{e}x_o^2y_o^2 + \bar{f}x_a^2y_a^2 + \bar{g}x_o^2y_a^2 + \bar{h}x_a^2y_o^2 + \bar{j}x_o^2x_a^2 + \bar{k}y_o^2y_a^2 \\
& + \bar{l}x_o^3x_a + \bar{m}y_o^3y_a + \bar{n}x_ox_a^3 + \bar{p}y_oy_a^3 + \bar{q}x_o^3y_a + \bar{r}x_o^3y_o \\
& + \bar{s}y_o^3x_a + \bar{t}y_o^3x_o + \bar{u}y_ox_a^3 + \bar{v}y_ax_a^3 + \bar{w}x_oy_a^3 + \bar{z}x_ay_a^3 \qquad (4.17) \\
& + \bar{\alpha}x_o^2x_ay_o + \bar{\beta}x_o^2x_ay_a + \bar{\gamma}x_o^2y_oy_a + \bar{\delta}x_a^2x_oy_o \\
& + \bar{\xi}x_a^2x_oy_a + \bar{\zeta}x_a^2y_oy_a + \bar{\lambda}y_o^2x_ox_a + \bar{\mu}y_o^2x_oy_a \\
& + \bar{\nu}y_o^2x_ay_a + \bar{\theta}y_a^2x_oy_o + \bar{\varphi}y_a^2x_ox_a + \bar{\varrho}y_a^2x_ay_o + \bar{\omega}x_oy_ox_ay_a
\end{aligned}
$$

The expressions for the coefficients $\bar{a}, \bar{b}, \ldots, \bar{\varrho}, \bar{\omega}$ are set out in full elsewhere (Hawkes, 1965b); here, we are concerned only with the aperture aberrations, which originate in the terms in $m^{(4)}$ of the form x_a^4, y_a^4, $x_a^3y_a$, $x_ay_a^3$, $x_a^2y_a^2$, $x_ox_a^3$, $y_oy_a^3$, $x_oy_a^3$, $y_ox_a^3$, $x_ox_ay_a^2$, $y_ox_ay_a^2$, $x_ox_a^2y_a$ and $y_ox_a^2y_a$. The associated coefficients are as follows:

$$
\bar{c}x_a^4: \qquad \bar{c} = X_1h_x^4 + X_2h_x^2h_x'^2 + Nh_x'^4
$$

$$
\bar{d}y_a^4: \qquad \bar{d} = Y_1h_y^4 + Y_2h_y^2h_y'^2 + Nh_y'^4
$$

$$
\begin{aligned}
\bar{f}x_a^2y_a^2: \qquad \bar{f} =\ & Vh_x^2h_y^2 + \tfrac{1}{2}R\{h_x^2(h_y^2)' - (h_x^2)'h_y^2\} \\
& + X_2h_x^2h_y'^2 + Y_2h_x'^2h_y^2 + 2Nh_x'^2h_y'^2
\end{aligned}
$$

$$
\bar{n}x_ox_a^3: \qquad \bar{n} = 4X_1g_xh_x^3 + X_2(h_x^2)'(g_xh_x)' + 4Ng_x'h_x'^3
$$

$$
\bar{p}y_oy_a^3: \qquad \bar{p} = 4Y_1g_yh_y^3 + Y_2(h_y^2)'(g_yh_y)' + 4Ng_y'h_y'^3
$$

$$
\begin{aligned}
\bar{\zeta}y_ox_a^2y_a: \qquad \bar{\zeta} =\ & 2Vh_x^2g_yh_y + R\{h_x^2(g_yh_y)' - (h_x^2)'g_yh_y\} \\
& + 2X_2h_x^2g_y'h_y' + 2Y_2h_x'^2g_yh_y + 4Nh_x'^2g_y'h_y'
\end{aligned}
$$

$$
\begin{aligned}
\bar{\varphi}x_ox_ay_a^2: \qquad \bar{\varphi} =\ & 2Vg_xh_y^2h_x + R\{g_xh_x(h_y^2)' - (g_xh_x)'h_y^2\} \\
& + 2X_2g_xh_xh_y'^2 + 2Y_2g_x'h_x'h_y^2 + 4Ng_x'h_x'h_y'^2
\end{aligned}
$$

$$ (4.18) $$

in which

$$
\begin{aligned}
X_1 =\ & \frac{1 + 2\varepsilon\Phi}{128[\Phi(1 + \varepsilon\Phi)]^{1/2}}\,\Phi^{(iv)} - \frac{(\Phi'' - D)^2}{128[\Phi(1 + \varepsilon\Phi)]^{3/2}} \\
& + \frac{1 + 2\varepsilon\Phi}{2[\Phi(1 + \varepsilon\Phi)]^{1/2}}\,D_1 - \eta Q_1 \\
& - \frac{1 + 2\varepsilon\Phi}{96[\Phi(1 + \varepsilon\Phi)]^{1/2}}\,D'' + \frac{\eta}{48}\,Q''
\end{aligned}
$$

$$ (4.19) $$

$$Y_1 = \frac{1 + 2\varepsilon\Phi}{128[\Phi(1 + \varepsilon\Phi)]^{1/2}} \Phi^{(iv)} - \frac{(\Phi'' + D)^2}{128[\Phi(1 + \varepsilon\Phi)]^{3/2}}$$

$$+ \frac{1 + 2\varepsilon\Phi}{2[\Phi(1 + \varepsilon\Phi)]^{1/2}} D_1 - \eta Q_1$$

$$+ \frac{1 + 2\varepsilon\Phi}{96[\Phi(1 + \varepsilon\Phi)]^{1/2}} D'' - \frac{\eta}{48} Q''$$

$$X_2 = \frac{1 + 2\varepsilon\Phi}{16[\Phi(1 + \varepsilon\Phi)]^{1/2}} (D - \Phi'')$$

$$Y_2 = - \frac{1 + 2\varepsilon\Phi}{16[\Phi(1 + \varepsilon\Phi)^{1/2}} (D + \Phi'')$$

$$N = - \frac{1}{8} [\Phi(1 + \varepsilon\Phi)]^{1/2} \qquad R = \frac{1}{4} \eta Q'$$

$$V = \frac{1 + 2\varepsilon\Phi}{64[\Phi(1 + \varepsilon\Phi)]^{1/2}} \Phi^{(iv)} - \frac{\Phi''^2 - D^2}{64[\Phi(1 + \varepsilon\Phi)]^{3/2}}$$

$$- 3 \frac{1 + 2\varepsilon\Phi}{[\Phi(1 + \varepsilon\Phi)]^{1/2}} D_1 + 6\eta Q_1$$

For the remaining coefficients, we find

$$\bar{u}y_o x_a{}^3: \qquad \bar{u} = Ih_x{}^3 g_y$$

$$\bar{v}x_a{}^3 y_a: \qquad \bar{v} = Ih_x{}^3 h_y$$

$$\bar{w}x_o y_a{}^3: \qquad \bar{w} = - Ig_x h_y{}^3$$

$$\bar{z}x_a y_a{}^3: \qquad \bar{z} = - Ih_x h_y{}^3 \qquad\qquad (4.20)$$

$$\bar{\varrho}y_o x_a y_a{}^2: \qquad \bar{\varrho} = - 3Ih_x g_y h_y{}^2$$

$$\bar{\xi}x_o x_a{}^2 y_a: \qquad \bar{\xi} = 3Ig_x h_x{}^2 h_y$$

in which the definition of I has simplified to

$$I = \frac{2(1 + 2\varepsilon\Phi)}{[\Phi(1 + \varepsilon\Phi)]^{1/2}} P_1 + 4\eta\varDelta_1 \qquad\qquad (4.21)$$

From Eqs. (2.23), we deduce that the aperture aberrations are as follows:[11]

[11] The aberration coefficients of quadrupole lens systems were first derived by Melkich in his Berlin Dissertation (1944, 1947); rectilinear orthogonal systems are considered in Part I. Melkich employed the trajectory method and deduced rather complicated formulas for all the aberration coefficients. Simpler formulas have been obtained by Hawkes (1965a,b) which are demonstrably equivalent to Melkich's expressions (1965b,

$$k_x x_c^{\mathrm{I}} = h_{xc} \left\{ x_a^3 \int_a^c \bar{n}\, dz + x_a^2 y_a \int_a^c \bar{\xi}\, dz + x_a y_a^2 \int_a^c \bar{\varphi}\, dz + y_a^3 \int_a^c \bar{w}\, dz \right\} \quad (4.22\text{a})$$

$$- g_{xc} \left\{ x_a^3 \int_0^c 4\bar{c}\, dz + x_a^2 y_a \int_0^c 3\bar{v}\, dz + x_a y_a^2 \int_0^c 2\bar{f}\, dz + y_a^3 \int_0^c \bar{z}\, dz \right\}$$

$$k_y y_c^{\mathrm{I}} = h_{yc} \left\{ x_a^3 \int_a^c \bar{u}\, dz + x_a^2 y_a \int_a^c \bar{\xi}\, dz + x_a y_a^2 \int_a^c \bar{\varrho}\, dz + y_a^3 \int_a^c \bar{p}\, dz \right\} \quad (4.22\text{b})$$

$$- g_{yc} \left\{ x_a^3 \int_0^c \bar{v}\, dz + x_a^2 y_a \int_0^c 2\bar{f}\, dz + x_a y_a^2 \int_0^c 3\bar{z}\, dz + y_a^3 \int_0^c 4\bar{d}\, dz \right\}$$

in which

$$k_x = [\Phi(1 + \varepsilon\Phi)]^{1/2}(g_x h_x' - g_x' h_x)$$
$$k_y = [\Phi(1 + \varepsilon\Phi)]^{1/2}(g_y h_y' - g_y' h_y)$$

$$(4.23)$$

or writing

$$x^{\mathrm{I}} = (30)_x x_a^3 + (21)_x x_a^2 y_a + (12)_x x_a y_a^2 + (03)_x y_a^3$$
$$y^{\mathrm{I}} = (03)_y y_a^3 + (12)_y x_a y_a^2 + (21)_y x_a^2 y_a + (30)_y x_a^3$$

$$(4.24)$$

we have

$$(30)_x = \frac{h_{xc}}{k_x} \int_a^c \bar{n}\, dz - 4 \frac{g_{xc}}{k_x} \int_0^c \bar{c}\, dz$$

$$(03)_y = \frac{h_{yc}}{k_y} \int_a^c \bar{p}\, dz - 4 \frac{g_{yc}}{k_y} \int_0^c \bar{d}\, dz$$

$$(12)_x = \frac{h_{xc}}{k_x} \int_a^c \bar{\varphi}\, dz - 2 \frac{g_{xc}}{k_x} \int_0^c \bar{f}\, dz$$

$$(21)_y = \frac{h_{yc}}{k_y} \int_a^c \bar{\xi}\, dz - 2 \frac{g_{yc}}{k_y} \int_0^c \bar{f}\, dz$$

$$(03)_x = - \frac{h_{xc}}{k_x} \int_a^c Ig_x h_y^3\, dz + \frac{g_{xc}}{k_x} \int_0^c Ih_x h_y^3\, dz$$

$$(4.25)$$

Appendix to Part I). Scherzer (1947) has calculated formulas for the aperture aberration coefficients in the course of his examination of the various ways of correcting the spherical aberration of round lenses. Formulas for these aberrations have also been obtained by Bernard and Hue (1956, 1957); in connection with their formulas, see also Septier and van Acker (1961), Strashkevich (1961, 1963, 1964), and Markovich and Tsukkerman (1960). Expressions for the aberration coefficients have also been derived by Meads (1963). *Note added in proof.* Many new forms of the formulae have been derived since this chapter was concluded (in December, 1964); see in particular Ovsyannikova and Yavor (1965) and Hawkes (1966/7) and for a general survey, see Hawkes (1966).

$$(30)_y = \frac{h_{yc}}{k_y} \int_a^c Ih_x{}^3 g_y \, dz - \frac{g_{yc}}{k_y} \int_o^c Ih_x{}^3 \, h_y \, dz$$

$$(21)_x = 3\frac{h_{xc}}{k_x} \int_a^c Ig_x h_x{}^2 h_y \, dz - 3\frac{g_{xc}}{k_x} \int_o^c Ih_x{}^3 h_y \, dz$$

$$(12)_y = -3\frac{h_{yc}}{k_y} \int_a^c Ih_x g_y h_y{}^2 \, dz + 3\frac{g_{yc}}{k_y} \int_o^c Ih_x h_y{}^3 \, dz$$

In the image plane, $z = z_i$, of a stigmatic system, $h_x(z_i) = h_y(z_i) = 0$, so that $g_{xi}/k_x = 1/[\Phi_i(1 + \varepsilon\Phi_i)]^{1/2}h'_{xi}$ and $g_{yi}/k_y = 1/[\Phi_i(1 + \varepsilon\Phi_i)]^{1/2}h'_{yi}$. If, therefore, the slopes of the rays $h_x(z)$ and $h_y(z)$ are the same at the image plane, the aberrations $(12)_x$ and $(21)_y$ will be equal. Alternatively, we may express the aberration in terms of the angle at which the outermost rays intersect the axis in the image plane (cf. Deltrap, 1964a): instead of

$$x_i{}^{\mathrm{I}} = (30)_x x_a{}^3 + (12)_x x_a y_a{}^2$$
$$y_i{}^{\mathrm{I}} = (03)_y y_a{}^3 + (21)_y x_a{}^2 y_a$$

we write

$$x_i{}^{\mathrm{I}} = [30]_x \alpha^3 + [12]_x \alpha\beta^2$$
$$y_i{}^{\mathrm{I}} = [03]_y \beta^3 + [21]_y \alpha^2\beta \tag{4.26}$$

(see Fig. 16). Clearly, $\alpha \simeq x_A h'_{xi}$ and $\beta \simeq y_A h'_{yi}$, so that provided the system is stigmatic, the coefficients $[12]_x$ and $[21]_y$ are always equal, whether or not the magnification is the same in the x and y directions.

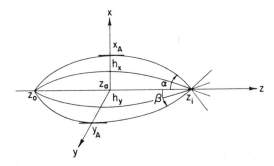

FIG. 16. The angles α and β which replace x_a and y_a in the alternative definition of the aperture aberration coefficients.

We shall see that the spherical aberration coefficients of magnetic and electrostatic round lenses can be cast into the simple form $\int f(z)t^4 dz$. This has the advantage that if the potential distribution can be satisfactorily replaced by an analytical model, no derivatives of computed functions appear; this is very convenient if the function $t(z)$ has to be obtained with the aid of a computer, which will be the case when the Gaussian equations of motion are not soluble in terms of tabulated functions. It is likewise possible to cast the quadrupole aperture aberration coefficients into simpler forms than those given in Eqs. (4.22), with (4.18) and (4.19), and we shall now list these simplified expressions, for·the general case of a mixed magnetic and electrostatic system, and for special cases of practical interest.

Of the terms which appear in Eqs. (4.22a) and (4.22b), the following can be simplified into the expressions set out below:

$$\int_o^c 4\bar{c}\,dz = \int_o^c Fh_x^4\,dz + c_{oc}$$

$$\int_o^c 4\bar{d}\,dz = \int_o^c \tilde{F}h_y^4\,dz + d_{oc}$$

$$\int_a^c \bar{n}\,dz = \int_a^c Fg_x h_x^3\,dz + k_x \int_a^c Eh_x^2\,dz + n_{ac}$$

$$\int_a^c \bar{p}\,dz = \int_a^c \tilde{F}g_y h_y^3\,dz + k_y \int_a^c \tilde{E}h_y^2\,dz + p_{ac}$$

$$\int_o^c 2\bar{f}\,dz = \int_o^c Kh_x^2 h_y^2\,dz + \int_o^c L\{h_x^2(h_y^2)' - (h_x^2)'h_y^2\}\,dz$$
$$+ \int_o^c M(h_x^2 h_y'^2 + h_x'^2 h_y^2)\,dz + f_{oc}$$

(4.27)

$$\int_a^c \bar{\xi}\,dz = \int_a^c Kh_x^2 g_y h_y\,dz + \int_a^c L\{h_x^2(g_y h_y)' - (h_x^2)'g_y h_y\}\,dz$$
$$+ \int_a^c M(h_x^2 g_y' h_y' + h_x'^2 g_y h_y)\,dz + \xi_{ac}$$

$$\int_a^c \bar{\varphi}\,dz = \int_a^c Kg_x h_x h_y^2\,dz + \int_a^c L\{g_x h_x(h_y^2)' - (g_x h_x)'h_y^2\}\,dz$$
$$+ \int_a^c M(g_x h_x h_y'^2 + g_x' h_x' h_y^2)\,dz + \varphi_{ac}$$

in which c_{oc}, d_{oc}, ..., ξ_{ac} and φ_{ac} denote definite integrals, listed below, and $E(z)$, $F(z)$, $K(z)$, $L(z)$ and $M(z)$ represent the following functions:

$$\frac{F(z)}{\varPhi^{1/2}} = 2\,\frac{D_1 - 2\eta Q_1}{\varPhi}\,\varPhi^{1/2} \qquad (octopole\ terms)$$

$$-\frac{5}{32}\left(\frac{\varPhi'}{\varPhi}\right)^4 + \frac{5}{32}\left(\frac{\varPhi'}{\varPhi}\right)^2 \frac{\varPhi''}{\varPhi} - \frac{1}{16}\left(\frac{\varPhi''}{\varPhi}\right)^2 + \frac{1}{64}\frac{\varPhi'''}{\varPhi}\frac{\varPhi'}{\varPhi}$$

$$(round\ lens)$$

$$+ \frac{5}{32} \left(\frac{\Phi'}{\Phi}\right)^2 \frac{D}{\Phi} - \frac{7}{64} \frac{\Phi'}{\Phi} \frac{D'}{\Phi} + \frac{1}{16} \frac{\Phi''}{\Phi} \frac{D}{\Phi}$$

(overlapping round lens and electrostatic quadrupole)

$$- \frac{5}{16} \eta \left(\frac{\Phi'}{\Phi}\right)^2 \frac{Q}{\Phi^{1/2}} + \frac{3}{16} \eta \frac{\Phi'}{\Phi} \frac{Q'}{\Phi^{1/2}} - \frac{1}{16} \eta \frac{\Phi''}{\Phi} \frac{Q}{\Phi^{1/2}}$$

(overlapping round lens and magnetic quadrupole)

$$- \frac{1}{12} \left(\frac{D}{\Phi}\right)^2 + \frac{1}{3} \eta \frac{D}{\Phi} \frac{Q}{\Phi^{1/2}} - \frac{1}{2} \eta^2 \frac{Q^2}{\Phi} + \frac{D'' - 4\eta Q'' \Phi^{1/2}}{\Phi}$$

(quadrupoles only)

(4.28a)

$$E(z) = \frac{1}{32} \frac{\Phi'''}{\Phi} - \frac{3}{16} \frac{\Phi''}{\Phi} \frac{\Phi'}{\Phi} + \frac{3}{16} \left(\frac{\Phi'}{\Phi}\right)^3$$

$$- \frac{1}{16} \frac{\Phi'}{\Phi} \frac{D - 3\eta Q \Phi^{1/2}}{\Phi} + \frac{1}{32} \frac{D' - 4\eta Q' \Phi^{1/2}}{\Phi}$$

(4.28b)

Replacing D by $-D$ and Q by $-Q$ converts $F(z)$ into $\tilde{F}(z)$ and $E(z)$ into $\tilde{E}(z)$. If the round electrostatic lens fields and the quadrupole fields overlap, the expressions for $K(z)$, $L(z)$, and $M(z)$ can be written in two different ways. Either

$$\frac{K(z)}{\Phi^{1/2}} = \frac{5}{64} \frac{\Phi'''}{\Phi} \frac{\Phi'}{\Phi} - \frac{3}{32} \frac{\Phi''}{\Phi} \left(\frac{\Phi'}{\Phi}\right)^2$$

$$+ \frac{1}{8} \left(\frac{D - 2\eta Q \Phi^{1/2}}{\Phi}\right) - 6 \frac{D_1 - 2\eta Q_1 \Phi^{1/2}}{\Phi}$$

(4.29a)

$$\frac{L(z)}{\Phi^{1/2}} = \frac{1}{16} \frac{D + \eta Q \Phi^{1/2}}{\Phi} \frac{\Phi'}{\Phi} - \frac{3}{32} \frac{D' - 4\eta Q' \Phi^{1/2}}{\Phi}$$

$$\frac{M(z)}{\Phi^{1/2}} = - \frac{3}{16} \left(\frac{\Phi'}{\Phi}\right)^2$$

or

$$\frac{K(z)}{\Phi^{1/2}} = \frac{1}{32} \frac{\Phi^{(iv)}}{\Phi} - \frac{1}{16} \frac{\Phi'''}{\Phi} \frac{\Phi'}{\Phi} + \frac{21}{64} \frac{\Phi''}{\Phi} \left(\frac{\Phi'}{\Phi}\right)^2$$

$$- \frac{3}{32} \left(\frac{\Phi''}{\Phi}\right)^2 - \frac{15}{64} \left(\frac{\Phi'}{\Phi}\right)^4$$

$$+ \frac{1}{8} \left(\frac{D - 2\eta Q \Phi^{1/2}}{\Phi}\right)^2 - 6 \frac{D_1 - 2\eta Q_1 \Phi^{1/2}}{\Phi}$$

$$\frac{L(z)}{\Phi^{1/2}} = \frac{1}{16} \frac{2D - 3\eta Q \Phi^{1/2}}{\Phi} \frac{\Phi'}{\Phi} - \frac{3}{32} \frac{D' - 4\eta Q' \Phi^{1/2}}{\Phi}$$

(4.29b)

$$\frac{M(z)}{\Phi^{1/2}} = \frac{3}{32} \left(\frac{\Phi'}{\Phi}\right)^2 - \frac{3}{16} \frac{\Phi''}{\Phi}$$

If, however, the system is designed in such a way that the quadrupole and round fields never overlap, the function $M(z)$ vanishes in the quadrupole regions, and the two sets of expressions for $K(z)$ and $L(z)$ reduce to

$$\frac{K(z)}{\Phi^{1/2}} = \frac{1}{8}\left(\frac{D - 2\eta Q\Phi^{1/2}}{\Phi}\right)^2 - 6\frac{D_1 - 2\eta Q_1 \Phi^{1/2}}{\Phi}$$

$$L(z) = -\frac{3}{32}\frac{D' - 4\eta Q'\Phi^{1/2}}{\Phi^{1/2}}$$

(4.29c)

The definite integrals $c_{oc}, \ldots, \varphi_{ac}$ are of the following forms, provided object and image lie in field-free space (see Hawkes, 1965a for the general case):

$$c_{oc} = -\tfrac{1}{2}\Phi_c^{1/2} h_{xc}'^3 h_{xc}$$

$$d_{oc} = -\tfrac{1}{2}\Phi_c^{1/2} h_{yc}'^3 h_{yc}$$

$$n_{ac} = -\tfrac{1}{2}\Phi_c^{1/2} g_{xc}' h_{xc} h_{xc}'^2 + \tfrac{1}{2}\Phi_a^{1/2} g_{xa}' h_{xa}'^2$$

$$+ \left(\frac{\Phi''}{8\Phi^{1/2}} + \frac{\Phi'^2}{8\Phi^{3/2}} - \frac{5D}{24\Phi^{1/2}} + \frac{1}{2}\eta Q\right)_a g_{xa}'$$

$$+ \left(\frac{\Phi'' - D}{8\Phi} + \frac{1}{4}\eta\frac{Q}{\Phi^{1/2}}\right)_a k_x$$

$$p_{ac} = -\tfrac{1}{2}\Phi_c^{1/2} g_{yc}' h_{yc} h_{yc}'^2 + \tfrac{1}{2}\Phi_a^{1/2} g_{ya}' h_{ya}'^2$$

$$+ \left(\frac{\Phi''}{8\Phi^{1/2}} + \frac{\Phi'^2}{8\Phi^{3/2}} + \frac{5D}{24\Phi^{1/2}} - \frac{1}{2}\eta Q\right)_a g_{ya}'$$

$$+ \left(\frac{\Phi'' + D}{8\Phi} - \frac{1}{4}\eta\frac{Q}{\Phi^{1/2}}\right)_a k_y$$

$$f_{oc} = -\tfrac{1}{4}\Phi_c^{1/2} h_{yc}' h_{xc}' (h_x h_y)_c'$$

$$\zeta_{ac} = -\tfrac{1}{8}\Phi_c^{1/2}\{(h_x^2)_c' g_{yc}' h_{yc}' + h_{xc}'^2 (g_y h_y)_c'\}$$

$$+ \tfrac{1}{8}\Phi_a^{1/2}(2h_{xa}' g_{ya}' h_{ya}' + h_{xa}'^2 g_{ya}')$$

$$+ \frac{1}{8}\frac{\Phi_a'}{\Phi_a^{1/2}} g_{ya}' h_{ya}' - \frac{3D_a - 4\eta Q_a \Phi_a^{1/2}}{32\,\Phi_a^{1/2}} g_{ya}' + \frac{1}{32}\frac{\Phi_a''}{\Phi_a^{1/2}} g_{ya}'$$

$$\varphi_{ac} = -\tfrac{1}{8}\Phi_c^{1/2}\{(h_y^2)_c' g_{xc}' h_{xc}' + h_{yc}'^2 (g_x h_x)_c'\}$$

$$+ \tfrac{1}{8}\Phi_a^{1/2}(2h_{ya}' g_{xa}' h_{xa}' + h_{ya}'^2 g_{xa}')$$

$$+ \frac{1}{8}\frac{\Phi_a'}{\Phi_a^{1/2}} g_{xa}' h_{xa}' + \frac{3D_a - 4\eta Q_a \Phi_a^{1/2}}{32\,\Phi_a^{1/2}} g_{xa}' + \frac{1}{32}\frac{\Phi_a''}{\Phi_a^{1/2}} g_{xa}'$$

(4.30)

When the current plane, $z = z_c$, is a stigmatic image plane, the definite integrals which remain, c_{oc}, d_{oc}, and f_{oc}, all vanish.

Special Cases

(i) *Systems containing electrostatic and magnetic quadrupoles and octopoles, but no round lenses.* In this situation, $\Phi(z)$ is constant, and with $x_c^I = (30)x_a^3 + (12)x_a y_a^2$, $y_c^I = (03)y_a^3 + (21)x_a^2 y_a$, we find

$$
(30) = \frac{h_{xc}}{k_x} \int_a^c \left(\frac{D'' - 4\eta Q'' \Phi^{1/2}}{96\,\Phi^{1/2}} + 2\frac{D_1 - 2\eta Q_1 \Phi^{1/2}}{\Phi^{1/2}} \right.
$$

$$
\left. - \frac{D^2 - 4\eta DQ\Phi^{1/2} + 6\eta^2 Q^2 \Phi}{12\Phi^{3/2}} \right) g_x h_x^3 \, dz
$$

$$
+ h_{xc} \int_a^c \frac{D' - 4\eta Q' \Phi^{1/2}}{32\Phi}\, h_x^2 \, dz
$$

$$
- \frac{g_{xc}}{k_x} \int_0^c \left(\frac{D'' - 4\eta Q'' \Phi^{1/2}}{96\Phi^{1/2}} + 2\frac{D_1 - 2\eta Q_1 \Phi^{1/2}}{\Phi^{1/2}} \right.
$$

$$
\left. - \frac{D^2 - 4\eta DQ\Phi^{1/2} + 6\eta^2 Q^2 \Phi}{12\Phi^{3/2}} \right) h_x^4 \, dz
$$

$$
+ [30] \tag{4.31a}
$$

$$
(12) = \frac{h_{xc}}{k_x} \int_a^c \left\{ \frac{(D - 2\eta Q\Phi^{1/2})^2}{8\Phi^{3/2}} - 6\frac{D_1 - 2\eta Q_1 \Phi^{1/2}}{\Phi^{1/2}} \right\} g_x h_x h_y^2 \, dz
$$

$$
- \frac{h_{xc}}{k_x} \int_a^c \frac{3}{32}\frac{D' - 4\eta Q' \Phi^{1/2}}{\Phi^{1/2}} \{g_x h_x (h_y^2)' - (g_x h_x)' h_y^2\}\, dz
$$

$$
- \frac{g_{xc}}{k_x} \int_0^c \left\{ \frac{(D - 2\eta Q\Phi^{1/2})^2}{8\Phi^{3/2}} - 6\frac{D_1 - 2\eta Q_1 \Phi^{1/2}}{\Phi^{1/2}} \right\} h_x^2 h_y^2 \, dz
$$

$$
+ \frac{g_{xc}}{k_x} \int_0^c \frac{3}{32}\frac{D' - 4\eta Q' \Phi^{1/2}}{\Phi^{1/2}} \{h_x^2 (h_y^2)' - (h_x^2)' h_y^2\}\, dz
$$

$$
+ [12] \tag{4.31b}
$$

To obtain (03) and (21) from (30) and (12) respectively, we replace D by $-D$, Q by $-Q$, h_x by h_y, and g_x by g_y. The quantities [30] and [12] denote the definite integrals, which are of the form

$$
[30] = \frac{g'_{xa} h_{xc}}{2k_x} \left(\Phi^{1/2} h'^2_{xa} - \frac{D_a - 3\eta Q\Phi^{1/2}}{6\Phi^{1/2}} \right) + \frac{1}{2} h_{xc} h'^2_{xc}
$$

$$
[12] = \frac{g'_{xa} h_{xc}}{k_x} \left(\frac{1}{4} \Phi^{1/2} h'_{xa} h'_{ya} + \frac{1}{8} \Phi^{1/2} h'^2_{ya} + \frac{3D_a - 4\eta Q_a \Phi^{1/2}}{32\Phi^{1/2}} \right)
$$

$$
+ \tfrac{1}{8} h'_{yc}(h_{xc} h'_{yc} + 2h'_{xc} h_{yc})
$$

and the same replacements convert [30] into [03] and [12] into [21].

(ii) *Systems containing only electrostatic quadrupoles and octopoles.* In practice, systems are usually constructed wholly from magnetic elements, or wholly from electrostatic units; the most important exceptions to this are achromatic systems, in which both types of element must be present [cf. Hawkes (1965c) in which all the earlier references are listed]. For the electrostatic case, we find

$$k_x(30) = h_{xc} \int_a^c \left(\frac{D''}{96\Phi^{1/2}} - \frac{D^2}{12\Phi^{3/2}} + 2\frac{D_1}{\Phi^{1/2}} \right) g_x h_x^3 \, dz$$

$$+ k_x h_{xc} \int_a^c \frac{D'}{32\Phi} h_x^2 \, dz$$

$$- g_{xc} \int_0^c \left(\frac{D''}{96\Phi^{1/2}} - \frac{D^2}{12\Phi^{3/2}} + 2\frac{D_1}{\Phi^{1/2}} \right) h_x^4 \, dz + k_x[30]$$

$$k_x(12) = h_{xc} \int_a^c \left[\left(\frac{D^2}{8\Phi^{3/2}} - 6\frac{D_1}{\Phi^{1/2}} \right) g_x h_x h_y^2 \right.$$

$$\left. - \frac{3}{32} \frac{D'}{\Phi^{1/2}} \{ g_x h_x (h_y^2)' - (g_x h_x)' h_y^2 \} \right] dz$$

$$- g_{xc} \int_0^c \left[\left(\frac{D^2}{8\Phi^{3/2}} - 6\frac{D_1}{\Phi^{1/2}} \right) h_x^2 h_y^2 \right.$$

$$\left. - \frac{3}{32} \frac{D'}{\Phi^{1/2}} \{ h_x^2 (h_y^2)' - (h_x^2)' h_y^2 \} \right] dz$$

$$+ k_x[12] \qquad\qquad (4.32)$$

$$k_y(03) = h_{yc} \int_a^c \left(-\frac{D''}{96\Phi^{1/2}} - \frac{D^2}{12\Phi^{3/2}} + 2\frac{D_1}{\Phi^{1/2}} \right) g_y h_y^3 \, dz$$

$$- k_y h_{yc} \int_a^c \frac{D'}{32\Phi} h_y^2 \, dz$$

$$- g_{yc} \int_0^c \left(-\frac{D''}{96\Phi^{1/2}} - \frac{D^2}{12\Phi^{3/2}} + 2\frac{D_1}{\Phi^{1/2}} \right) h_y^4 \, dz + k_y[03]$$

$$k_y(21) = h_{yc} \int_a^c \left[\left(\frac{D^2}{8\Phi^{1/2}} - 6\frac{D_1}{\Phi^{1/2}} \right) h_x^2 g_y h_y \right.$$

$$\left. - \frac{3}{32} \frac{D'}{\Phi^{1/2}} \{ h_x^2 (g_y h_y)' - g_y h_y (h_x^2)' \} \right] dz$$

$$- g_{yc} \int_0^c \left[\left(\frac{D^2}{8\Phi^{3/2}} - 6\frac{D_1}{\Phi^{1/2}} \right) h_x^2 h_y^2 \right.$$

$$\left. - \frac{3}{32} \frac{D'}{\Phi^{1/2}} \{ h_x^2 (h_y^2)' - (h_x^2)' h_y^2 \} \right] dz + k_y[21]$$

(iii) *Systems consisting of magnetic quadrupoles and octopoles only.* We now find

$$k_x(30) = h_{xc} \int_a^c \left(-\frac{\eta}{24} Q'' - \frac{\eta^2}{2} \frac{Q^2}{\Phi^{1/2}} - 4\eta Q_1 \right) g_x h_x{}^3 \, dz$$

$$- k_x h_{xc} \int_a^c \frac{\eta}{8} \frac{Q'}{\Phi^{1/2}} h_x{}^2 \, dz$$

$$- g_{xc} \int_o^c \left(-\frac{\eta}{24} Q'' - \frac{\eta^2}{2} \frac{Q^2}{\Phi^{1/2}} - 4\eta Q_1 \right) h_x{}^4 \, dz$$

$$+ k_x[30]$$

$$k_x(12) = h_{xc} \int_a^c \left[\left(\frac{\eta^2}{2} \frac{Q^2}{\Phi^{1/2}} + 12\eta Q_1 \right) g_x h_x h_y{}^2 \right.$$

$$\left. + \frac{3\eta}{8} Q' \{ g_x h_x (h_y{}^2)' - (g_x h_x)' h_y{}^2 \} \right] dz$$

$$- g_{xc} \int_o^c \left[\left(\frac{\eta^2}{2} \frac{Q^2}{\Phi^{1/2}} + 12\eta Q_1 \right) h_x{}^2 h_y{}^2 \right.$$

$$\left. + \frac{3\eta}{8} Q' \{ h_x{}^2 (h_y{}^2)' - (h_x{}^2)' h_y{}^2 \} \right] dz$$

$$+ k_x[12]$$

$$k_y(03) = h_{yc} \int_a^c \left(\frac{\eta}{24} Q'' - \frac{\eta^2}{2} \frac{Q^2}{\Phi^{1/2}} - 4\eta Q_1 \right) g_y h_y{}^3 \, dz$$

$$+ k_y h_{yc} \int_o^c \frac{\eta}{8} \frac{Q'}{\Phi^{1/2}} h_y{}^2 \, dz$$

$$- g_y \int_o^c \left(\frac{\eta}{24} Q'' - \frac{\eta^2}{2} \frac{Q^2}{\Phi^{1/2}} - 4\eta Q_1 \right) h_y{}^4 \, dz$$

$$+ k_y[03]$$

$$k_y(21) = h_{yc} \int_a^c \left[\left(\frac{\eta^2}{2} \frac{Q^2}{\Phi^{1/2}} + 12\eta Q_1 \right) h_x{}^2 g_y h_y \right.$$

$$\left. + \frac{3\eta}{8} Q' \{ h_x{}^2 (g_y h_y)' - (h_x{}^2)' g_y h_y \} \right] dz$$

$$- g_{yc} \int_o^c \left[\left(\frac{\eta^2}{2} \frac{Q^2}{\Phi^{1/2}} + 12\eta Q_1 \right) h_x{}^2 h_y{}^2 \right.$$

$$\left. + \frac{3\eta}{8} Q' \{ h_x{}^2 (h_y{}^2)' - (h_x{}^2)' h_y{}^2 \} \right] dz$$

$$+ k_y[21]$$

(4.33)

These are the most important special cases; any others can be straight-forwardly derived from the general formulas. In order to compute the aberrations in any given case, the form of the functions $\Phi(z)$, $D(z)$, $D_1(z)$, ... which describe the potential distribution must first be ascertained, either by measurement or computation; when is it permissible to express these potential functions as analytic functions, the remaining analysis will be considerably simplified. The first-order equations of motion must then be solved, and again, it is very convenient if the solutions can be expressed in terms of tabulated functions. From the general solution, the particular trajectories $g_x(z)$, $h_x(z)$, $g_y(z)$, and $h_y(z)$ can be deduced, and substituted into the appropriate formulas for the aberration coefficients. This involves integration, which will usually have to be performed with the aid of an electronic computer, even if the explicit expressions for the first-order trajectories have been obtained.

Line foci. We mentioned earlier that in some applications, the line foci of astigmatic quadrupole systems are used directly. The line foci are formed in the planes in which $h_x(z)$ and $h_y(z)$ vanish, and the aberrations of these lines can be deduced from the preceding formulas by substituting either

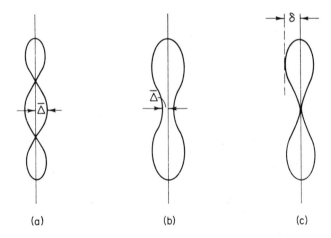

(a)   (b)   (c)

FIG. 17. The effect of the aperture aberration in the neighborhood of a line focus. (a) The line converted into three loops; (b) The line converted into an hour-glass figure; (c) The figure-of-eight pattern, formed in a plane close to the line focus plane.

$h_{xc} = 0$ or $h_{yc} = 0$. The effect of the aperture aberrations is to convert the line into three loops (Fig. 17a) or an hour-glass figure (Fig. 17b); near the line focus plane is a plane, $z = z_s$, in which the central loop or narrow neck disappears, and the maximum width of the outer loops in this plane (2δ in Fig. 17c) has been used by Septier as a measure of the aperture aberration.[12]

If we consider the line focus at which $h_x(z)$ vanishes, in the plane $z = z_L$, say, we have

$$\bar{x} = x_L - g_{xL}x_0 = x_a\{(30)x_a^2 + (12)y_a^2\}$$

$$\bar{y} = y_L - g_{yL}y_0 = y_a\{h_{yL} + (21)x_a^2 + (03)y_a^2\}$$

$\bar{x}$ vanishes when $x_a \neq 0$ if $\tan\theta_a = \pm[-(30)/(12)]^{1/2}$, so that the aberration figure is of the three-looped form (Fig. 17a) if (30) and (12) have opposite signs, and of the hour-glass form (Fig. 17b) if they have the same sign. The half-width of the central loop (or of the narrowest part of the hour-glass), $\bar{A}$, is given by

$$\bar{A} = (30)r_a^3$$

In an adjacent plane, and considering only the modifications due to the terms in $h_x x_a$ and $h_y y_a$, we find

$$\bar{x} = x_a\{h'_{xL}\delta z + (30)x_a^2 + (12)y_a^2\}$$

$$\bar{y} = y_a\{h_{yL} + h'_{yL}\delta z + (21)x_a^2 + (03)y_a^2\}$$

so that the half-width of the central loop (or of the neck of the hour-glass) vanishes if

$$\delta z = -\frac{(30)}{h_{xL}}r_a^2 \tag{4.34}$$

In this plane, $z = z_s$, we have

$$\bar{x} = x_a\{(12) - (30)\,y_a^2\}$$

which has a maximum value, δ, given by

$$\delta = \frac{2}{3\sqrt{3}}r_a^3\{(12) - (30)\} \tag{4.35}$$

Septier's measure, δ, is thus proportional to the difference between the aperture aberration coefficients in the x direction, and the order in which the aberration figures are seen as we examine planes in the vicinity of a line focus depends upon the sign of (30). Near the other line focus, ($z = z_{L'}$),

---

[12] This line focus aberration has been extensively studied by Septier; a coherent account of all his work is to be found in Grivet and Septier (1960) and in Septier (1961).

where $h_y(z)$ vanishes, the aberrations have a similar effect; the half-width of the central loop or neck is now $(03)r_a^3$, and vanishes in a plane distant $-(03)r_a^2/h'_{yL'}$ from the line focus plane. The aberration δ is now given by

$$\delta = \frac{2}{3\sqrt{3}} r_a^3 \{(21) - (03)\}$$

Round Systems. The potential distribution is now completely characterized by $\Phi(z)$ and $\Omega(z)$:

$$m^{(0)} = [\Phi(1 + \varepsilon\Phi)]^{1/2}$$

$$m^{(2)} = -\frac{1 + 2\varepsilon\Phi}{8[\Phi(1 + \varepsilon\Phi)]^{1/2}} \Phi''(X^2 + Y^2)$$

$$+ \tfrac{1}{2}[\Phi(1 + \varepsilon\Phi)]^{1/2}(X'^2 + Y'^2) - \tfrac{1}{2}\eta\Omega(XY' - X'Y) \qquad (4.36)$$

$$= -\frac{1}{8[\Phi(1 + \varepsilon\Phi)]^{1/2}} \{(1 + 2\varepsilon\Phi)\Phi'' + \eta^2\Omega^2\}(x^2 + y^2)$$

$$+ \tfrac{1}{2}[\Phi(1 + \varepsilon\Phi)]^{1/2} (x'^2 + y'^2)$$

$$m^{(4)} = \theta(x^2 + y^2)^2 + \varkappa(x'^2 + y'^2)^2 + \lambda(xy' - x'y)^2$$

$$+ \mu(x^2 + y^2)(x'^2 + y'^2) + \nu(x^2 + y^2)(xy' - x'y) \qquad (4.37)$$

$$+ \xi(x'^2 + y'^2)(xy' - x'y)$$

in which

$$\theta = -\frac{1}{128[\Phi(1 + \varepsilon\Phi)]^{1/2}} \left\{ \frac{1}{\Phi(1 + \varepsilon\Phi)} \times \right.$$

$$(\Phi''^2 + 2(1 + 2\varepsilon\Phi)\Phi''\eta^2\Omega^2 + \eta^4\Omega^4) - (1 + 2\varepsilon\Phi)\Phi^{(iv)} - 4\eta^2\Omega\Omega'' \Big\}$$

$$\varkappa = -\frac{1}{8}[\Phi(1 + \varepsilon\Phi)]^{1/2}$$

$$\lambda = -\frac{1}{8}\eta^2 \frac{\Omega^2}{[\Phi(1 + \varepsilon\Phi)]^{1/2}} \qquad (4.38)$$

$$\mu = -\frac{1}{16[\Phi(1 + \varepsilon\Phi)]^{1/2}} \{(1 + 2\varepsilon\Phi)\Phi'' + \eta^2\Omega^2\}$$

$$\nu = -\frac{1}{16}\eta \left\{ \frac{1 + 2\varepsilon\Phi}{\Phi(1 + \varepsilon\Phi)} \Phi''\Omega + \frac{\eta^2\Omega^3}{\Phi(1 + \varepsilon\Phi)} - \Omega'' \right\}$$

$$\xi = -\frac{1}{4}\eta\Omega$$

Substituting $x = x_o g + x_a h$ and $y = y_o g + y_a h$ into $m^{(4)}$, and writing $r_o{}^2 = x_o{}^2 + y_o{}^2$, $r_a{}^2 = x_a{}^2 + y_a{}^2$, $\zeta = x_o x_a + y_o y_a$ and $\sigma = x_o y_a - x_a y_o$, we obtain

$$m^{(4)} = \tfrac{1}{4}\bar{A}r_o{}^4 + \tfrac{1}{4}\bar{B}r_a{}^4 + \bar{C}\zeta^2 + \bar{D}r_o{}^2 r_a{}^2 + \bar{E}r_o{}^2\zeta + \bar{F}r_a{}^2\zeta + \bar{e}r_o{}^2\sigma$$
$$+ \bar{f}r_a{}^2\sigma + \bar{c}\zeta\sigma \tag{4.39}$$

The coefficients $\bar{A}$, $\bar{B}, \ldots, \bar{c}$ are given in full by Glaser (1956, p. 223). Only $\bar{B}$, $\bar{F}$, and $\bar{f}$ contribute to the aperture aberration

$$\bar{B} = 4\theta h^4 + 4\varkappa h'^4 + 4\mu h^2 h'^2$$
$$\bar{F} = 4\theta g h^3 + 4\varkappa g' h'^3 + \mu(gh)'(h^2)' \tag{4.40}$$
$$\bar{f} = (\nu g^2 + \xi g'^2)\,(gh' - g'h)$$

With the aid of Eqs. (2.18), we find

$$kx_c{}^{\mathrm{I}} = h_c\left\{r_a{}^2 x_a \int_a^c \bar{F}\,dz + r_a{}^2 y_a \int_a^c \bar{f}\,dz\right\}$$
$$\qquad - g_c r_a{}^2 x_a \int_o^c \bar{B}\,dz \tag{4.41}$$

$$ky_c{}^{\mathrm{I}} = h_c\left\{r_a{}^2 y_a \int_a^c \bar{F}\,dz - r_a{}^2 x_a \int_a^c \bar{f}\,dz\right\}$$
$$\qquad - g_c r_a{}^2 y_a \int_o^c \bar{B}\,dz$$

In the Gaussian image plane, $h_i = 0$, and

$$x_i{}^{\mathrm{I}} = Br_a{}^2 x_a \qquad y_i{}^{\mathrm{I}} = Br_a{}^2 y_a$$

in which

$$B = -\frac{1}{[\Phi_i(1 + \varepsilon\Phi_i)]^{1/2}h_i{}'} \int_o^i \bar{B}\,dz \tag{4.42}$$

The aberration curve in the image plane is a circle, radius $Br_a{}^3$; it is usual to define the spherical aberration coefficient C_s in terms of the radius of the circle in the object plane of which the spherical aberration disc is the Gaussian image. Since the transverse magnification is g_i, the radius of this circle in the object plane, ϱ_o, will be given by $g_i\varrho_o = Br_a{}^3$; if the slope of the outermost ray from the object point is α (see Fig. 18a), we have $\alpha \simeq h_o' r_a$, so that

$$\varrho_o = \frac{B}{h_o'^3 g_i}\,\alpha^3$$

The coefficient of spherical aberration, C_s, is defined by

$$\varrho_0 = C_s \alpha^3$$

so that

$$C_s = \frac{B}{g_i h_o'^3} = - \frac{1}{h_o'^4 [\Phi_o (1 + \varepsilon \Phi_o)]^{1/2}} \int_0^i \bar{B} \, dz \qquad (4.43)$$

If the lens or optical system is used in the reverse sense, with the same pair of conjugate planes, the spherical aberration coefficient will not be the same.

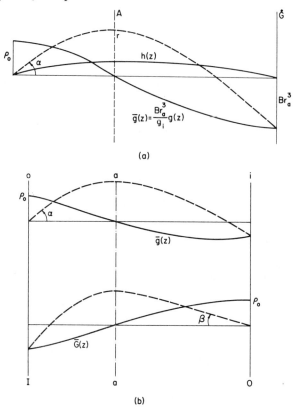

(a)

(b)

FIG. 18. (a) The spherical aberration and the definition of C_s. A and G are the aperture and Gaussian image planes, respectively; – – – – Marginal ray affected by spherical aberration ——— Ray satisfying the Gaussian equations of motion; (b) The "forward" and "backward" spherical aberration coefficients.

[Cf. Archard (1958) for a discussion of this point in very different terms.] Comparing the two situations, we find (Fig. 18b)

Left-to-right   *Right-to-left*

$$B = -\frac{g_i}{k}\int_o^i \bar{B}\,dz \qquad B' = -\frac{G_I}{K}\int_o^I \bar{B}\,dz$$

$$C_s = \frac{B}{g_i h_o'^3} = -\frac{1}{k h_o'^3}\int_o^i \bar{B}\,dz \qquad C_s' = \frac{B'}{G_I h_o'^3} = -\frac{1}{k' h_o'^3}\int_o^I \bar{B}\,dz$$

$$\varrho_0 = C_s a^3 \qquad\qquad \varrho_0 = C_s' \beta^3$$

Thus

$$C_s/C_s' = \frac{k'}{k}\frac{h_o'^3}{h_o'^3}$$

and since $h_o' = h_i'$ and $k' = \Phi_o^{1/2} G_I h_o'$, we find[13]

$$\frac{C_s}{C_s'} = \frac{1}{g_i^4} \tag{4.44}$$

For a *magnetic* round lens, or system of magnetic lenses, we have

$$C_s = \frac{1}{[\Phi(1+\varepsilon\Phi)]^{1/2}}\int_o^i \left[\left\{\frac{\eta^4}{32}\frac{\Omega^4}{[\Phi(1+\varepsilon\Phi)]^{3/2}} - \frac{\eta^2}{8}\frac{\Omega\Omega''}{[\Phi(1+\varepsilon\Phi)]^{1/2}}\right\}t^4\right.$$
$$\left. + \frac{1}{2}[\Phi(1+\varepsilon\Phi)]^{1/2}t'^4 + \frac{\eta^2}{4}\frac{\Omega^2}{[\Phi(1+\varepsilon\Phi)]^{1/2}}t^2t'^2\right]dz \tag{4.45a}$$

in which $t(z)$ denotes the ray that satisfies the boundary conditions $t(z_o) = 0$, $t'(z_o) = 1$, so that $t(z) = h(z)/h_o'$. The terms in t'^4 and $t^2t'^2$ can be eliminated by partial integration, yielding

$$C_s = \frac{\eta^2}{48\Phi(1+\varepsilon\Phi)}\int_o^i\left(\frac{4\eta\Omega^4}{[\Phi(1+\varepsilon\Phi)]^{1/2}} + 5\Omega'^2 - \Omega\Omega''\right)t^4\,dz \tag{4.45b}$$

For an *electrostatic* round lens, or system of lenses, the corresponding expression is

$$C_s = \frac{1}{[\Phi_o(1+\varepsilon\Phi_o)]^{1/2}}\int_o^i\left[\left\{\frac{\Phi''^2}{32[\Phi(1+\varepsilon\Phi)]^{3/2}} - \frac{1+2\varepsilon\Phi}{32[\Phi(1+\varepsilon\Phi)]^{1/2}}\Phi^{(iv)}\right\}t^4\right.$$
$$\left. + \frac{1}{2}[\Phi(1+\varepsilon\Phi)]^{1/2}t'^4 + \frac{1+2\varepsilon\Phi}{4[\Phi(1+\varepsilon\Phi)]^{1/2}}\Phi''t^2t'^2\right]dz \tag{4.46a}$$

[13] This result is an obvious consequence of Eq. (4.43), but since confusion can arise over these definitions, this list of relationships between the "forward" and "backward" properties of a system will perhaps be found helpful.

Eliminating all derivatives of $t(z)$, and considering only the nonrelativistic approximation, formula (4.46a) becomes

$$C_s = \frac{1}{64\Phi_0^{1/2}} \int_0^i \Phi^{1/2}(4T'^2 + 3T^4 - 5T^2T' - TT'')t^4 \, dz \tag{4.46b}$$

in which $T(z)$ denotes $\Phi'(z)/\Phi(z)$.

Scherzer's Theorem. It has been shown by Scherzer (1936) that subject to limitations set out below, the spherical aberration coefficient can never change sign. Scherzer states that the general formula for C_s when both electrostatic and magnetic round lenses are present, their fields overlapping, can be cast into the form:

$$
\begin{aligned}
C_s = \frac{1}{16\Phi_0^{1/2}} \int_0^i \Phi^{1/2} &\left\{ \frac{5}{4} \left(\frac{\Phi''}{\Phi} t + \frac{\Phi'}{\Phi} t' - \frac{\Phi'^2}{\Phi^2} t \right)^2 \right. \\
&+ \frac{\Phi'^2}{\Phi^2} \left(t' + \frac{7}{8} \frac{\Phi'}{\Phi} t \right)^2 + \frac{2}{\Phi} \eta^2 \left(\Omega't + \Omega t' - \frac{5}{4} \Omega \frac{\Phi'}{\Phi} t \right)^2 \\
&+ 2\eta^2 \frac{\Omega^2}{\Phi} \left(t' + \frac{1}{4} \frac{\Phi'}{\Phi} t \right)^2 + \frac{1}{64} \frac{\Phi'^4}{\Phi^4} t^2 + \eta^2 \frac{\Omega^4}{\Phi^2} t^2 \\
&+ \left. \frac{1}{16} \eta^2 \Omega^2 \frac{\Phi'^2}{\Phi^3} t^2 \right\} t^2 \, dz
\end{aligned}
\tag{4.47}
$$

As Φ is essentially positive, C_s is itself positive and can only vanish if $\Phi(z)$ is constant and $\Omega(z)$ vanishes everywhere or if the conditions subject to which the proof is valid are not fulfilled. It is this important result that is known as "Scherzer's Theorem." The theorem is valid provided that all the fields possess rotational symmetry, and are static, that $\Phi'(z)/\Phi(z)$ and $\Phi(z)$ are continuous functions of z, that no appreciable space-charge distribution is present, and that a real image of an object placed at z_0 will be formed at z_i. Scherzer himself (1947) subsequently suggested methods of designing systems free of spherical aberration in which one or other of these necessary conditions is not satisfied. [A detailed account of the ensuing attempts to correct spherical aberration is given in Septier (1966).]

The fact that C_s is always positive implies that marginal rays intersect the axis nearer the optical system than peripheral rays. This is a consequence of the definition of C_s, for the longitudinal spherical aberration, δz_A, which is measured from the Gaussian image plane towards the system, is given by $\delta z_A = (0021)r_A^2/h_i'$ and substituting $C_s = (0021)/h_0'^3 g_i$ and $h_i' = \Phi_0^{1/2}h_0'/\Phi_i^{1/2}g_i$, we find $\delta z_A = [\Phi_1/\Phi_0]^{1/2}g_i^2h_0'^2r_A^2C_s$. The sign of δz_A is thus always the same as that of C_s.

REFERENCES

Amboss, K. (1959). Electron optics: aberrations of air-cored magnetic lenses. Thesis, London.

Archard, G. D. (1958). *Rev. Sci. Instr.* **29**, 1049–1050.

Bernard, M.-Y., and Hue, J. (1956). *Compt. Rend.* **243**, 1852–1854.

Bernard, M.-Y., and Hue, J. (1957). *Compt. Rend.* **244**, 732–735.

Bok, A. B., Kramer, J. and Le Poole, J. B. (1964). *Proc. 3rd European Regional Conf. Electron Microscopy, Prague, 1964* Appendix p. 9. Publ. House Czech. Acad. Sci., Prague.

Burfoot, J. C. (1954). *Proc. Phys. Soc.* B67, 523–528.

Burfoot, J. C. (1956). *Proc. 3rd Intern. Conf. Electron Microscopy, London, 1954* pp. 105–109. Roy. Microscop. Soc., London.

Carathéodory, C. (1937). *Ergeb. Math. Grenzgeb.* **4**, Pt. 5, 104 pp.

Chako, N. (1957). *Trans. Chalmers Univ. Technol. Gothenburg* No. **191**, 50 pp.

Cotte, M. (1938). Recherches sur l'optique électronique. Thesis, Paris and *Ann. Phys. (Paris)* [11] **10**, 333–405.

Deltrap, J. H. M. (1964a). Correction of spherical aberration of electron lenses. Thesis, Cambridge.

Deltrap, J. H. M. (1964b). *Proc. 3rd European Regional Conf. Electron Microscopy, Prague, 1964.* A45–46. Publ. House Czech. Acad. Sci., Prague.

Dušek, H. (1958). Über die elektronenoptische Abbildung durch Orthogonalsysteme unter Verwendung eines neuen Modellfeldes mit streng angebbaren Paraxialbahnen. Dissertation, Vienna.

Dušek, H. (1959). *Optik* **16**, 419–445.

Glaser, W. (1952). "Grundlagen der Elektronenoptik", Springer, Vienna.

Glaser, W. (1956). *In* "Handbuch der Physik" (S. Flügge, ed.), Vol. 33, pp. 123–395.

Grivet, P., and Septier, A. (1960). *Nucl. Instr. Methods* **6**, 125–156 and 243–275.

Hawkes, P. W. (1965a). *Optik* **22**, 349–368.

Hawkes, P. W. (1965b). *Phil. Trans. Roy Soc.* A **257**, 479–552.

Hawkes, P. W. (1965c). *Optik* **22**, 543–551.

Hawkes, P. W. (1965/6). *Optik* **23**, 244–250.

Hawkes, P. W. (1966). *Springer Tracts Mod. Phys.* **42**.

Hawkes, P. W. (1966/7). *Optik* **24**, 252–262 and 275–282.

Hawkes, P. W. and Cosslett, V. E. (1962). *Brit. J. Appl. Phys.* **13**, 272–279.

Klemperer, O. (1953). "Electron Optics," p. 135. Cambridge Univ. Press, London and New York.

Le Poole, J. B. (1964). *Proc. 3rd European Regional Conf. Electron Microscopy, Prague, 1964* Appendix, p. 8. Publ. House Czech. Acad. Sci., Prague.

Luneburg, R. K. (1964). "Mathematical Theory of Optics," pp. 234–239. Univ. of California Press, Berkeley, California.

Markovich, M. G., and Tsukkerman, I. I. (1960). *Zh. Tekhn. Fiz.* **30**, 1362–1368; *Soviet Phys.–Tech. Phys.* **5**, 1292–1298.

Meads, P. F. (1963). The theory of aberrations of quadrupole focusing arrays. Pt. I of a thesis, California; reproduced as UCRL–10807.

Melkich, A. (1944). Ausgezeichnete astigmatische Systeme der Elektronenoptik. Dissertation, Berlin.

Melkich, A. (1947). *Sitzber. Akad. Wiss. Wien, Math. Naturwiss. Kl., Abt. IIa* **155**, 393–471.

Ovsyannikova, L. P. and Yavor, S. Ya. (1965). *Zh. Tekhn. Fiz.* **35**, 940–946, *Soviet Phys.—Tech. Phys.* **10**, 723–726.

Picht, J. (1963) "Einführung in die Theorie der Elektronenoptik," 3rd ed., p. 81 ff. Barth, Leipzig.

Rose, H. (1966/7). *Optik* **24**, 36–59 and 108–121.

Scherzer, O. (1936). *Z. Physik* **101**, 593–603.

Scherzer, O. (1947). *Optik* **2**, 114–132.

Septier, A. (1961). *Advan. Electron. Electron Phys.* **14**, 85–205.

Septier, A. (1966). *Advan. Opt. Electron Microscopy* **1**, 204–274.

Septier, A., and van Acker, J. (1961). *Nucl. Instr. Methods* **13**, 335–355.

Strashkevich, A. M. (1961). *Radiotekhn. i Elektron.* **6**, 1562–1565; *Radio Eng. Electron.* (*USSR*) **6**, 1392–1396.

Strashkevich, A. M. (1963). *Zh. Tekhn. Fiz.* **33**, 512–517; *Soviet Phys.—Tech. Phys.* **8**, 380–383.

Strashkevich, A. M. (1964). *Zh. Tekhn. Fiz.* **34**, 1401–1408; *Soviet Phys.—Tech. Phys.* **9**, 1082–1086.

Sturrock, P. A. (1951). *Proc. Roy. Soc.* **A210**, 269–289.

Sturrock, P. A. (1955). "Static and Dynamic Electron Optics," Cambridge Univ. Press, London and New York.

CHAPTER 2.6

ELECTRON MICROPROBES

T. Mulvey

DEPARTMENT OF PHYSICS, UNIVERSITY OF ASTON IN
BIRMINGHAM, ENGLAND

2.6.1. Introduction

Focused electron beams or " probes " have been used from the earliest days of electron optics when the chief interest was to provide a fine, intense electron spot for the high-voltage oscillograph. This was generally done by forming an electron-optical image of a suitable source, such as a gas discharge tube or later a hairpin filament, by means of a magnetic solenoid. The solenoid was later provided with an iron shroud and became the basis of present day magnetic electron lenses. Spot sizes of about 100 μ could be produced in this way. The electron-optical techniques that were developed for the high-voltage oscillograph later proved useful in the design of the more sophisticated lenses needed for the electron microscope. The subsequent intensive study of the electron optics of the electron microscope in the last twenty-five years has led, in turn, to a considerable improvement in the performance of many allied instruments such as the scanning electron microscope (von Ardenne, 1938a), and the shadow electron microscope (Boersch, 1939). More recently the electron probe x-ray microanalyzer (Castaing, 1956) and electron beam machining equipment (Steigerwald, 1953), have become common laboratory tools. In all these instruments a finely focused electron beam, which may vary in size from a few Angstroms units (A.U.) to a few microns, has to be accurately positioned on a specimen and controlled in intensity. In the last few years it has also been realized that the performance of an electron microscope itself can be greatly improved by reducing the size of the focused illuminating beam to less than a micron by means of a double or triple condenser lens. Riecke (1960) succeeded in building such a three-stage illuminating system for the electron microscope in which a probe diameter of some 100 A.U. was obtained at the specimen. This enabled selected area diffraction patterns to be obtained from correspondingly small areas of the specimen, a technique that was first introduced by von Ardenne (1941). Duncumb (1962) has designed a special electron-probe illuminating system for a combined electron probe analyzer–electron microscope in which a probe of 1000 A.U. in diameter allows x-ray microanalysis to be carried out on selected areas of the specimen. Many difficulties are encountered in the production of such fine electron probes.

2.6.2. General Principles

This chapter is concerned with the electron-optical problems that have to be solved in order to obtain maximum intensity in the smallest possible probe bearing in mind the special applications of probes of different diameter.

Although the general principles underlying the design of electron-optical systems for the production of fine electron probes are similar, there are nevertheless important differences in the requirements to be fulfilled and these have important consequences for the electron-optical design. For example, the objective lens of a high-resolution electron microscope capable of resolving a few Angstrom units would not be suitable as the final lens in an electron probe x-ray microanalyzer or even a scanning electron microscope, since the different operating conditions in an electron probe analyzer would introduce excessive aberrations.

A. ELECTRON SOURCES

The overall design of any electron–probe system depends greatly on the type of electron source that is employed. A full account of this subject is given in Chapter 2.1 (Haine and Linder). It may be mentioned that the most common electron source is the triode gun with hairpin filament, which is generally suitable if the total probe current required is less than one milliampere. This type of source produces a spot with a Gaussian distribution; source size is therefore generally specified in terms of the halfwidth of this distribution. In special cases the sloping sides of the distribution can be inconvenient; this can often be improved by placing a suitable aperture below the electron gun (Haine and Mulvey, 1952) but usually such precautions are not necessary. The halfwidth of the electron source is usually in the region of 40-50 microns; since the final probe size may vary between one micron and a few Angstroms a demagnifying system must be provided with a demagnification in the range 100–100,000. For example, a shadow electron microscope may call for a demagnification of 100,000 ×, a scanning electron microscope 10,000 × and an electron probe microanalyzer perhaps 250 ×.

B. THE DEMAGNIFYING SYSTEM

Figure 1 shows the general arrangement of a two-stage demagnifying system. The first projector lens[1] produces a demagnified image I_1 of the source. This is further demagnified by the second lens which forms the final image I_2 of diameter d_2 on the specimen or workpiece. The specimen plane is usually fixed and cannot easily be changed. It is generally important that the final image should be physically accessible, that is, well clear of the lens body, to allow for specimen manipulation, the extraction of secondary electrons or x-rays from the specimen itself and sometimes for visual in-

---

[1] Often referred to as the first condenser lens. Electron-optically it corresponds to the projector lens in an electron microscope.

spection in an optical microscope. The condition of accessibility is difficult to fulfil without allowing the lens aberrations to increase to the point where they seriously enlarge the diameter of the electron probe. This can be countered by restricting the maximum lens aperture, but this reduces the maximum available probe current. The successful design is one which reconciles these conflicting requirements with the minimum sacrifice of performance.

It will be shown later that the condition of physical accessibility precludes the use of lenses of intrinsically low aberrations. No such restriction exists in the first projector lens so that its aberrations can be considerably smaller than those of the final lens. Hence the first projector lens need not contribute to the overall aberrations of the system even if the final lens provides no demagnification, a fact which simplifies calculations.

Before discussing the design of the electron–optical system in detail, it may be useful to discuss some fundamental limitations that are relevant to the production of fine electron probes.

C. THEORETICAL LIMITATIONS IN PROBE-FORMING SYSTEMS

In the absence of aberrations in the final lens (Gaussian image formation) the current I_p in a probe of cross-sectional area $\pi d_g^2/4$ and (small) semi-angle α_2 (Fig. 1) is given by the product of the area of the image, the solid

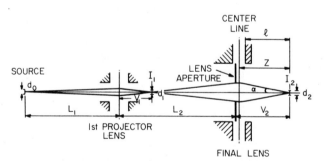

FIG. 1. Schematic arrangement of a two-stage demagnifying system.

angle that contains the impinging electrons and the " brightness " β of the source. Hence

$$I_p = \beta(\pi\alpha^2)\,\pi d_g^2/4 \tag{1}$$

where $\beta = (\varrho_c e V/\pi KT)$ is the current density per unit solid angle emitted by the source (see Chapter 2.1). ϱ_c is the current density at the cathode, e the electronic charge, K Boltzmann's constant, T the absolute temperature of the cathode, and V the accelerating voltage. The quantity β varies directly with accelerating voltage; a typical value for a hairpin filament would be 10^5 A/cm²/sr at 2700°K and 50 kV.

Equation (1) is adequate as it stands for calculating the expected probe current where lens aberrations are unimportant, for example, in the illuminating system of a conventional electron microscope in which the minimum spot size is several microns and the illuminating angle small ($< 10^{-3}$ radian). In general, however, lens aberrations are a controlling factor in the design.

D. EFFECT OF LENS ABERRATIONS

Aberrations enlarge the Gaussian image without increasing the number of electrons in it, thereby altering unfavorably the current-density distribution and, therefore, the effective probe size. Under these conditions the aperture angle and the demagnification have to be adjusted to give the most satisfactory compromise between probe size, probe current, and current distribution. The principal aberrations to be considered are spherical aberration, diffraction at the aperture, chromatic aberration caused by inadequate stabilization of electrical supplies, astigmatism caused by lens asymmetry, and space-charge aberrations. The effect of space charge will be referred to subsequently; fortunately, it need not be a serious problem for electron probes. Astigmatism can be corrected by means of suitable quadrupole lenses (Bertein, 1947, 1948; Scherzer, 1947) and chromatic aberration can be made negligible by using suitably stabilized supplies for the lens currents and accelerating voltage.

The problem then remains of combining spherical aberration, diffraction and the Gaussian image itself in the optimum proportions. A rigorous wave–optical computation for an arbitrary current distribution of the source would be difficult and has not yet been attempted.

A wave–optical solution is available (cf. Haine, 1954) for the limiting case in which the Gaussian image size is small compared with the central diffraction disk whose diameter d_d is given by

$$d_d = 1.22\,\lambda/a \tag{2}$$

Here λ is the electron wavelength and a the aperture angle.

It can be shown (Conrady, 1919) that if the lens is always focused to give the smallest disk, the diffraction disk is not significantly affected by spherical aberration provided that the path difference $\frac{1}{4}\,C_s a^4$ introduced by spherical aberration does not exceed the Rayleigh limit of $\lambda/4$. Under these conditions

$$C_s a^4/4 = \lambda/4 \tag{3}$$

that is,

$$a^4 = \lambda/C_s \tag{4}$$

that is,

$$a^3 = 0.82 \, d_d/C_s \tag{5}$$

$$a = 0.94(d_d/C_s)^{1/3} \tag{6}$$

and

$$d_{min} = 1.22 \, C_s^{1/4} \lambda^{3/4} \tag{7}$$

at 25 kV $\lambda = 0.077$ A.U. and the minimum possible value of C_s for an immersed specimen is 0.04 cm (Liebmann and Grad, 1951) so that $d_{min} = 8.0$ A.U.

In practice, for a probe formed just outside the lens body, C_s would be about 1 cm, giving $d_{min} = 15$ A.U.

Under these limiting conditions, the probe size is not greatly dependent on C_s but the probe current is negligible since it has been tacitly assumed that the Gaussian image size is negligible compared with the Airy diffraction disk. Probes in which a readily detectable current is required, for example, in a scanning electron microscope, must be appreciably larger than this.

When diffraction is negligible an exact geometrical solution can be given for the case of a source of uniform current density and a probe whose diameter is large compared with that of the Airy disk. This is a useful approximation for electron-probe analyzers and electron beam machining equipment.

In the Gaussian image plane the probe current I_p is related to the Gaussian image diameter by Eq. (1).

$$I_p = \beta(\pi^2/4)d_g^2 a^2$$

Each point on the source is imaged as a disk of confusion whose diameter in the Gaussian image plane is given by

$$d_s = 2C_s a^3 \tag{8}$$

Note that the current density in the image is then no longer uniform, but falls to zero toward the edge.

The overall diameter of the image d is thus

$$d = d_g + d_s \tag{9}$$

so that

$$I_p = \beta(\pi^2/4)(d - 2C_s\alpha^3)^2\alpha^2 \tag{10}$$

This has a maximum if

$$\alpha = (d/8C_s)^{1/3} \tag{11}$$

and

$$d_g = 3d/4 \tag{12}$$

giving a maximum current

$$I_p = \frac{9\pi^2}{256} \beta \frac{d^{8/3}}{(C_s)^{2/3}} \tag{13}$$

According to this calculation, the maximum probe current is obtained if the Gaussian size of the probe is set to three quarters of the finally desired probe size and the aperture angle adjusted in accordance with Eq. (11).

Equation (13) shows that the available probe current is directly proportional to the gun brightness β and to the 8/3 power of probe diameter. Spherical aberration enters as the 2/3 power of C_s; it is therefore an advantage to reduce C_s as much as possible.

In practice it is permissible to use a larger angle than that given by Eq. (11) provided that the lens focus is readjusted to give best results. A number of approximate solutions appropriate to particular applications have been given by Castaing (1960), Mulvey (1959), and Einstein et al. (1963). The " optimum " combination of Gaussian image size and lens aperture chosen by these authors differs in detail according to the various assumptions made and so leads to different values for the final probe current. It should be remembered that the choice of operating conditions is largely dictated by the application. For example, in electron beam machining it may be more important to have a well-defined cutting beam of nearly uniform intensity even if this means sacrificing total probe current, whereas for electron probe x-ray microanalysis in which the x-ray resolution limit is not set directly by the diameter of the electron probe, a nonuniform distribution, with the benefit of a larger probe current, may be quite acceptable.

In practice it is often preferable to follow the procedure used by Smith (1956; see also Pease and Nixon, 1965) in which, following von Ardenne (1938b), the aberration disks and Gaussian image are assumed to add in quadrature. Although arbitrary, this procedure yields results that correspond closely to optimum values derived by more complicated methods. The final probe diameter d is then given by

$$d^2 = d_g^2 + d_s^2 + d_d^2 + \text{etc.} \tag{14}$$

where d_g is the diameter of the Gaussian image, d_s the minimum disk of confusion because of spherical aberration measured in the plane of least confusion and d_d that due to diffraction. Hence

$$d^2 = \frac{4I_p}{\pi^2 \beta a^2} + \frac{C_s^2 a^6}{4} + \frac{(1.22\lambda)^2}{a^2} \tag{15}$$

that is,

$$I_p = \frac{\pi^2 \beta}{4} \left[d^2 - \frac{C_s^2 a^6}{4} - \frac{(1.22\lambda)^2}{a^2} \right] a^2 \tag{16}$$

This has a maximum value that is independent of λ when $a = a_{\text{opt}}$, where

$$a_{\text{opt}} = (d/C_s)^{1/3} \tag{17}$$

a value only slightly different from that given by Eq. (6).

The corresponding value of the current is given by

$$I_{p\max} = \frac{3\pi^2 \beta}{16} \left[\frac{d^{8/3}}{C_s^{2/3}} - \frac{4}{3} (1.22\lambda)^2 \right] \tag{18}$$

Equation (18) shows that the effect of diffraction is thus to reduce the maximum current; in the limit when $I_{p\max} = 0$, $d = 1.29\, C_s^{1/4}\lambda^{3/4}$, the smallest probe size obtainable, in reasonable agreement with the more rigorous result of Eq. (7).

Diffraction is not important if the first term in the brackets is appreciably larger than the second, namely, if

$$d^{8/3}/C_s^{2/3} > \tfrac{4}{3} (1.22\lambda)^2$$

or if

$$d > 1.29\, C_s^{1/4}\, \lambda^{3/4} \tag{19}$$

Diffraction is therefore important only for probes in the 5–20 A.U. range. In practice Eq. (18) may usually be simplified to

$$I_{p\max} = \frac{3\pi^2 \beta}{16} \frac{d^{8/3}}{C_s^{2/3}} \tag{20}$$

which corresponds to $d_s = d/2$ and $d_g = (\sqrt{3}/2)d$.

A good working rule in all probe systems is therefore to make an initial setting in which the demagnification is adjusted to produce a Gaussian image of about $0.9d$ and an aperture $a = (d/C_s)^{1/3}$. Further changes can then be made, if necessary, to suit particular requirements.

E. ELECTRON–OPTICAL DESIGN OF THE DEMAGNIFYING SYSTEM

Since the position of the final image is usually fixed (Fig. 1), the size of the Gaussian image has to be varied by altering the size of the intermediate image, simply by varying the current in the first lens. As the demagnification must be carefully chosen to give maximum probe current, it is generally preferable to use two lenses in an electron probe system even where it might be feasible to realize the necessary demagnification with one lens. A one-lens system can of course be constructed for a specific task but is usually far too inflexible for general use.

F. OPTIMUM ARRANGEMENT OF THE LENSES

Figure 1 shows the position of the intermediate image I_1. In the absence of aberrations the diameter d_1 of this image I_1 is given by

$$d_1 = d_0 \, v_1 / L_1 \tag{21}$$

where d_0 is the diameter of the source and L_1 and v_1 are the object and image distances, respectively, as shown in Fig. 1. The final probe diameter d_2 is similarly given by

$$d_2 = d_0 v_1 v_2 / L_1 \cdot (L_2 - v_1) \tag{22}$$

where $(L_2 - v_1)$ and v_2 are the object and image distance, respectively, of the second lens. Since v_1 is usually small compared with L_2 :

$$d_2 = d_0 v_1 v_2 / L_1 L_2 \tag{23}$$

The minimum value of v_2 is generally fixed by mechanical considerations of accessibility. For example, in an electron microscope it might be necessary to provide room for a specimen airlock; in a scanning electron microscope space must be provided for the secondary electron detector. The distance v_1 is approximately equal to the minimum projector focal length and as explained below this is usually of the order of a few millimeters. For a given overall length therefore the demagnification is controlled primarily by the product $L_1 L_2$ and will be a maximum if $L_1 = L_2$. It should be noted however, that if v_2 is greater than L_2 the final lens will *magnify* the intermediate image. This has two disadvantages; image defects in the first lens will be magnified and may appear in the final image plane. More serious, however, is the fact that the aperture angle of the intermediate lens and the convergence of the final lens is larger than it need be; thereby increasing the aberrations of the final lens, as discussed below. Under these circumstances it may be better to make L_2 larger than L_1 even though this may mean increasing the overall length of the instrument. In general, every effort should be made to see that each lens, and especially the last one, demagnifies the preceding image; this is best done by keeping the distance v_2 to an absolute minimum even if this causes some difficulties in the mechanical design.

G. Special Features of a Demagnifying System

In electron optical *magnifying* systems the specimen is usually more or less deeply immersed in the magnetic field of the objective lens. This gravely restricts the size of the specimen and its ease of handling. Consequently, only that part of the field that lies between specimen and the image plane contributes to the imaging process. In the subsequent magnifying (projector) lenses the whole of the magnetic field contributes to the focusing properties and this leads to a sharp distinction between the electron-optical properties of objective and projector lenses. In addition the magnification of both the objective and the projector lens is large and this simplifies the calculation of the aberrations. The situation is usually quite different in electron probe instruments. Here the specimen is generally outside the magnetic field of its associated lens so that the whole of the field contributes to the imaging process. One is therefore primarily concerned with the properties of " projector " lenses or at least with the range of excitation where the properties of objective or projector lenses do not differ from each other. The properties of electron lenses are dealt with in Chapter 2.1. It is probably true to say, however, that the extensive literature on the focal properties of projector and objective lenses has concentrated on finding simplified methods of presenting large amounts of data in the form of universal curves. This apparent simplicity has sometimes been obtained at the price of obscuring some important physical limitations in actual lenses. For example the flux density B_p in the parallel part of the air gap is a primary design parameter since the saturation properties of iron set a limit to the possible combinations of lens geometry and excitation. The flux-density limitation in fact determines the absolute minimum focal length that can be obtained in practice. In the usual representation of lens properties the focal properties are expressed in terms of an arbitrary unit such as the sum of the lens bore D and lens gap S. Thus it can be shown that $f/(S/D)$ has a minimum value, where f is the projector focal length. However, the minimum absolute focal length f_{min} is quite different from the relative minimum focal length $[f/(S + D)]_{min}$

and occurs at a much lower excitation. It is possible to replot the lens data
of Liebmann and Grad (1951) in such a way as to provide the necessary
design data for electron probe systems directly (Mulvey and Wallington,
1965). This has been done in Fig. 2 in which the quantity $f_{\text{proj}}B_p/V_r^{1/2}$ is
plotted as a function of lens excitation NI/NI_A. Here f_{proj} is the actual

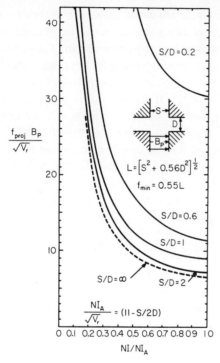

FIG. 2. Absolute projector focal length as a function of lens excitation.

projector focal length in cm, B_p is the flux density in the parallel part of
the air gap, in gauss, and V_r is the relativistically corrected accelerating
voltage. The excitation parameter NI/NI_A is the ratio of the number of
ampere-turns NI actually applied to the lens and the number of ampere-
turns NI_A that would be required to produce the absolute minimum focal
length possible. In the range of S/D from 0 to 2 as shown in Fig. 2, NI_A may
be found simply from the following:

$$NI_A/V_r^{1/2} = (11 - S/2D) \tag{24}$$

It can also be shown (Mulvey and Wallington, 1965) that the absolute minimum focal length $f_{\min}$ is given by

$$f_{\min} = 0.55\,L \tag{25}$$

where

$$L = (S^2 + 0.56\,D^2)^{1/2} \tag{26}$$

Equations (24) and (25) are useful inasmuch as they enable the minimum projector focal length of any given lens to be ascertained quickly, and therefore serve as a useful check on any given design. The curves of Fig. 2, which enable a suitable design to be chosen in the first place, show that as the excitation is increased the projector focal length falls steadily and reaches a minimum when $NI = NI_A$. If the excitation is increased further the projector focal length begins to rise once more (not shown in Fig. 2). The absolute projector focal length falls rapidly as the ratio S/D is increased from 0.2 to 1 then leveling off so that very little is gained by making S/D greater than 2. A convenient practical value for the first lens is $S/D = 1$. Table I summarizes the minimum focal length obtainable with such a lens, assuming $B_p = 20$ kG. The focal length at any other field strength is obtained by scaling the focal length inversely as the field strength.

TABLE I

MINIMUM PROJECTOR LENGTH FOR $S/D = 1$ AND $B_p = 20$ KG

V (kV):	10	20	30	50	75	100
V_r (kV):	10.1	20.4	30.9	52.5	80.6	110
NI_A (amp-turn):	1060	1490	1850	2410	2980	3480
$f_{\min}$ (cm) ($B_p = 20$ kG):	0.045	0.067	0.079	0.103	0.128	0.15

For example, at an accelerating voltage of 50,000 V the minimum projector focal length obtainable is about 1 mm at 20 kG and 2 mm at 10 kG; a corresponding excitation of 2,400 ampere-turns has to be provided irrespective of the value of B_p. It should be remembered that this figure represents the ampere turns in the air gap itself; a further allowance of some 5% should be made for losses in the iron circuit. A demagnification of about 100 × is easily possible in a projector stage such as this. If a larger demagnification is needed further stages, designed in a similar way, may be added.

H. DESIGN OF THE FINAL LENS

Figure 1 shows the important electron optical parameters of the final lens. The image I_2 is located at a distance Z from the center line of the lens. The working distance l is defined as the distance between the *inside* of the final poleface and the focused image. This is the maximum possible working distance; in practice, an allowance has to be made for the thickness of the final polepiece. In designing the final lens it is important to secure the greatest working distance for a given amount of spherical aberration. Figure 3 shows l/C_s plotted as a function of NI/NI_A for a series of symmetrical lenses for various ratios of S/D. Clearly, the ratio l/C_s improves as S/D is reduced from 2 to 0.2. A further slight improvement can be obtained by using even smaller S/D ratios but this is not usually advisable as the increase in flux density in the gap might cause difficulties with the magnetic circuit. Figure 3 shows that for the best S/D ratio a lens excitation $NI/NI_A \simeq 0.9$, that is, just less than for minimum focal length, is required.

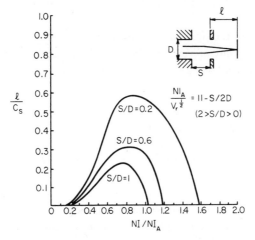

FIG. 3. Ratio of working distance (l) to spherical-aberration coefficient (C_s) as a function of the excitation parameter NI/NI_A for various S/D.

Figure 3 also shows that l cannot exceed $0.6\ C_s$. Having thus determined NI/NI_A, and S/D it remains to fix the actual size of the pole pieces. This can be done by noting (Fig. 1) that

$$l = Z - S/2 \qquad (27)$$

The distance l is determined by the specimen handling arrangements. Table II gives a series of values of (l/S) over a useful range of excitations for

TABLE II

$S/D = 0.2$

NI/NI_A:[a]	0.7	0.8	0.9	1.0
$NI/V_r^{1/2}$:[b]	7.63	8.74	9.8	10.9
l/S:	2.7	2.0	1.5	1.1

[a] I (Amperes).   [b] V (volts).

$S/D = 0.2$, from which the required value of S may be deduced. In particular, the lens with $S = 2\, l/3$ gives the best ratio of working distance to spherical aberration coefficient, corresponding to the peak to the curve for $S/D = 0.2$ in Fig. 3.

A lens designed on these lines will have fairly large bores, usually several centimeters, whereas in a projector lens, or an electron microscope objective lens where l can be negative the bore and gap would be not more than a few millimeters.

This means that the optimum lenses for electron probes are relatively poor lenses from the point of view of aberrations. It can generally be reckoned that such lenses will have a value of C_s ten times greater than that of a good electron microscope objective.

I. MAGNETIC FIELD AT THE SPECIMEN

The chief practical disadvantage of the optimum symmetrical lens for an electron probe system is that the axial field distribution is broad; an appreciable field strength can occur at the specimen and this is usually undesirable. This field can be reduced by reducing the bore of the final pole piece which causes the axial field to fall off much more rapidly than before. If at the same time, the other bore is increased so as to maintain the average bore size constant the first-order focal properties are hardly changed. This comes about because the maximum field on the axis is hardly affected as shown in Fig. 4 and first-order focal properties are not greatly influenced by the form of the axial field distribution.

Liebmann (1955) has treated the case of a highly asymmetrical lens by regarding the smaller bore as a weak pinhole lens. This approximation is entirely adequate for first order properties such as focal length, chromatic aberration coefficient, and so forth, but is liable to considerable error, however, when applied to the calculation of C_s since this aberration is very

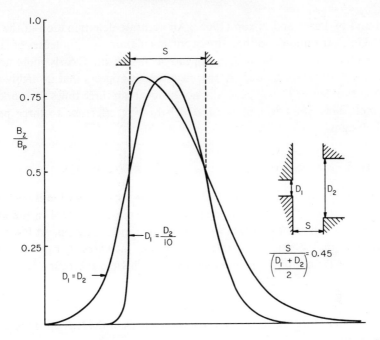

FIG. 4. Axial field distribution for a symmetrical and an unsymmetrical lens of the same $S/D = 0.45$.

sensitive to the precise form of the rapidly changing axial field in the gap region near the pinhole as shown in Fig. 5. Experiment indicates that this approximation is adequate for small amounts of asymmetry but consistently underestimates C_s for lenses of appreciable asymmetry; for ratios of the bores approaching ten, the error could well be of the order of 100%

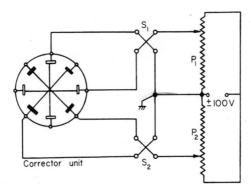

FIG. 5. Arrangement of an eight-pole astigmatism corrector to correct second order astigmatism.

as found by Pease and Nixon (1965). An accurate determination of the actual field distribution and the subsequent calculation of C_s in lenses of large asymmetry is more difficult than appears at first sight. Clearly more accurate calculations are needed; in the meantime it appears that in highly unsymmetrical lenses C_s will generally not be less than three times the working distance. This point will be illustrated below by reference to three practical designs.

J. Effect of Lens Conjugates on Spherical Aberration

Another factor that can cause an increase in lens aberration is sufficient demagnification. The spherical aberration coefficient C_s quoted in the literature is generally that of a lens working at high magnification. At low magnifications the spherical aberration coefficient increases. It can be shown (Petrie, 1962) that for a *thin lens* the effective spherical aberration coefficient

$$C_s' = C_s(1 + V/U) \tag{28}$$

where V and U are the image and object distances of the lens or in the notation of Fig. 1:

$$C_s' = C_s \left[1 + \frac{V_2}{L_2 - V_1} \right] \tag{29}$$

$$\simeq C_o \left[1 + \frac{V_2}{L_2} \right] \tag{30}$$

Thus if L_2 is reduced from a value large compared with V_2 to a value equal to V_2 the spherical aberration will be doubled. Archard (1958) has shown that for strong lenses Equation (28) overestimates the value of C_s' typically by about 40% at unit magnification but is a useful first approximation. Where optimum performance is required, it is clearly desirable to arrange that the ratio L_2/V_2 should be appreciably greater than one.

K. Correction of Astigmatism

In the previous analysis of the probe–forming system it was assumed that astigmatism was negligible.

Astigmatism in the final lens arises from lack of axial symmetry in the lens construction. The principal causes of this are

(1) Lack of roundness of the lens bores and corrugation of the pole faces;

(2) Unsymmetrical charging up of apertures;

(3) Lack of symmetry in the coil winding itself;

(4) Inhomogeneity of the pole piece material.

Astigmatism makes its appearance in the form of two line foci separated by an axial distance z_A ; between these two line foci there is a disk of minimum confusion whose diameter $d_A = z_A \alpha$ where α is the semiaperture angle. Fortunately, astigmatism can be corrected by means of a suitably oriented cylindrical lens or quadrupole (Bertein, 1947; Scherzer, 1947). Astigmatism will have to be corrected if d_A becomes comparable with d. From Equation (2)

$$d_A = z_A(d/C_s)^{1/3} \tag{31}$$

that is, correction is needed if

$$d \leq (z_A^3/C_s)^{1/2} \tag{32}$$

With care in machining the bores and pole faces, z_A can be kept down to 1–$2\,\mu$.

So for example, with $C_s = 2.5$ cm, $z_A = 2 \times 10^{-4}$ cm, correction will be needed if d is 200 A.U. or less. It is useful to have an astigmatism corrector even for larger probe diameters, as astigmatism arising from unsymmetrical charging of the aperture can then easily be corrected.

With fine electron probes lack of symmetry in the coil winding itself can give rise to astigmatism. In certain cases (cf. Riecke, 1962) a form of astigmatism of threefold symmetry has been observed whose correction calls for a corrector of threefold symmetry. This type of aberration will tend to arise in lenses with a large bore and gap resulting in poor magnetic screening between coil winding and lens axis, thereby allowing the nonuniform axial field from the coil to perturb the field created by the pole pieces. For lenses with small S/D ratios, this problem is less severe and can be overcome by keeping the windings well clear of the pole gap.

Astigmatism is conveniently corrected by creating a magnetic or electrostatic field of twofold or threefold symmetry in the path of the beam in the region of maximum radius, that is, as near to the lens as possible. A simple electrostatic octupole corrector for use in electron probe systems has been described by Mulvey (1959); this consists of eight insulated electrodes placed around the axis. For the correction of twofold astigmatism opposite electrodes are connected together as shown in Fig. 5. An analogous magnetic system consisting of eight coils is also useful.

This type of corrector may be regarded as two quadrupoles producing astigmatic components at right angles to each other and whose amplitudes are proportional to the voltages V_x and V_y in Fig. 5. In operation V_y is set to zero and the astigmatism reduced to a minimum by V_x ; the remaining astigmatism is then removed by adjusting V_y . The same electrode system can also be used to correct both second- and third-order astigmatism by superimposing additional potentials on the electrodes (see Kawakatsu and Kanaya, 1961). By using specially designed circular potentiometers with insulated wiper arms, it is possible to devise circuits in which the control of amplitude is separated from that of changing the direction of the correction. The increased complexity of the circuitry is usually justified by the increased ease of applying the correction.

It should be noted, however, that in order to avoid interaction between the controls it may be better in practice to provide a separate electrode or coil system for correcting third-order astigmatism; a nine or twelve electrode system is excellent for this purpose.

L. Stability Requirements

If the voltage and current vary with time, a point in the image will be enlarged to a disk of diameter d_c where

$$d_c = C_c \, \alpha(dV/V - 2dI/I), \tag{33}$$

where C_c is the chromatic aberration coefficient, dV/V and dI/I are the changes in voltage and current, respectively; α is the semiangle as before. For lenses with a positive working distance $C_c \backsimeq f$, where f is the focal length.

Chromatic aberration becomes serious if d_c becomes comparable with d; that is, if

$$dV/V - 2 \, dI/I \leq (d^2 C_s)^{1/3}/f \tag{34}$$

for example, if $f = 1$ cm, $C_s = 2$ cm, $d = 25$ A.U.:

$$(dV/V - 2 \, dI/I) = 5 \times 10^{-5} \tag{35}$$

that is, for a 25-A.U. probe the stability of the accelerating voltage and the lens current supplies should preferably be better than one or two parts in 10^5.

2.6.3. Practical Applications

Besides the theoretical principles outlined above there are many other factors to be considered in the design of a practical probe forming system. These are possibly best illustrated by reference to three specific electron probes covering between them a range of probe sizes from $5\,\mu$ to 50 A.U.

A. AN ELECTRON PROBE FOR THE MICRON RANGE

Figure 6 shows the construction of a fairly simple electron–optical system (Mulvey, 1959) for producing a high-intensity probe in the range 5–$0.5\,\mu$. The electron gun (1) produces a beam of electrons which passes through a fixed aperture into a Faraday cage (4) which enables a check to be made on gun brightness (β). This part of the column also acts as a vacuum pumping section (3). The first projector lens (5) has a bore of 1.2 cm and gap of 1.0 cm ($S/D = 0.6$) and demagnifies the source by a maximum of 15.5 times, giving a total demagnification of some $100 \times$, sufficient for a probe of one half a micron diameter. Greater demagnifications could be obtained, if needed, by reducing the lens gap and bore. Larger probes sizes are obtainable of course, simply by reducing the excitation of the first projector lens.

The final demagnifying lens (7) has asymmetrical polepieces to reduce the magnetic field at the specimen (12) as much as possible. The larger bore is 6 cm in diameter and the small bore is 0.6 cm diameter, with a gap of 1.7 cm giving a focal length of 2.8 cm with $C_s \simeq 3.5$ cm. An excitation of some 1300 At (ampere-turns) at 30 kV is needed. The lenses axes are prealigned by accurately machined spigots and bolted together rigidly. The employment of large bores helps considerably in this respect. The only alignment mechanisms needed are the adjustable anode plate which enables the electron beam to be tilted until it is parallel to the optical axis and the translation of the gun as a whole (2) which allows the beam to be brought onto the lens axes. The aperture (9) of the final lens is mounted in a cone that can be inserted or removed through the small bore. The wide bore of the final lens allows plenty of room for an 8 pole astigmatism corrector located at position (8) (detail not shown in Fig. 6) and an electrostatic beam deflector system (10) which enables the electron probe to be moved electrically over an area of the sample of approximately $120 \times 120\,\mu$. As shown in the figure the electron–probe system is incorporated in an electron probe x-ray microanalyzer but as it is a self-contained unit, it can easily be detached and used for example as a fine focus x-ray source, an x-ray projection microscope (cf. Brümmer *et al.*, 1963), or for simple electron beam machining applications.

B. 1000-A.U. Electron–probe System for a Combined Electron Microscope–Microanalyzer

Here a finer electron probe was required for the examination of details observed in the electron microscope. Figure 7 (Duncumb, 1962) shows an electron probe of 1000 A.U. in diameter for the examination of thin specimens as prepared for normal electron microscopy. The separation between the probe forming lenses must now be greater than that shown in Fig. 6 in order to produce the necessary demagnification of about 500 times.

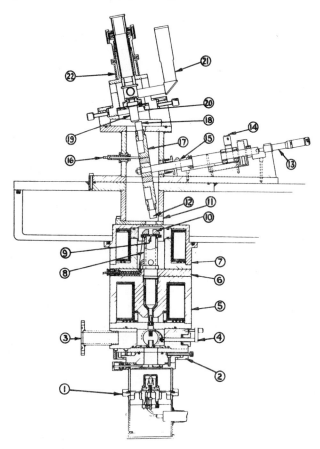

FIG. 6. Two-stage electron probe forming system (Mulvey, 1959) for producing probes in the 5–0.5 μ diameter range. Working distance 1.0 cm.

Referring to Fig. 7, the electron gun that operates up to 50 kV is provided with a filament than can be centered under vacuum. The source is demagnified up to 25 × by the *condenser lens* ($S/D = 1$) whose minimum focal length is about 1.0 cm. The pole pieces are removable so that different minimum focal lengths could be obtained if needed by changing the pole-piece bores. The intermediate lens can be traversed horizontally to bring it

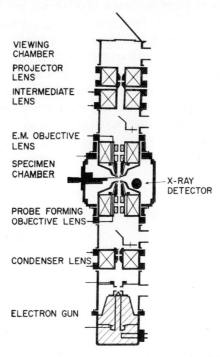

FIG. 7. Two-stage electron probe forming system (Duncumb, 1962) for producing a probe of 1000 A.U. in diameter. Working distance 1.1 cm.

onto the axis which is defined by two fixed reference points, namely, the center of the objective lens and the center of the anode of the electron gun.

The *probe-forming objective lens* (larger bore 1.9 cm diameter, smaller bore 0.32 cm, gap 0.32 cm) provides a further demagnification of 20 with a working distance of 0.95 cm, and a focal length of 1.3 cm. Probe currents of 10^{-9} A can be obtained at 30 kV for a probe diameter of 1000 A.U. A special feature of this lens are the extended pole pieces that provide extra space in the specimen chamber.

An accurate calculation of focal properties is difficult with such pole pieces; measurements by Duncumb indicate that an excitation of 1200 At (ampere-turns) at 30 kV is needed, some 30% more than calculated according to Liebmann's data for conventional pole pieces. The measured value of C_s (3–4 cm) is somewhat lower than that calculated (5.6 cm) but this could well be accounted for by the effect of the magnetic circuit in altering the slope of the axial field distribution and possible experimental errors in determining the value of C_s

The probe can be magnified some 12,000 times in a three stage electron microscope mounted above the probe forming system; this greatly helps in adjusting the electron optical system. Both the *probe forming objective lens* and the *electron microscope objective* lens are provided with magnetic deflector coils which enable the position of the scanning beam and the magnified image in the electron microscope to be adjusted electrically. The objective lens is similar in construction to the probe forming lens but operates at a shorter focal length (0.9 cm) and higher excitation (1620 At, 30 kV).

Since the probe forming lens has to operate at comparatively large angles (10^{-2} rad) for electron probe analysis and at small angles ($< 10^{-3}$ rad) when used as the illuminating system of an electron microscope it is necessary to change the aperture sizes frequently. This can be done under vacuum by an aperture-changing mechanism.

C. A. 50-A.U. Probe for a Scanning Electron Microscope

Good electron–optical design is of decisive importance in the production of probes smaller than 100 A.U.

Figure 8 shows a three-stage electron probe system (Pease and Nixon, 1965) capable of forming a 50 A.U. probe, for scanning electron microscopy. The first and second pole piece lenses ($S/D = 2$) provide a combined maximum demagnification of 3500 × which is increased to 14,000 by the last lens which was designed from Liebmann's (1955) data. The upper bore is 0.4 cm in diameter, the lower bore is 3.6 cm in diameter with a gap of 0.3 cm giving a calculated working distance l of 0.5 cm and $C_c = 0.8$ cm in good agreement with the measured values. The calculated value of C_s was 1.0 cm but experimental measurement gave $C_s = 2.0$ cm ($\pm$ 0.05 cm).

The calculated minimum probe diameter for these conditions is $d_{min} = 52$ A.U. for $I = 10^{-12}$ A at 15 kV.

The best results so far obtained at this voltage was $d_{min} = 75$ *A.U.* for $i = 4 \times 10^{-13}A$, but a probe size of 50 *A.U.*, diameter was obtained for $i = 8 \times 10^{-13}$ A at 30 kV.

From Fig. 8 it may be seen that a rigid construction has been adopted, with generously designed magnetic circuits in order to minimize undesirable fields on the axis which can cause difficulty in alignment. The *first* and *second lenses* can be moved horizontally in an enclosing framework so that their axes coincide with that of the *final lens.*

The corrugation of the pole faces was less than 2.5 μ and the bores had an ellipticity ($r_{max} - r_{min}$) of less than 0.5 μ which according to the calcu-

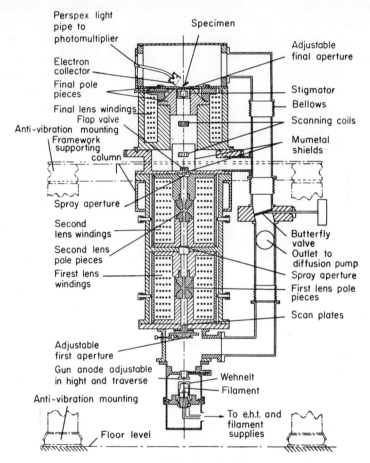

Perspex light pipe to photomultiplier

Specimen

Electron collector

Adjustable final aperture

Final pole pieces

Stigmator

Final lens windings

Bellows

Flap valve

Scanning coils

Anti-vibration mounting

Framework supporting column

Mumetal shields

Spray aperture

Second lens windings

Butterfly valve

Second lens pole pieces

Outlet to diffusion pump

Firest lens windings

Spray aperture

First lens pole pieces

Scan plates

Adjustable first aperture

Gun anode adjustable in hight and traverse

Wehnelt

Filament

Anti-vibration mounting

To e.h.t. and filament supplies

Floor level

FIG. 8. Three-stage electron probe forming system (Pease and Nixon, 1965) for producing a probe of some 50 A.U. in diameter. Working distance 0.5 cm.

lations of Archard (1953) makes Z_A less than 1 μ. Accordingly, from Eq. (32) no correction should be needed for astigmatism caused by bore ellipticity. This was borne out by experiment. An eight-pole electrostatic astigmatism corrector was fitted, but was found to be unnecessary when the column was clean.

D. ELECTRON PROBES OF HIGH POWER

In the preceding examples, the probe current was fairly low—less than a few microamperes.

In electron equipment designed for welding and cutting massive samples probe currents can be several orders of magnitude higher. Whereas this gives rise to engineering problems in the manipulation of the workpiece under vacuum or the shielding of the electron optical system (see Bas *et al.*, 1962) from vaporized metal, the general design principles remain unchanged. Generally speaking in such equipment, the working distance and consequently the spherical aberration of the final lens will be an order of magnitude greater than that, for example, of an electron probe analyzer.

Space-Charge Effects

When a beam of electrons is converged to a focus by a lens, collisions between electrons, together with the Coulomb forces of repulsion tend to diverge the beam, and introduce aberrations.

In practice space-charge effects are partly neutralized by the presence of a large number of positive ions generated by collisions between electrons and residual gas molecules in the vacuum, since working pressures are usually in the range $10^{-4} - 10^{-5}$ torr.

It can be shown (Glaser, 1956) that in the absence of positive ions a perfect lens would bring the electrons to a minimum disk of confusion d_m given by

$$d_m = 2af \exp(-\alpha^2/\sigma) \tag{36}$$

where α is the semiaperture angle, f the focal length, and σ a space-charge factor given by

$$\sigma = \frac{1}{2\pi\varepsilon_0} \left(\frac{m}{2e}\right)^{1/2} \frac{I}{V^{3/2}} = 3.04 \times 10^4 \frac{I}{V^{3/2}} \tag{37}$$

where I is the probe current, V is the accelerating voltage, ε_0 the permittivity of free space, and m/e the ratio of mass to charge of the electron.

Substitution of typical values for the various constants in Eq. (36) shows that the space-charge limit is not nearly as serious as that set by spherical aberration, even for the highest probe currents in present use.

However, Dolder and Klemperer (1955) have pointed out that Eq. (36) is based on the assumption of a continuous static distribution of space charge and this is generally not applicable to electron-optical instruments.

They showed experimentally that defocusing effects can be observed at lower values of current than those given by Eq. (36); more important they measured by means of experimental ray tracing the longitudinal spherical aberration caused by space charge. In the range of probe currents of interest in the present discussion the sign of the spherical-aberration coefficient was found to be negative. Unfortunately, their experimental arrangement did not allow the focused spot to be observed directly.

A similar result has been obtained theoretically by Kanaya et al. (1965) by adapting the eiconal method of evaluating lens aberrations to the problem of space charge. Their results show that the effect of space charge due to the beam itself is to defocus the probe and to introduce aberrations analogous to the five Seidel aberrations of classical optical and electron–optical systems. In particular, the sign of the aberration coefficients are negative; that is, they tend to cancel the geometrical aberrations of the final lens. In particular, the only important aberration is that of defocusing; this can be corrected simply by strengthening the excitation of the final lens.

Space-charge effects are magnified considerably if ions are used instead of electrons. It has been shown experimentally by Kanaya and his co-workers (1964) that even with argon ions, defocusing of the probe is still the only serious aberration, and the maximum probe current is limited by the spherical aberration of the final lens.

Further experiments by Hamisch et al. (1964) who used an electron microscope to observe directly a focused probe of some 43 μ in diameter have confirmed that even for currents as low as one microampere appreciable defocusing takes place, which is easily corrected by increasing the lens strength. In addition aberrations of magnitude about 3000 A.U. were observed at these currents. The authors explain these effects in terms of the collision of electrons in the probe-forming system resulting in an increase both in the chromatic energy spread of the electrons and a lateral spread in position.

It may therefore be concluded that appreciable space-charge effects take place even in probes carrying a current as low as one microampere, but with suitable adjustment the ultimate performance of an electron probe forming system is limited fundamentally by the brightness of the source and the aberrations of the final lens.

REFERENCES

Archard, G. D. (1958). *Rev. Sci. (Instr.* 29, 1049.
Archard, G. D. (1953). *J. Sci. (Instr.* 30, 352.

Bas, E. B., Cremosnik, G., and Lerch, H. (1962). *Trans. 8thVacuum Symp. 2nd Intern. Congr.* pp. 817–829. Macmillan (Pergamon), New York.

Bertein, F. (1947). *Ann. Radioelec. Compagn. Franc. Assoc. T.S.F.* 2, 379–408.

Bertein, F. (1948). *Ann. Radioelec. Compagn. Franc. Assoc. T.S.F.* 3, 49–62.

Boersch, H. (1939). *Naturwissenschaften* 27, 418.

Brümmer, O., Brauer, K. H., and Suwalski, G. (1963). *Z. Angew. Phys.* 16, 27–37.

Castaing. R. (1956). *Proc. Int. Conf. Elec. Micros.* London 1954, p. 301–304. Royal Microscopical Society, London.

Castaing, R. (1960). *Advan. Electron. Electron Phys.* 13; 317–386.

Conrady, A. E. (1919). *Monthly Notices Roy. Astron. Soc.* June, p. 575.

Dolder, K. T., Klemperer, O. (1955). *J. Appl. Phys.* 26, 146–147.

Duncumb, P. (1962). *Proc. Electron Microscopy 5th Intern. Conf. Electron Microscopy Philadelphia, 1962.* Vol. I, KK4. Academic Press, New York, 1962.

Einstein, P. A., Harvey, D. R., and Simmonds, P. (1963). *J. Sci. Instr.* 40, 562–567.

Glaser, W. (1956). *In* "Handbuch der Physik" (S. Flügge, ed.), Vol. 33, p. 379. Springer, Berlin.

Haine, M. E. (1954). *Advan. Electron. Electron Phys.* 6, 295–370.

Haine, M. E., and Mulvey, T. (1952). *J. Opt. Soc. Am.* 42, 763–773.

Hamisch, K., Loeffler, K. H., and Kaisler, H. J. (1964). *Proc. Intern. Conf. Electron Microscopy, Prague, 1964* p. 11, Czechoslovak Acad. Sci., Prague.

Kanaya, K., Kawakatsu, H., Matsui, S., Yamazaki, H., Okazaki, I., and Tanaka, T. (1964). *Optik* 21, 399–422.

Kanaya, K., Kawakatsu, H., and Yamazaki, H. (1965). *Brit. J. Appl. Phys.* 16, 355.

Kawakatsu, H., and Kanaya, K. (1961). *Bull. Electrotech. Lab. (Tokyo)* 25, 641–656 & 801–814.

Liebmann, G. and Grad E. M. (1951). *Proc. Phys. Soc.* B 64, 956–971.

Liebmann, G. (1955). *Proc. Phys. Soc.* B 68, 682–685.

Mulvey, T. (1959). *J. Sci. Inst.* 36, 350–355.

Mulvey, T., and Wallington, M. (1965). Unpublished material.

Pease, R. F. W. and Nixon, W. C. (1965). *J. Sci. Instr.* 42, 81–85.

Petrie, D.P.R. (1962). *Proc. Int. Congr. Elec. Micros.* Philadelphia, 1, p. kk–2. Academic Press (London and New York).

Riecke, W. D. (1960). *Proc. 4th Intern. Conf. Electron Microscopy, Berlin, 1958.* Vol. 1, p. 189. Springer, Berlin.

Riecke, W. D. (1962). *Optik* 19, 81–116.

Scherzer, O. (1947). *Optik* 2, 114–132.

Smith, K.C.A. (1956). Ph.D. dissertation. Cambridge University.

Steigerwald, K. H. (1953). *Physik. Verhandl.* 4, 123.

von Ardenne, M. (1938a). *Z. Physik.* 109, 553–572.

von Ardenne, M. (1938b). *Z. Physik* 108, 338.

von Ardenne, M. (1941). *Z. Physik* 117, 515.

AUTHOR INDEX

Number in parentheses are references unmbers and are included to assist in locating references in which authors' names are not mentioned in the text. Numbers in italics refer to pages on which the references are listed.

A

Adler, S. E., 185, 186, *226*
Agar, A. W., 236, *249*
Alekseev, A. C., 217, *226*
Alfvén, H., 33, *44*
Allen, J. E., 215, *229*
Alon, G. I., 185, 186, 207, *226*
Amboss, K., 420, 437, *467*
Andryushchenko-Lutsenko, N. I., 193, 217, *229*
Arnal, R., 302, 303, *305*
Archard, G. D., 273, 292, *305*, 464, *467*, 484, 491, *494*
Arnaud, M., 185, 221, *226*

B

Barber, E., 164, *226*
Bardin, B. M., 208, *228*
Barthere, J., 319, 326, 327, 328, *352*
Bass, E. B., 299, *305*, 492, *494*
Batskikh, G. J., 193, 217, *229*
Bayh, W., 300, *305*
Bellman, R., 62, *98*
Berglund, S., 215, *226*
Bernard, M. Y., 289, *305*, 452, *467*
Bernstein, H. J., 165, *229*
Bertein, F., 473, 485, *494*
Bertram, S., 12, *44*
Birkhoff, G., 68, *98*
Birss, R. R., 212, *226*
Bleeker, J. J., 219, *226*
Blewett, J. P., 191, *226*, *408*
Bloomer, R. W., 238, 239, *249*

Bogart, L., 200, *226*
Bok, A. B., 449, *467*
Boersch, H., 235, *249*, 292, 302, 303, 304, *305*, *306*, 470, *494*
Borchards, P. H., 246, *250*
Borisov, V. S., 217, 223, 226, *226*
Bouvier, P., 302, *305*
Bowman-Manifold, M., 115, *161*
Brack, K., 293, *306*
Bradley, D., 237, 238, *249*
Brauer, K. H., 488, *494*
Braunersreuther, E., 174, 203, 204, 212, 214, *226*, *408*
Brewer, G. R., 103, 115, 117, 119, 120, 121, 122, *162*
Bricka, M., 244, *249*
Brown, D. E., 191, *228*
Brown, R. A., 202, *226*
Brown, W., 181, *226*
Bruck, H., 244, *249*
Brümmer, O., 488, *494*
Buckey, C. R., 103, 115, 121, 122, *162*
Burfoot, J. C., 433, 434, 435, 436, 437, *467*
Burson, S. B., 186, *226*
Busheva, G. K., 197, *229*
Buss, L., 200, *226*

C

Cahen, O., 185, 221, *226*
Caldecourt, V. J., 185, 186, *226*
Carathéodary, C., 428, *467*
Carré, B. A., 73, *98*
Castaing, R., 470, 475, *494*
Chako, N., 420, *467*

SUBJECT INDEX

A

Aberrations, 411-468
  analyses, 412
  aperture, 412, 434-435
  chromatic, 411
    immersion lens, 295
    paraxial, 255-258
    single lens, 279-282
  coefficients, 339, 440-466, 481
  electrostatic lenses, 251-260
    third-order, 261-265
  Fermat's principle, 413-419
  geometrical, 252, 411-412, 415-417
  magnetic electron lens, 317
  mechanical, 411-412
  and probe size, 473-476
  relativistic, 411
  rosette, 436, 439
  space-charge, 412
  spherical, 259-260, 283-285, 287, 422-423, 473
    constants, 263-264
    determination, 267
    disk, 261-262, 267
    lens conjugates, 484
    strong magnification, 262-263
  star, 436
  symmetry, 420-439
Absorption, 166
Acceptance, 401-402
Air gap, 24
Ampere's
  law, 22, 25
  theorem, 21, 26, 312
Amplifier, dc, 191-193
Analyzer, electron probe, 492
Anisotropy, 155-156
Anticoma, 430-432

Aperture
  aberrations, 412, 434-435, 463
    coefficients, 450-452
    and line focus, 460-462
  quadrupole, 360
  single lens, 296
Araldite, 103, 178
Astigmatism, 424-425, 430-432, 473
  axial, 301
  bore ellipticity, 491
  correction, 484-486
  eight-pole corrector, 483
  off-axis, 260

B

Barium oxide, 236
Beamlets, Gaussian. 88-92
  annular, 92-93
  space charge, 91
Beams
  isogenous, 41
  particle, 37-43
Biot and Savart equation, 18
Boundaries, insulating, 104
Brass, 112
Brightness, electron gun, 234-236

C

Capacitor, flat, 16
Cardinal points
  electrostatic lens, 252-254, 265-267
  magnetic lens, 269
  unipotential lens, 277-279
Cathode
  design, 236-240
  materials, 236-237
  point and ground, 237-238

501

concentration, 111
deionized water, 112
Electron
beams, 83-98
calculation, 92-98
microprobes, 469-494
design, 471
mirror, 293
-optical systems, 46-56
sources, 471
trajectories, 76-82
paraxial, 83-88
Picht's equation, 78
Ellipse, 403-405
Ellipticity, bore, 491
Emittance, 401-402
Energy
analyzer, 291-292
dispersion, 251-252
equation, 85
Equipotentials, 113, 125
Euclidian matrices, 34
Euler-Lagrange equations, 31
Excitation, square-wave, 108-112
equivalent circuiti, 109
Extremum method, 6

F

Fermat's principle, 413-419
Field
curvature, 424-425
electrical
perpendicular, 30-31
uniform, 29, 30-31
electrostatic, 4-17
equations, 5
potential, 10-14
with space charge, 115-120
without space charge, 103-115
in vacuum, 5-8
variables, 8-10
magnetic, 17-26, 120-129, 482-484
boundary conditions, 121
caused by poles, 24-25
focusing, 17-18
measurement, 163-229
perpendicular, 30-31

polar pieces, 122-123
rectilinear wire, 19
resistance networks for, 153-161
scalar potential, 120-121, 153-156
uniform, 29, 30-31
multipole, 17
F-F problem, 385-387
Filaments
hairpin, 237-238, 240, 470-471
life, 238-239
pointed, 238
supplies, 239-240
Filter lenses, 252
Finite difference equations, 6, 129-131, 139-140
mesh points, 130
solution, 131-132
Flux magnetic, 158-161
Fluxmeter, 209-212, 222
Grassot, 187-188
photoelectric, 211
Focal distance,
doublet, 370
quadrupole, 363-364
triplet, 391, 393
Focusing, 353-410
beam
astigmatic, 388-389
divergent, 383-385
parallel, 381
doublet, 368-389
field, 359-360
ideal and practical, 358-359
matching, 400-405
multiplet, 396-400, 405-408
structure, magnetic, 17-18
systems, 354-359
matrix, 355
optics, 355-358
triplet, 389-396
F-P problem, 382-383
Frenet trihedral, 39
Fourier integral formula, 11

G

Galvanometer, 209-212, 222
ballistic, 187-189